Emerging Trends in Biomedical Science and Health

Emerging Trends in
Biomedical Science and Health

Editors

Prof. D.V. Rai
Department of Biophysics
Basic Medical Sciences Block
Panjab University, Chandigarh, India
e-mail: dvrai@pu.ac.in, dvrai@yahoo.com

Prof. R.C. Sobti
Vice Chancellor
Department of Biotechnology
Panjab University, Chandigarh, India
e-mail: rcsobti@pu.ac.in

Prof. Raj Bahadur
Director Principal
Government Medical College and Hospital
Chandigarh, India
email: dpgmcc@yahoo.com, rajbahadur9@gmail.com

I.K. International Publishing House Pvt. Ltd.

NEW DELHI • BANGALORE

Published by
I.K. International Publishing House Pvt. Ltd.
S-25, Green Park Extension
Uphaar Cinema Market
New Delhi 110 016 (India)
E-mail: info@ikinternational.com

ISBN 987-93-80026-22-0

Published by Krishan Makhijani for I.K. International Publishing House Pvt. Ltd. S-25, Green Park Extension, Uphaar Cinema Market, New Delhi 110 016. Printed by Rekha Printers Pvt. Ltd., Okhla Industrial Area, Phase II, New Delhi 110 020.

"The university stands for humanism and tolerance, for reason, for adventure of ideas and for the search of truth. It stands for the forward march of the human race towards even higher objectives. If the universities discharge their duties adequately then it is well with the nation and the people".

Pt. Jawaharlal Nehru

Biomedical science without human ethics is lame, human health without consciousness is blind.

— D.V. Rai

Foreword

"Every great advance in Science has issued
from a new audacity of imagination"
John Dwey

Health as defined in the World Health Organization's constitution is "a state of complete physical, mental and social well-being and not merely the absence of disease or infirmity." The state of health of an individual or population depends upon complex interactions of the physical, biological and social domains. Over the past century, human activities and behavior have played a major role in global environmental change, in turn; these and other global lifestyle affect the health and welfare of human populations at remarkable rates. Biomedical science and health education have always played an important role to protect human health with these drastic changing situations either on the virtue of its advanced technological interventions or by arousing global consciousness among populations regarding adoption or copying of unaware lifestyle patterns. Apart from these two strategies there is a need of amalgamatory approach to keep pace with these ever changing challenges which in turn rquire the knowledge of other multidisciplinary fields. Health research builds primarily on the basic sciences of biology, chemistry, and physics and needs other primarily research-oriented fields that make exceptionally significant contributions to health science like biomedical engineering, biotechnology, nursing, nutrition, pharmacology, pharmacy, psychology, physical therapy, medicine biochemistry, epidemiology, genetics, and biophysics to improve, prevent and cure diseases.

The present book entitled "*Emerging trends in Biomedical Science and Health*" reveals an excellent compilation of summary of information of contemporary issues and solutions related to human health. The book broadly covers the topics like oxidative stresses and their solution with antioxidants, role of biological rhythms in pharmacology and toxicology, impact of bone physiology on human health, prevention and treatment of cancer by the process of biotransformation of carcinogens, control of epilepsy by herbs and their products, role of immunotoxicology in maintenance of human health. It also discusses the perspectives and prospects of stem cells in biomedicine, diversified role of microorganisms in human helath and disease, effects of shilajit in maintaining human health and vigour, new health prospets of vitamins and minerals, impact and risks of genetically engineered foods of human health and society. The present book elaborates upon the effects and consequences of bioaerosols and various energy sources on long-term human health. It also concludes with the introduction of a new and challenging field of medical geology and biomedical geography and their role in assessing, diagnosing, preventing the epidemiology of various human ailments and diseases to improve the human health and their surrounding environment.

This book is a vital addition to the knowledge base of all professionals, academics and students of graduate and postgraduate courses working in the field of biomedical sciences, disease management and worldwide development of human health.

I wish the endavor of authors and editors all the success.

(Prof. N.K. Ganguly)
MD, Dsc(hc), FMedSci (London), FRC Path (London), FAMS, FNA, FASc.
FNASc. FTWAS (Italy), FIACS (Canada), FIMSA

– Adviser to the Hon'be Health Minister
– Distinguished Biotechnology Fellow & Advisor
– Translational Health Science & Technology Institute

Preface

Recent developments in biomedical science and technology fascinated the attention of scientific, academic as well as general community because of several reasons. One of them is the impact of different factors related to lifestyle, environment and nutrition on human health. This book is specially meant for graduate and postgraduate students, teachers, researchers, administrators, planners/policy-makers and non-governmental organizations concerned with education as well as research and development in biomedical sciences and awareness as well as protection against diseases for maintaining health.

The book broadly deals with the lifestyles related to health diseases, human rhythms in toxicology, cancer prevention, stem cells in biomedicine, health effects of exposure to electromagnetic energy sources, microorganisms and bioaerosols in environment, health impacts of biotechnological food products and medical geology and geography.

We extend our profound gratitude to Prof. K.N. Pathak, (Former) Vice Chancellor, Panjab University for being a source of encouragement to complete this compilation in the form of book. We acknowledge with thanks the consistent efforts of our guest authors who helped us to complete this academic assignment. To appreciate the faculty members, staff and students of the Department of Biophysics and other sister departments of the Panjab University and other universities in India for their help and support.

D.V. Rai
R.C. Sobti
Raj Bahadur

Contents

Contributors

Anant Singh University Institute of Pharmaceutical Sciences, Panjab University Chandigarh-160014	—	M. Pharma
Anil Kumar University Institute of Pharmaceutical Sciences, Panjab University Chandigarh-160014 e-mail: kumaruips@yahoo.com	Reader	M. Pharma, Ph.D.
Ashish Dhir University Institute of Pharmaceutical Sciences, Panjab University Chandigarh-160014	—	M. Pharma
Ashwani Koul Department of Biophysics Panjab University, Basic Medical Sciences Block, Chandigarh-160014 e-mail: ashwanik@pu.ac.in	Reader	M.Sc., Ph.D.
D.K. Rahi Department of Microbiology, Basic Medical Sciences Block, Panjab University Chandigarh-160014 e-mail: dkrio@pu.ac.in	Lecturer	M.Sc., Ph.D.

D.V. Rai Department of Biophysics, Basic Medical Sciences Block, Panjab University Chandigarh-160014 e-mail: dvrai@pu.ac.in	Professor	M.Sc., M.Tech., Ph.D.
Gaurav Kumar Department of Biophysics, Basic Medical Sciences Block, Panjab University Chandigarh-160014 e-mail: gauravcheppa@yahoo.com	—	M.Sc.
Geeta Shukla Department of Microbiology, Basic Medical Sciences Block, Panjab University Chandigarh-160014 e-mail: geeta_shukla@pu.ac.in	Senior Lecturer	M.Sc., Ph.D.
Harkiranpreet Kaur Department of Biophysics, Basic Medical Sciences Block, Panjab University Chandigarh-160014	—	M.Sc.
H.D. Khanna Department of Biophysics Institute of Medical Sciences, Banaras Hindu University, Varanasi-221 005 UP e-mail: hdkhanna@yahoo.com	Professor	M.Sc., Ph.D.
Krishna Mohan Department of Geography Panjab University Chandigarh-160014 e-mail: kmohan@pu.ac.in	Reader	M.A., Ph.D.

L.M. Aggarwal Department of Radiotherapy Institute of Medical Sciences Banaras Hindu University, Varanasi- 221005, U.P.	Radiological Safety Officer	M.Sc., Ph.D.
Mehak Goel Department of Biophysics Panjab University. Chandigarh-160014	—	M.Sc.
Naresh Tuli Advanced Centre of Geology Panjab University Chandigarh-160014 e-mail: nareshtuli@lycos.com	Reader	M.Sc., Ph.D.
Raj Bahadur Government Medical College and Hospital, Chandigarh Sector- 32, Chandigarh 160047 e-mail: zajhahadurg@gmail.com	Director Principal Professor	M.B.B.S., M.S., FAMS, FIMSA, FICA, FAOI, FIHE.
Raman Soni Department of Biotechnology D.A.V. College, Sector 11 Chandigarh-160011 e-mail: ramansoni@yahoo.com	Lecturer	M.Sc., M.Phil. Ph.D.
Ranjana Khanna Department of Chemistry BHU, Varanasi, 221 005	Reader	Ph.D.
Richa Goyal University Institute of Pharmaceutical Sciences, Panjab University Chandigarh-160014	—	M.Pharma

R.C. Sobti Department of Biotechnology Panjab University Chandigarh-160014 Chandigarh e-mail: rcsobti@pu.ac.in	Vice Chancellor Professor	M.Sc., Ph.D. FNASc, FAMS FZS, FPAS, FSCG
R.K. Sharma SAIF Panjab University Chandigarh-160014 e-mail: ramesh@pu.ac.in	Technical Officer	M.Sc., M.Tech.
R.S.Verma Associate Professor Deptt. of biotechnology Indian Institute of Technology, Chennai-600036. Madras – (T.N.) e-mail: vermars@iitm.ac.in	—	M.Sc., Ph.D.
Sachin Gupta Department of Immunopathology Post Graduate Institute of Medical Education & Research, Chandigarh-160012.	—	M.Sc.
Sandeep Thind Department of Geography Panjab University Chandigarh-160014	—	M.A.
Shruti Apollo Hospital, Varanasi-221 005 UP (India)	Resident Doctor,	BAMS
S.C. Gangar Department of Biophysics, Basic Medical Sciences Block, Panjab University Chandigarh-160014 e-mail:scgangar@yahoo.com	—	M.Sc., Ph.D.

S.K. Arora Department of Immunopathology Post Graduate Institute of Medical Education & Research, Chandigarh-160012 e-mail: skarora_in@yahoo.com	Professor	M.Sc., Ph.D.
S.K. Kulkarni University Institute of Pharmaceutical Sciences Panjab University, Chandigarh-160014. e-mail: skpu@yahoo.com	Dean University Instructions Professor	M. Pharma, Ph.D. FAMS, FNASc. FPAS, FIPA
S.K. Soni, Department of Microbiology Basic Medical Sciences Block, Panjab University Chandigarh-160014 e-mail: sonisk@pu.ac.in	Reader	M.Sc., Ph.D.

I

Oxidative Stress and Antioxidant Protection in Disease States

H.D. Khanna, Ranjana Khanna and Shruti

INTRODUCTION

The body tissues require continuous nourishment for their normal functioning. Healthful state is directly related to the defence system of the body to fight the diseases. The immunity is the ultimate strength of the body and is the power to fight the diseases. The quality of immune response is dependent on the quality of food and the metabolic state of the body. The whole process is interrelated. As organic beings, what comes out of us generally reflects what goes into us. Our energy and expression mirrors what we take into ourselves via nutrition. This occurs on three primary levels. The first is food that builds up the physical body or 'physical nutrition'. The second is our sensory impression that builds up the mind or 'psychological nutrition'. The third is our associations, which nourish the heart and the soul or what could be called 'spiritual nutrition'. If these three factors of right diet consisting of fresh fruits or freshly cooked vegetables rich in terms of calories; secondly, the right diet for mind consisting of wholesome impressions drawn from the world of nature, art or temples and right association consisting of healthy relationship with friends and family; thirdly, of the spiritual level from the gurus and teachers occur—then we are happy and healthy and are bound to grow spiritually. If the process of conversion of food is defective, the end products in the form of metabolites shall also be defective. The defective metabolic disorder contains the precursors of highly reactive electrochemically unstable molecules i.e. free radicals, the initiators of destructive chain reactions. If the defence mechanism of the body fails to combat these free radicals or they are not used properly in the body, these silent killers pose a threat by injuring tissues, DNA, proteins or lipid components of cells causing cell damage that could lead to cell death.

FREE RADICAL

Free radicals are atoms or group of atoms with an odd (unpaired) number of electrons and can be formed when oxygen interacts with certain molecules. Once formed, these highly reactive radicals can start a chain reaction. Their chief danger comes from the damage they can do when they react with important cellular components such as DNA, or the cell membrane. Cells may function poorly or die if this occurs. To prevent free radical damage, the body has a defence system of *antioxidants.* Free radicals play an important role, in both health and disease, and have been implicated in countless human disease processes. Free radicals are vital to human health. They are essential for life; we need them, and they fight infection. Our cells give a burst of these free radicals to kill the germs when we become infected with bacteria fungus and parasites.

Oxygen is what is known as a free radical; it is highly unstable (Fig.1). Oxygen is life-saving but if you give 100% oxygen, it is highly toxic. Why does oxygen become toxic? It is a molecule which consists of two atoms of oxygen, and when they are combined, they balance each other out. But it is unstable, and if the two atoms are separated, they become two free radicals. But free radicals are not limited to oxygen. Any molecule, any atom in the body can become a free radical, and when it does it will try to balance itself by picking up an electron from the neighboring molecule, which in turn becomes a free radical and the vicious cycle continues until it is checked and controlled. If it is not controlled, damage occurs in different parts of the body.

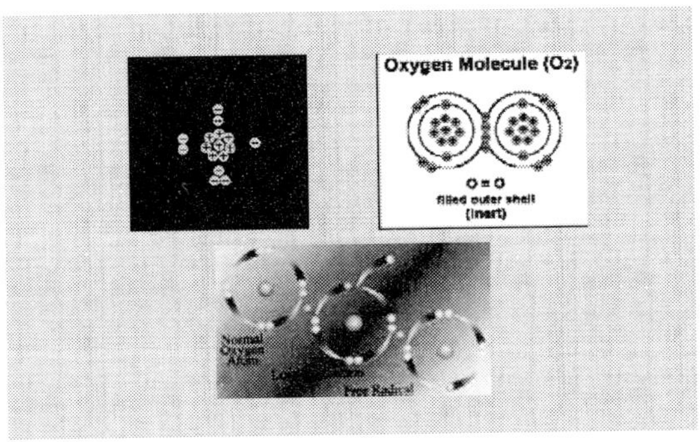

Fig. 1. Free radical generation.

Free radicals "steal" electrons from other molecules in order to restore their own balance. In the process, they create another free radical that can participate in a chain reaction with resultant damage to the involved body tissue. When DNA or genetic material is involved in free radical reactions, mutations or genetic alterations can result. Free radical chain reactions are stopped by antioxidants. Under normal conditions, every cell in an individual encounters about 10,000 free radical "hits" on a daily basis. When our antioxidant defences are adequate, this free radical damage is repaired without much consequence.

Also of importance is the fact that different antioxidants are active against different free radicals and active in different parts of the cell. For optimal health, a broad array of antioxidants is important to provide protection against free radicals.

REACTIVE OXYGEN SPECIES

The term 'reactive oxygen species' (ROS) and 'reactive nitrogen species' (RNS) is a collective term that includes not only the radicals but also the non-radicals. Free radicals and reactive oxygen/

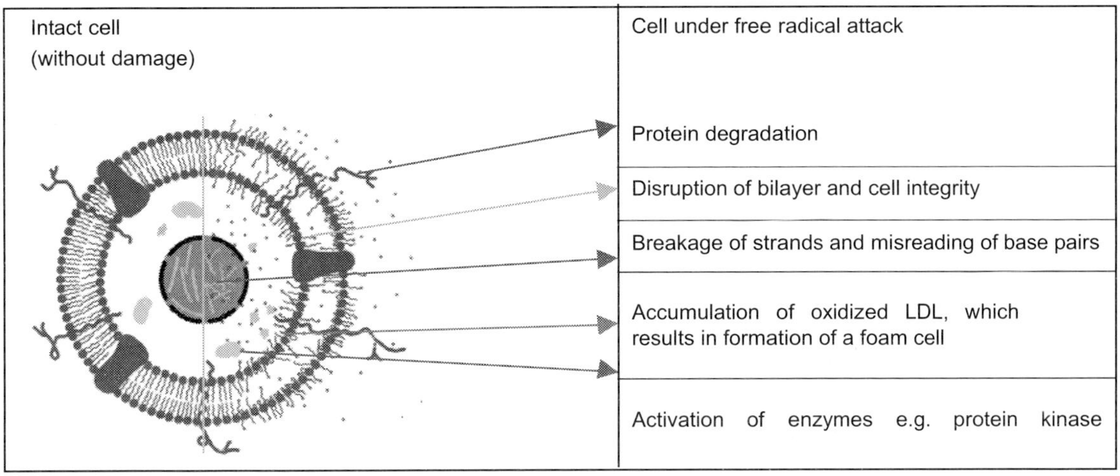

Fig. 2. Cell under free radical attack.

nitrogen species of importance in living organisms include hydroxyl ($OH^{\cdot}$), superoxide ($O_2^{\cdot -}$), nitric oxide ($NO^{\cdot}$), nitrogen dioxide ($NO_2^{\cdot}$) and peroxyl ($ROO^{\cdot}$). Peroxynitrite ($OONO^-$), hypochlorous acid (HOCl), hydrogen peroxide (H_2O_2), singlet oxygen ($^1O_2$), ozone (O_3), nitrous acid (HNO_2) and dinitrogen trioxide (N_2O_3) are not free radicals but can easily lead to free radical reactions in living organisms.

These reactive oxygen intermediates may participate in reactions, which give rise to free radical species. Unstable free radical species attack cellular components, causing damage to lipids, proteins and DNA, which can initiate a chain of events, which results in the onset of disease (Fig. 2).

A poor diet, cigarette smoking, stress, alcohol, carcinogens, environmental pollutants and ultraviolet (UV) radiation from the sun also increase the free-radical load, creating a situation known as "oxidative stress".

A free radical has three stages: the initiation stage, propagation and finally, termination. They are terminated or neutralized by nutrient antioxidants, enzymatic mechanisms, or by recombining with each other. The aim is to attain a delicate balance between free radical activity and optimum antioxidant activity, thereby achieving a state of balance (homeostasis).

Defense System Against Free Radicals

As soon as these free radicals are created, they are checked by enzymes in the body, which are known as *antioxidant enzymes*, or nutrients in our food that are known as *antioxidant nutrients* like *vitamin C, vitamin E, beta-carotene and bioflavonoids*. These are present in the plants, herbs, fruits and vegetables and are very powerful. If the generation of free radicals is much more than can be handled by the enzymes or antioxidant nutrients, then trouble arises, and they create damage in the cell membrane. In a joint, it ultimately results in arthritis, in lungs we have emphysema and bronchitis, in the blood vessels we have atherosclerosis or heart disease, in the stomach we have peptic ulcer, in the skin, aging and wrinkling. These radicals can produce damage in each area. In the nucleus, if free radicals are produced in the DNA, the DNA becomes transformed and it becomes mutated. Mutated transformed DNA will produce cancer, leukemia, and lymphoma. Diabetes, kidney problems, liver problems, almost every disease is related to the damage caused by these free radicals.

ANTIOXIDANT

An antioxidant is a molecule capable of slowing or preventing the oxidation of other molecules. Oxidation is a chemical reaction that transfers electrons from a substance to an oxidizing agent. Oxidation reactions can produce free radicals, which start chain reactions that damage cells. Antioxidants terminate these chain reactions by removing free radical intermediates and inhibit other oxidation reactions by being oxidized themselves. As a result, antioxidants are often reducing agents.

Oxidation occurs when high levels of oxygen react with exposed surfaces during respiration. It happens every time we take in a breath of air. This process has byproducts called free radicals. Free radicals being unstable molecules react with normal, healthy body cells and damage them. This repetitive damage leads to degenerative diseases like cancer, ischemic heart disease, diabetes, arthritis, cataract, aging and many others.

Oxidation reactions are crucial for life, they can also be damaging; hence, plants and animals maintain complex systems of multiple types of antioxidants, such as glutathione, vitamin C and vitamin E as well as enzymes such as catalase, superoxide dismutase and various peroxidases. Low levels of antioxidants, or inhibition of the antioxidant enzymes, causes oxidative stress and may damage or kill cells.

OXIDATIVE STRESS

Oxidative stress is caused by an imbalance between the production of reactive oxygen and a biological system's ability to readily detoxify the reactive intermediates or easily repair the resulting damage. All forms of life maintain a reducing environment within their cells. This reducing environment is preserved by enzymes that maintain the reduced state through a constant input of metabolic energy. Disturbances in this normal redox state can cause toxic effects through the production of peroxides and free radicals that damage all components of the cell, including proteins, lipids and DNA.

As oxidative stress has been associated with the pathogenesis of many human diseases, the use of antioxidants in pharmacology is intensively studied, particularly as treatments for stroke and neurodegenerative diseases. However, it is unknown whether oxidative stress is the cause or the consequence of such diseases. Antioxidants are also widely used as ingredients in dietary supplements in the hope of maintaining health and preventing diseases such as cancer and coronary heart disease.

In humans, oxidative stress is involved in many diseases, such as atherosclerosis, Parkinson's disease and Alzheimer's disease and it may also be important in ageing. However, reactive oxygen species can be beneficial, as they are used by the immune system as a way to attack and kill pathogens.

ANTIOXIDANT SYSTEM

Antioxidants are molecules, which can safely interact with free radicals and terminate the chain reaction before vital molecules are damaged. Although there are several enzyme systems within the body that scavenge free radicals, the principal micronutrient (vitamin) antioxidants are vitamin E, beta-carotene and vitamin C. Additionally, selenium, a trace metal that is required for proper function of one of the body's antioxidant enzyme systems, is sometimes included in this category. The body cannot manufacture these micronutrients so they must be supplied in the diet.

Fortunately, the body has its own antioxidants for damage limitation. One of the most effective of these is the substance tocopherol (vitamin E). This vitamin dissolves in fat and that is especially important because by far the most significant free radical damage in the body is damage to the membranes of cells and to low-density lipoproteins, and these are made of fat molecules. Vitamin C is also a powerful antioxidant, but is soluble in water, not in fat. This means that it gets distributed to all parts of the body. The two vitamins are both highly efficient at mopping up free radicals and sometimes even co-operate in so doing.

Other natural body antioxidants include compounds such as *cysteine, glutathione* and *D-penicillamine,* and blood constituents such as the iron-containing molecule *transferrin* and the protein *ceruloplasmin.* These act either by preventing free radicals from being produced or by mopping them up.

The body also contains a number of important antioxidant *enzymes.* An enzyme is a highly active protein that accelerates a chemical reaction. Most of what goes on in the body is promoted by thousands of different enzymes. The most interesting antioxidant enzyme is *superoxide dismutase.* This excited enormous interest when it was discovered, as it has no other function than to change the dangerous superoxide free radical to the safer hydrogen peroxide. This made the scientists think and really got the doctors interested in free radicals. Hydrogen peroxide (H_2O_2), although not a free radical, is not particularly pleasant stuff to have around. The extra oxygen atom is readily available to cause oxidation, making it an active compound (Fig. 3). So the body has two other enzymes, *catalase* and *glutathione peroxidase* that break down hydrogen peroxide to water and oxygen.

Table 1 shows the important enzymatic and non-enzymatic physiological antioxidants.

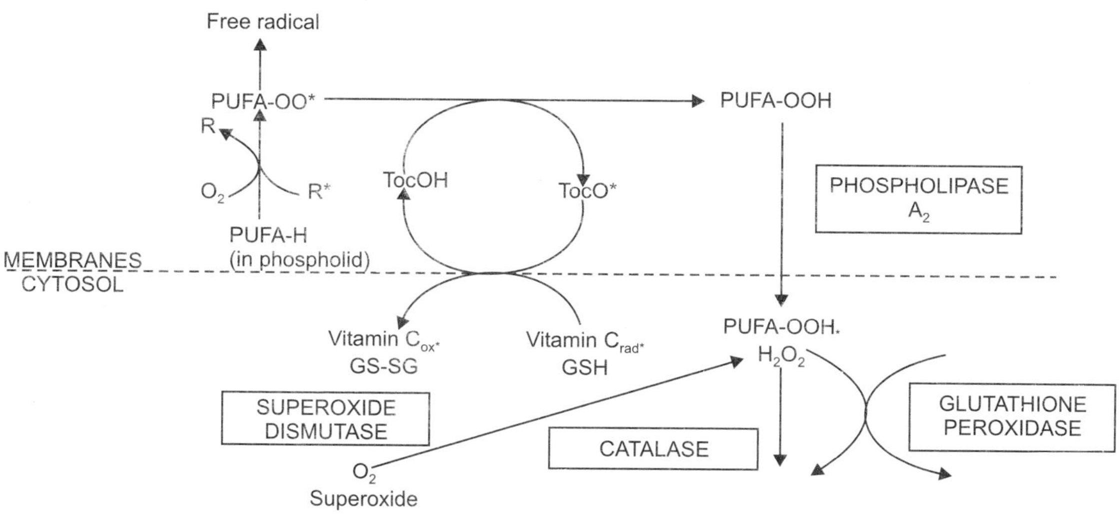

Fig. 3. Antioxidant defence system.

Table 1. Important enzymatic and non-enzymatic physiological antioxidants

Enzymatic antioxidants	Location	Properties
Superoxide dismutase (SOD)	Mitochondria cytosol	Dismutase superoxide radicals
Glutathione peroxidase (GSH))	Mitochondria and cytosol	Removes hydrogen peroxide and organic hydroperoxide
Catalase (CAT)	Mitochondria and cytosol	Removes hydrogen peroxide
Non-enzymatic antioxidants	**Location**	**Properties**
Vitamin C	Aqueous phase of cell	Acts as free radical scavenger and recycles vitamin E
Vitamin E	Cell membrane	Major chain-breaking antioxidant in cell membrane
Uric acid	Product of purine metabolism	Scavenger of OH radicals
Glutathione	Nonprotein thiol in cell	Serves multiple roles in the cellular antioxidant defence
δ-lipoic acid	Endogenous thiol	Effective in recycling vitamin C, may also be an effective glutathione substitute
Carotenoids	Lipid soluble antioxidants, located in membrane tissue	Scavengers of reactive oxygen species, singlet oxygen quencher
Bilirubin	Product of heme metabolism in blood	Extracellular antioxidant
Ubiquinones	Mitochondria	Reduced forms are efficient antioxidants
Metal ions sequestration: transferrin, ferritin, lactoferrin,		Chelating of metal ions, responsible for Fenton reactions
Nitric oxide		Free radical scavenger, inhibitor of LP

Important antioxidants present in the food:

- **Vitamin C** is plentiful in fruits and vegetables, especially dark green leafy vegetables, citrus fruits, strawberries, red peppers, kiwi, papaya, broccoli, cabbage and cauliflower. If you eat six servings or more of fruits and vegetables each day, you may not need vitamin C supplement.
- **Carotenoids** are found in orange fruits and vegetables and in red and dark green vegetables. Apricots, carrots, cantaloupe, sweet potatoes and winter squash are good sources of beta-carotene.
- Lycopene is found in tomatoes.
- Lutein is found in dark green leafy vegetables and red peppers.
- Alpha carotene is found in pumpkin, carrots, yellow peppers and winter squash.
- Vegetable oils, nuts and seeds, wheat germ, and dark leafy greens all contain **vitamin E** But it is impossible to get therapeutic amounts of the vitamin from diet alone. For example, you need to eat 25 pounds of almonds or consume nearly 9 cups of canola oil to get 400 IU of vitamin E.
- **Flavonoids** are found in a wide array of fruits and vegetables. In particular, beets contain anthocyanidins, green tea contains polyphenols, and apples and onions contain quercetin. Other good flavonoid sources include citrus fruits, berries and red wine.

Table 2 shows the concentration of antioxidant metabolites in human serum and liver tissue.

Table 2. Concentration of antioxidant metabolites in human serum and liver tissue

Antioxidant metabolite	Solubility	Concentration in human serum (μM)	Concentration in liver tissue (μmol/kg)
Ascorbic acid (vitamin C)	Water	50 – 60	260 (human)
Glutathione	Water	325 – 650	6,400 (human)
Lipoic acid	Water	0.1 – 0.7	4 – 5 (rat)
Uric acid	Water	200 – 400	1,600 (human)
Carotenes	Lipid	β-carotene: 0.5 – 1 retinol (vitamin A): 1–3	5 (human, total carotenoids)
α-tocopherol (vitamin E)	Lipid	10 – 40	50 (human)
Ubiquinol (coenzyme Q)	Lipid	5	200 (human)

OTHERS ANTIOXIDANTS

Curcumin

A substance called curcumin (from the herb turmeric) significantly improves brain functions in animals with memory deficits, so is true for preparations that include alpha-lipoic acid and vitamins C and E.

Coenzyme Q10

Coenzyme Q10 (CoQ10) or ubiquinone is an essential cofactor of the electron transport chain as well as a potent free radical scavenger in lipid and mitochondrial membranes. Because of its strong antioxidant properties, it is being tested in patients with degenerative neurological diseases, Parkinson's disease and multiple sclerosis. The hope is that coenzyme Q10 can slow down progression of the diseases. Many studies suggest that coenzyme Q10 may be better strategy than vitamin E in treatment of neuro-degenerative diseases in increasing survival time. Oral administration of coenzyme Q10 increases both brain and brain mitochondrial concentrations.

Astaxanthin

It is a carotenoid compound. Some forms are insoluble (many antioxidants are fat or water soluble). Beta carotene (a preliminary stage of vitamin A) is sometimes recommended as an antioxidant and astaxanthin supplies carotenoids.

Astaxanthin has been claimed to be 10 times more effective than beta-carotene and 100 times more effective than vitamin E in preventing lipid peroxidation. Astaxanthin works better in combination with vitamin E.

Carotenoids

They are a group of antioxidant nutrients found in many fruits and vegetables. They are more easily absorbed in supplement form and even more effective if taken with highly coloured fruits and vegetables such as carrots, tomatoes, watermelon and beets high in carotenoids. Like all antioxidants, carotenoids work better in combination than if taken alone. For example, beta carotene boosts the antioxidant activity of lycopene. Carotenoids include beta carotene, alpha carotene, cryptoxanthin, lycopene, lutein and zeaxanthin. Each performs a slightly different function at a different level and is dependant on other carotenoids for its effectiveness. Synthetic carotenoids have little biological effect and should be avoided. Naturally derived carotenoids suspended in an oil base benefit effectively.

Phenolics

They are naturally occurring antioxidant substances found in the skins of many fruits, vegetables and herbs. Fruits and vegetables with brightly coloured skins are generally considered to be relatively high in phenolics and the skins and seeds of the grapes (vitus vinifera) are used to make red wine.

Drinking green tea may help prevent heart attacks and sudden cardiac death.

IMPLICATION OF OXIDATIVE STRESS IN DISEASE STATES

In recent years, several studies have provided enough evidence to show the role of oxidative stress and involvement of various reactive oxygen species (ROS) in the pathogenesis of several degenerative diseases (Table 3). This has led to considerable interest to study the role of various antioxidants in the prevention and treatment of various diseases. As a result of these studies, a large number of dietary as well as plant derived components with antioxidant properties have been identified. Majority of these agents possess phenolic groups, which are responsible for their antioxidant properties. These compounds have been studied to assess their therapeutic potential in the treatment as well as prevention of various diseases and it has been found that they exhibit biological activities through different mechanisms. Based on these observations, many of these compounds have been used for structural

Table 3. Diseases in which oxidative free radicals have been implicated

Adult respiratory distress syndrome	Emphysema
Aging and premature aging	Hepatic injury (xenobiotics)
Atherosclerosis	Intestinal ischaemic injury
Alzheimer's disease	Ionizing radiation injury
Alcohol toxicity	Keloids
Carcinogenicity	Mucocutaneous syndrome
Cataract	Organ transplant rejection
Cerebral ischaemia	Eclampsia
Chronic bronchitis	Paraquat toxicity
Crohn's disease	Progressive systemic sclerosis
Dermatitis herpetiformis	Pulmonary fibrosis
Diabetes mellitus	Retrolental fibroplasias
Drug-induced toxicity	Rheumatoid arthritis
Pancreatitis	Schizophrenia
Ulcerative colitis	Sickle cell anaemia
Vitamin deficiency	Haemochromatosis

modification to optimize their therapeutic activities. Antioxidants can act at different levels. These are broadly classified into five levels. Classification of anti-oxidants is shown in Table 4.

Table 4. Levels of antioxidant action

Level 1	Prevention of radical formation: mainly by enhancing the levels of SOD and catalase, or by sequestering transient metals like iron.
Level 2	Scavenging of primary radicals (OH^0, O_2, $^{1}O_2$) or by breaking of chain initiation.
Level 3	Scavenging of secondary radical (ROC, LOO) or breaking of chain propagation (conventionally studied as inhibition of lipid peroxidation)
Level 4	Repair and reconstruction of lipid membrane aided by glutathione peroxidase or by increase in level of glutathione
Level 5	Repair of DNA and other cellular constituents

OXIDATIVE STRESS AND HYPERTENSION

The term 'Hypertension' denotes systolic blood pressure (SBP) of 140 mm Hg or greater, diastolic blood pressure (DBP) of 90 mmHg or greater, or taking antihypertensive medication. In most of the cases, hypertension is idiopathic and is referred to as essential or primary hypertension. Hypertension may lead to serious complications, which include heart failure, coronary artery disease, valvular heart disease, cardiac arrhythmia, cardiomyopathy and abnormal renal function. Pathogenesis of essential hypertension may include a heterogenicity of complex series of causes, including heredity, increased fluid volume, renal sodium transport deficiency, increased sympathetic activity and increased vascular tone, involvement of the renin – angiotensin – aldosterone system, chronic stress, diminished vasodepressor hormones (prostaglandins, atrial natriuretic factor) activity etc. In addition, factors like obesity, physical inactivity, occupation, smoking and alcoholism may also contribute. In less than 5% cases, hypertension is secondary to an identifiable cause including renal, endocrine and neurological abnormalities, or congenital cardiovascular anatomical defects or iatrogenic causes.

Hypertension is a well-established risk factor for cardiovascular disease, in part because it is proatherogenic. It presumably facilitates injury to the vascular endothelium by increasing stress caused by flowing blood. The pathogenesis of hypertension can involve both genetic predisposition and exposure to toxins, including cigarette smoke. Normal blood pressure is maintained by a complex web of systems affecting cardiac output, sodium balance, vasodilation and renal function. One of the most important of these is the renin/angiotensin system. Inhibitors of angiotensin-converting enzyme, which converts angiotensin I to the hypertensive peptide angiotensin II, are frequently used to treat hypertension.

It has been suggested that one mechanism contributing to hypertension is increased generation of $O_2^{\bullet -}$ in the vessel wall, inactivating $NO^{\bullet}$ and causing $ONOO^-$ formation. Angiotensin II may increase levels of O_2 in the vessel wall. Impaired endothelial response to vasodilators such as acetylcholine occurs in atherosclerosis and has been described in hypercholesterolaemic patients.

Dietary supplementation with antioxidant vitamins such as vitamin C, vitamin E and lipoic acid has the potential to lower blood pressure in both genetic and acquired hypertension. These nutritional supplements may act to prevent one of the underlying causes of essential hypertension i.e. elevated tissue aldehydes. In essential hypertension, excess endogenous aldehydes bind sulphahydryl groups of membrane proteins, altering membrane Ca^{2+} channels and increasing cytosolic free calcium and blood pressure. Increased oxidative stress can also lead to a further increase in reactive aldehydes through increased lipid peroxidation. Control of excess metabolic aldehydes and oxidative stress may be critical to the control of essential hypertension.

OXIDATIVE STRESS IN ATHEROSCLEROSIS

Endothelial lining of normal blood vessels are smooth and flexible, and blood flows through them to carry oxygen and other nutrients to different parts of the body. With aging, lipids and substances in the blood are deposited in the vessel walls and are called plaque. This is known as atherosclerosis or hardening of the arteries (Fig. 4). As the deposits become larger, the lumen of vessels becomes narrowed or blocked.

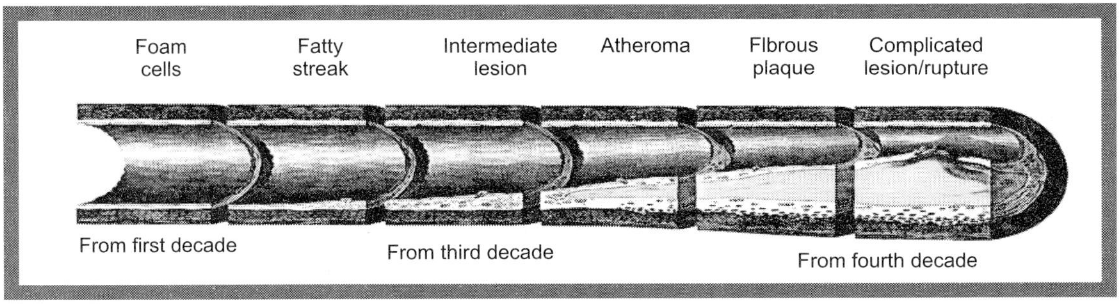

Fig. 4. Atherosclerosis.

Atherosclerosis is a slow progressive disease that may begin in childhood (Fig. 5). It often takes years for the plaque to build up inside any artery in the body. Atherosclerosis can lead to many types of circulatory problems.

Two of the major problems due to atherosclerosis are ischaemic heart disease and cerebrovascular events. Atherosclerosis is especially likely to be a major problem when it develops in the coronary arteries and affects myocardial oxygen blood supply and causes ischaemic heart disease and various acute coronary syndrome.

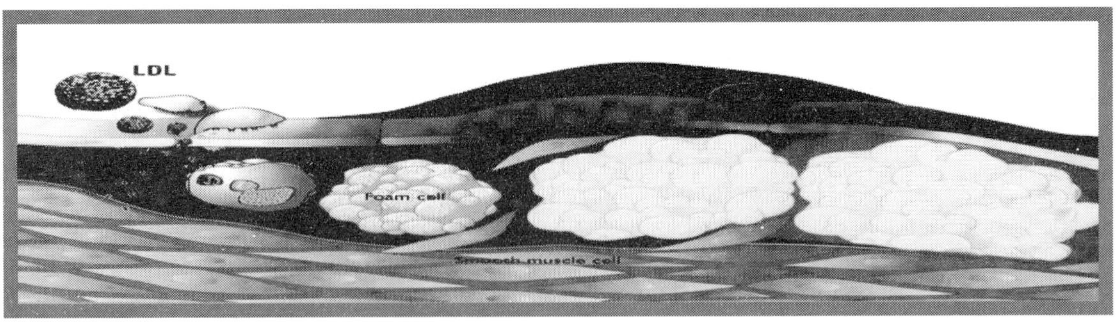

Fig. 5. Progression of atherosclerosis.

Atherosclerosis in Hypertensives

Elevation of blood pressure is associated with increased incidence of atherosclerosis. The metabolic correlates of human hypertensive disease are diabetes mellitus, obesity, hyperuricaemia, etc. Peripheral resistance to insulin has been identified in hypertensives and the level of insulin resistance/defect in whole body. Glucose utilization is also positively correlated with the severity of hypertension.

There are two main hypotheses that could explain the initiation of atherosclerosis: oxidation of LDL and an inappropriate immune response to vascular injury. Human low-density lipid protein (LDL) in its native state is not so atherogenic as is the oxidized LDL, which is formed due to damage to LDL. This oxidized-LDL is chemotactic for monocytes and macrophages, which engulf it to become foam cells. If the blood LDL level is too elevated, cholesterol derived from the LDL accumulates in and among the foam cells. The accumulated cholesterol, cells and debris constitute an atheroma, which in time can narrow the channel of the artery and so lead to thrombosis. To prevent this oxidation, LDL contains various antioxidant molecules such as alpha tocopherol and beta carotene, out of which alpha tocopherol is predominant. Ascorbic acid in plasma prevents LDL from getting oxidized. Probocol, a synthetic antioxidant also exerts strong protective effect. The oxidized LDL (ox LDL) is present in large amounts in atherosclerotic plaques where it stimulates T-cell migration, is immunogenic and induces antibody production. Although considerable evidence supports the involvement of inflammation and immunity in atherogenesis, the role of immune system has been questioned. Much of the research on the origin of atherosclerosis has concentrated on the interplay between lipid metabolism, cytokines and cellular activity within arterial wall. Consequently, the genes coding for LDL receptor (LDL-R); LXR-a; CD36; PPARs (a,g) and C-myc were recognized to play crucial role on the interplay of these processes responsible for atherogenesis. It is thus reasonable to assume that "Cross Talk" between such genes can define the genomics of atherosclerosis. Although polyphenols, derived from various herbs, have been shown to possess antioxidant properties, no heed has so far been paid to explore their effects on the genomics of atherosclerosis.

Increased production of free radicals is a feature of many human diseases including coronary vascular disease and dietary antioxidants, which may be important in protecting against these diseases associated with free radical injury. Oxidative modification of lipoproteins, particularly low-density lipoprotein (LDL) is thought to increase atherogenicity. And there is evidence for the presence of oxidized LDL and lipid peroxidation products in human atherosclerotic lesions. LDL oxidation appears to enhance macrophage uptake via scavenger receptors and thus contributing to atherosclerotic process.

The important implication of the oxidative hypothesis of atherosclerotic coronary heart disease is that antioxidants that can protect LDL against oxidation may limit the spread of degenerative illness and may act as antiatherosclerotic agents. The various antioxidant defense mechanisms present within the cell and outside the cell should be adequate to protect against oxidative damage. When pro-oxidant increases or antioxidant decreases, a situation of oxidative stress occurs that leads to excessive molecular damage and tissue injury. Countless stresses from smoking to aging can accelerate oxidation reactions. It is generally accepted that the micronutrients can promote optimal health and prevent chronic diseases by neutralizing the free radicals. However, the preventive potential of antioxidants is hardly restricted to isolated specific components and is rather due to various synergistic interactions of all antioxidant micronutrients.

NEUROLOGICAL DISORDERS

Oxidative stress in the brain is likely to occur as the brain uses up to 20% of the body's inspired oxygen, yet only accounts for 2% body weight. The brain also houses large concentrations of polyunsaturated fatty acids (PUFA), which may undergo lipid peroxidation in such an oxygen-rich environment. High iron and ascorbate content of the brain provides additional factor for providing favourable environment for generation of oxygen-free radicals. The concentration of antioxidative enzymes like SOD, glutathione peroxidase and reductase, and catalase is low in brain. The interplay of all these factors contributes to enhance the oxidative stress in the brain.

Neurons and glia undergo oxidative stress damage by: (a) increased lipid peroxidation, (b) oxidative damage to DNA, causing base modification and strand breakage, (c) damage to proteins, and (d) induction of apoptosis and necrosis.

Involvement of free radical reactions has been found in several diseases of the CNS such as chronic organic brain syndrome, Alzheimer's disease, Parkinson's disease, in some psychiatric syndromes (schizophrenia, paranoid psychosis, Korsakoff's syndrome), in cerebrovascular diseases, trauma of the central nervous system and in cerebral lipofuscinosis.

Antioxidants and Neurological Disease

Reduced levels of antioxidants such as vitamins E and C have been reported in many neurological conditions. Vitamin E supplementation in deficient individuals has been reported to either prevent or at least halt the progression of many neurological features except in patients of Parkinson's disease. Total antioxidant levels are elevated in Alzheimer's disease patients. Superoxide dismutase (SOD) levels are elevated in patients with many different neurological pathologies including Down's syndrome resulting in accumulation of hydrogen peroxide, which generates free radicals that damage brain tissues. Similar changes occur during ageing, this may explain the premature ageing observed in Down's syndrome patients. Glutathione peroxidase levels are elevated in patients with neurological pathologies including Down's syndrome and Parkinson's disease.

OXIDATIVE STRESS AND AGEING

Aging is a progressive deterioration of various functions that progressively converts physiologically and cognitively fit healthy adults into less fit individuals with increasing vulnerability to injury and illness. Theories of aging are of three types; genetically programmed aging changes; stochastic biological aging changes—free radical damage (DNA, proteins, lipids) or non-oxidative changes in protein structures and function; and decline in functions of physiological systems—immune system, endocrine system, muscle function, neurodegenerative changes. Mitochondria are a major in vivo source of oxygen-free radicals and cellular ATP, and play a role in initiating apoptosis of damaged cells. Ageing is thought to be caused by free radical reactions generated in the mitochondria, which accumulate with age. Free radicals cause the progressive oxidation of protein and lipid components in the cellular membranes and also activate phospholipases, proteases and endonucleases. Increased lipid peroxidation has been implicated in the ageing process followed by induction of detoxification enzymes. Wrinkle formation in the skin has also been attributed to long-term exposure to oxidative damage from UV light, resulting in the production of free radicals. Intervention with the new generation of antioxidant drugs, rasayanas and immunomodulators has been successful in controlling the action of the reactive oxygen species and free radical reactions.

CANCER AND HEART DISEASE—DO ANTIOXIDANTS HELP?

Strong evidence for the involvement of free radicals in the free radicals-induced DNA damage and consequent carcinogenesis comes from the fact that these processes can be abolished by free radical scavengers/ antioxidants. Some of the antioxidants appear to inhibit carcinogenesis by inhibiting stem cell origin and multistage nature of carcinogenesis and metabolism of a procarcinogen to its ultimate carcinogenic form. The antioxidant has been observed to inhibit chemically induced lung, mammary, fore stomach, colon and liver cancer in experimental animals. Polyphenols (antioxidant present in green tea) have been observed, likewise, to inhibit lung tumorigenesis. These findings indicate the involvement of free radicals in the process of tumor initiation. It can be deemed due to one or many of the following observations:

1. Free radicals may be involved in metabolism required for the activation of procarcinogen.
2. Reactions involving metabolic activation and/or degradation of carcinogen may release free radicals that can attack DNA.
3. Metabolites of carcinogen may themselves become free radicals and may interact directly with DNA.
4. Following binding of ultimate carcinogen to DNA, DNA-carcinogen adducts may produce free radicals in vicinity of DNA (Fig. 6).

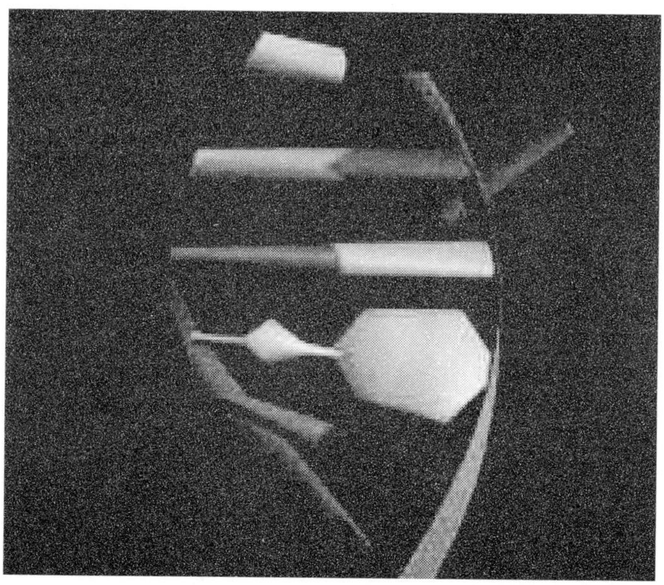

Fig. 6. Primary site of free radical damage in the DNA found in mitochondria.

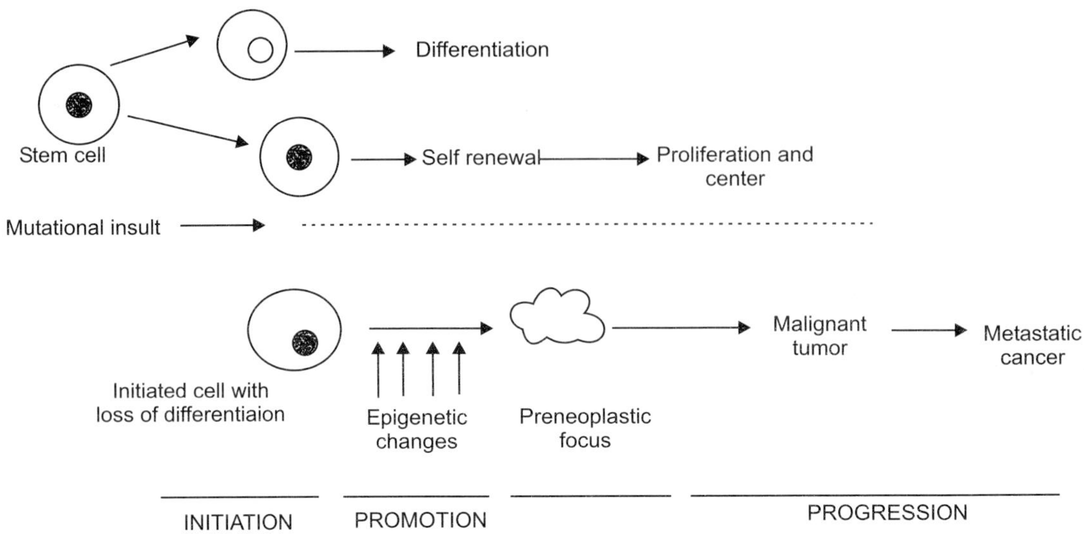

Fig. 7. Stem cell origin and multistage nature of carcinogenesis.

DIET AND CANCER

Epidemiologic observations show lower cancer rates in people whose diets are rich in fruits and vegetables. This has led to the theory that these diets contain substances, possibly antioxidants, which protect against the development of cancer. Fig. 7 shows the origin and different stages of carcinogenesis.

No single food removes cancer but experts believe that consuming plant-based foods helps, because they are low in saturated fats (linked to an increased risk of cancer, heart and other diseases) and high in fibre (believed to lower the risk of colon cancer). They are the best source of phytochemicals – natural substances in fruits and vegetables that contribute to the colour and flavour of vegetables and also appear to prevent the formation of certain tumors. One phytochemical widely used is the antioxidant beta carotene. But researchers believe that other carotenoids such as lutein – found in spinach, and other green leafy vegetables—are more effective than beta carotene. Other phytochemicals that help prevent cancer are limonene and phenols, found in citrus fruits, allyl sulfide in garlic and onions, sulforaphane, indoles and isothiocynates in broccoli, cauliflower, and other cruciferous vegetables.

Anti-cancer effects of green tea, turmeric and other agents will eventually add years to the human life. Several studies have suggested green tea might ward off heart disease and certain cancers (of stomach or esophagus) possibly due to potent antioxidant substances called polyphenols, which are also found in foods like grapes and wine. Antioxidants help protect the body from cell-damaging forms of oxygen that occur naturally in the body and are believed to contribute to a range of diseases.

CONCLUSION

Oxidative stress is an underlying factor in health and disease. A proper balance between oxidants and antioxidants is involved in maintaining health and longevity, and altering this balance in favour of oxidants may result in pathological responses causing functional disorders and diseases.

The foundation of good health is exercise and a diet rich in fruits and vegetables. Dark green leafy vegetables, spinach, fenugreek (methi), mustard green broccoli (sarson sag), yellow and orange coloured vegetables, carrots, pumpkins, sweet potato, fresh seasonal fruits, wheat and gram are rich in antioxidants. Vitamin E, citrus fruits, strawberries, green pepper, cauliflower and cabbage are rich in antioxidants/vitamin C. Sprouts from pulses, gram are full of vitamins, A, E and C (antioxidants) besides fiber and other nutrients including minerals and tracer elements like calcium, iron and zinc. They should be used by all for good health, preventing and fighting of disease.

SUGGESTED READING

1. *The Effect of Vitamin E and Beta Carotene on the Incidence of Lung Cancer and Other Cancers in Male Smokers*, New England Journal of Medicine (NEJM). 330 (15), pp 1029–1035, 1994, Apr. 14.
2. *A Clinical Trial of Antioxidant Vitamins to Prevent Colorectal Adenoma* NEJM, 331 (3), pp. 141–147, 1994, July 21.
3. *Antioxidant Vitamins—Benefits Not Yet Proved* (editorial), NEJM 330 (15), p. 1080–1081, 1994, Apr. 14.
4. *Antioxidants and Physical Performance* (review) Critical Reviews in Food Science and Nutrition, 35(1&2): 131–141, 1995.
5. *Increased blood antioxidant systems of runners in response to training load.* Clinical Science. 80, 611–618, 1991.
6. *Exercise, Oxidative Damage and Effects of Antioxidant Manipulation* (review). Journal of Nutrition 122(3 suppl): 766–73, 1992.
7. *Antioxidants: role of supplementation to prevent exercise-induced oxidative stress* (review). Medicine and Science in Sports and Exercise. 25(2): 232–6, 1993 Feb.
8. *Prospects for the use of antioxidant therapies.* (Review) Drugs 49(3): 345–61, 1995.

2

Biological Rhythms in Pharmacology and Toxicology

S.K. Kulkarni and Ashish Dhir

INTRODUCTION

Jean Baptiste Dortous de Mairin, in 1729, noticed that *Mimosa* and other plants cultivated in total darkness continue to open and close their leaves each day as if the sun were present, following the signals of their internal clocks. This was the first clue for the cyclic behavior that persists even in the absence of external signals. Since then, endogenous daily rhythms have been observed in a wide range of organisms at all levels of complexity and organization, including man himself.

Chronobiology is the branch of science, which objectively explores and quantifies mechanism of biological and medical time structure, including the important rhythmic manifestations of life, the internal clocks. "Chronobiology", generally is considered by many as a somewhat esoteric discipline, which deals with very minute variations, which mainly occur during the regulation processes to maintain the "steady state" of an organism. Not only pharmacological but also toxicological effects can be influenced by rhythms. The present document attempts to briefly discuss both the pharmacological and toxicological aspects of chronobiology and its therapeutic implications.

BIOLOGICAL RHYTHMS OR BIORHYTHMS

Biological rhythms are defined as regular variations in a biophysical or biochemical process with respect to time occurring in a predictable manner. Almost every physiological or behavioral variable is rhythmic at least in the healthy organism. The spectrum of rhythms is broad, ranging from cycles per second to cycles per year. The rhythm, that has received the greatest attention, is the *circadian* having a period, which corresponds approximately to the 24-hour rotation of the earth. One of the most fundamental rhythms of the life is the rhythm of generation or reproductive rhythm, namely estronuous or menstrual cycle. It is believed that the menstrual cycle in women is regulated by a time clock in the brain. The pituitary and the hypothalamus react cyclically to the ovarian steroid. Women's

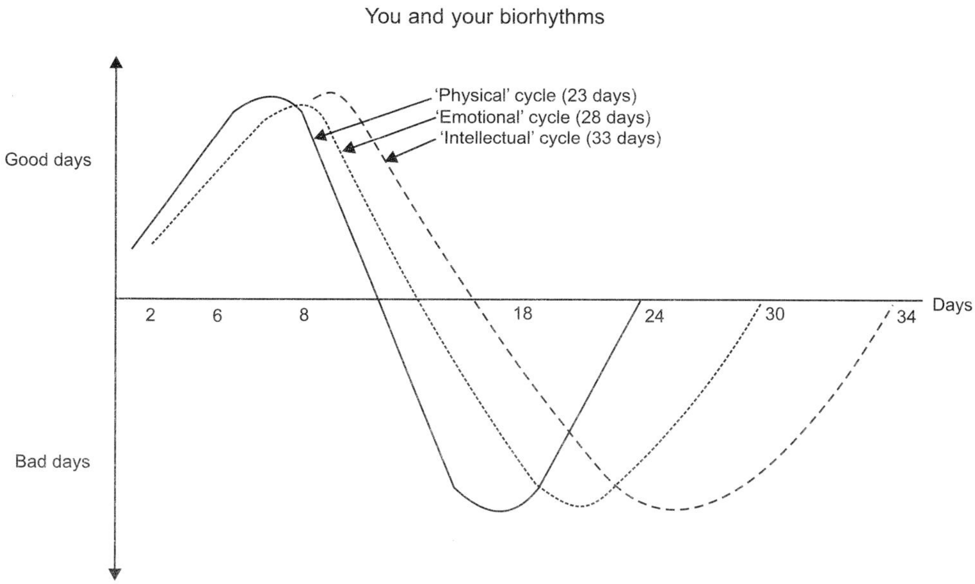

Fig. 1. Biorhythm cycle: Locate on the horizontal line the number of days you are into any of the three cycles then find the position of the appropriate curve above or below it. The curves above the horizontal lines indicate "good" days and below it indicate "bad" days, respectively. The cross-over points are "critical" days.

mood swings are often correlated with the monthly cycles with changing hormonal levels. Although there is no such defined periodicity defined for man, it is now believed that both men and women experience alike cyclic changes in their mental, physical and emotional performances (Fig .1).

The classification of various biological rhythms based on their periods are shown in Table 1. The most studied rhythm is the one which has a period of approximately 24 hours and which is referred to as circadian (from the Latin circa, meaning approximately and die, meaning day).

Table 1. Special domains of biological rhythms

Period (t)	**Nomenclature**	**Example**
t < 20 h	Ultradian	Cardiac rhythm
20 h < t < 28 h	Circadian	Activity and resting period
28 h < t < 2.5 days	Infradian	–
2.5 days < t < 7 days	Circaseptian	–
7 days < t < 20 days	Circavingintan	–
20 days < t < 30 days	Circatrigintan	Female menstruation in humans
t ≅ 1 year ± 2 months	Circannual	Sexual behaviour in some animals

The spectrum analysis of time-series data is determined by using various techniques such as chronogram or periodogram, the cosinor plot. Fig. 2 shows a chronogram of the circadian rhythm of the body temperature of a man determined at different times. This type of graph is a plot of estimated values or of the amplitude of a biological variable against a sequence of periods of time. This figure

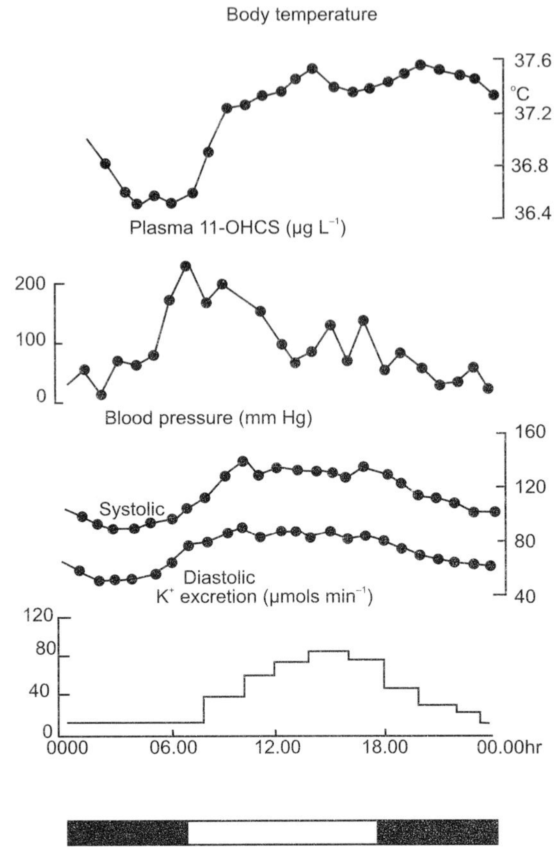

Fig. 2: The normal circadian variations in deep body (rectal) temperature, plasma 11-hydroxycorticosteroids (11-OHCS), blood pressure and urinary excretion of potassium in healthy subjects.

shows that during the sleep time, the activities of the body are minor, as a result of which the heart rate, blood pressure, body temperature and urinary excretion of potassium are low during these hours; by contrast, during the daytime we are active and take meals leading to rise in all the variables. Careful inspection of the data shows that blood pressure and body temperature continue to fall for some time after sleep onsets (23.30 h) and start to rise before waking takes place (07.30 h). These observations suggest that some factor in addition to direct influence of the sleep-wakefulness cycle is involved.

Some terminology used in chronogram include: acrophase, which is the time during the period when the particular biological activity is at its highest value; mesor is the rhythm-adjusted mean value of a biological function during a period. The mesor is the same value as the arithmetic mean between the maximal and minimal values of the temporal variation.

The statistical analysis based on least-squares method and the use of computers has provided another means of characterization of a rhythm by fitting the data to the best wave function of the form:

$$f(t) = M + A \cos(\omega t + \varphi)$$

where the value at time t of the function $f(t)$ is calculated from the values of the amplitude (A), mesor (M), acrophase (φ) and the angular frequency (ω, expressed in degrees/unit/time with 360° representing a complete cycle). This method is referred to as cosinor analysis. This method is helpful if the data are better described by a cosine curve than by linear relationship with time. The cosinor plot represents the period studied as a full circle in degrees (360°) and hours (24 h) so that 1 hour is equivalent to 15°. Midnight (00.00) and 0° are the starting reference of the period. Each vector represents a parameter studied. The direction of the vector expressed in negative degrees from 0° indicates the acrophase, and the corresponding clock hours are shown on the outside of the plot. The length of the vector indicates the amplitude, and the circle or ellipse around its end represents the 95% joint confidence region for the acrophase and the amplitude. The polar cosinor plot is shown in Fig. 3.

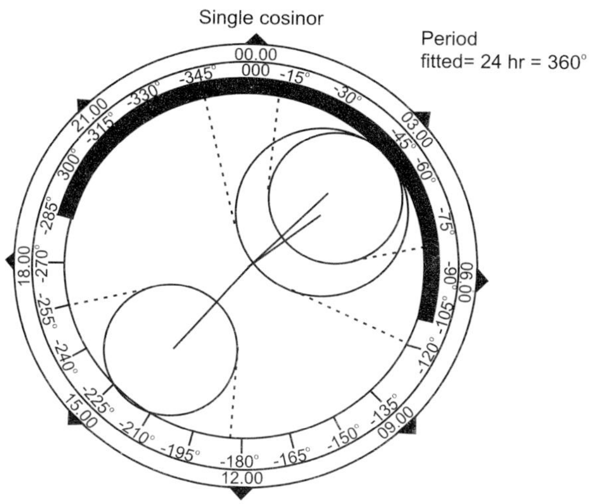

Fig 3. Polar cosinor plot of the circadian rhythm.

Very well studied is the circadian rhythm of the cholesterol-7α-hydroxylase, the rate limiting enzyme in bile acid synthesis in rats. This rhythm could be suppressed by hypophysectomy and by adrenalectomy. Therefore, it is assumed that this circadian rhythm might be controlled by an endocrine system. Melatonin is the neuroendocrine hormone produced by the pineal gland in the body and is involved in the maintenance of rhythms in the body. In other words, melatonin resets the biological clock. Therefore, melatonin is also known as "chronobiotic".

CHRONOPHARMACOLOGY

Chronopharmacology is the study of effects of drugs varying with biological timing and endogenous periodicities. It became recognized as a scientific domain of investigations only in the early 1970's. The goal is to improve our understanding of periodic and thus predictable (e.g. circadian) changes in both desired effects (chronoeffective) and undesired effects (chronotoxicity) of medication. Virey, as early as 1814, described that an individual who eats once a day in the evening had different temperament as compared to the one who eats once only in the morning. If Virey's early observations are extended to dosage schedules, it is quite appropriate to say that any given drug is not indicated equally at all the hours of the day. This questions the very validity of conventional dosage schedule such as three times or four times a day. A recent report also indicates that the survival of young mice can depend upon the timing of the food availability during the 24-hour-cycle. When the animals were housed individually at room temperature on a 12-hour-day/night cycle and were given food only for 4 hours at the beginning of the daily light cycle—a simulated late dinner, most of the mice died. On the other hand, if the food was made available for 4 hours at the beginning of the dark cycle—a simulated breakfast, most of the mice survived. Dosing-time dependent changes also include quantification of parameters characterizing endogenous circadian rhythms (CR), in terms of pharmacological effects.

CHRONOAVAILABILITY OF DRUGS

It is interesting to note that there is a tremendous variation in the bioavailability of drugs in the body depending on the time of the day when they are administered. For example, when animals are injected with a fixed dose of pentobarbital, they wake up with statistically significant different concentrations of drugs in the blood and the brain as a function of the time (of the day; 24 hr scale) of its administration. Fig. 4 shows how same dose of pentobarbitone at one time of the day produces hypnosis but the same dose given at other time of the day does not produce sleep. The same plasma concentration of pentobarbitone at one circadian time produces hypnosis while it fails to do so at other time.

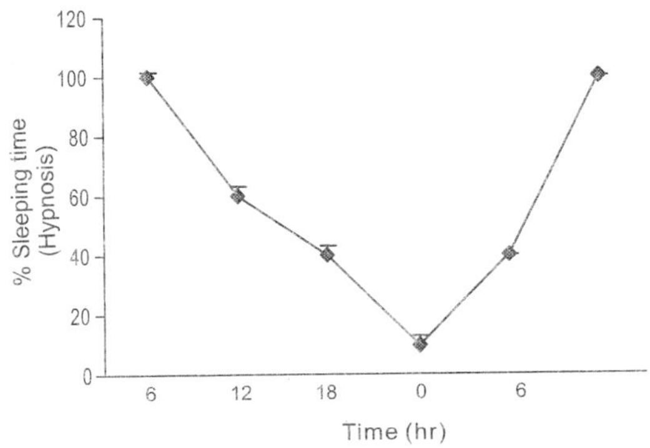

Fig. 4. Chronoavailability of pentobarbitone.

A circadian periodicity exists in the plasma levels of cortisol and the urinary excretion of 17-hydroxycorticosteroid in conscious man. The levels of circulating cortisol is reported to be maximum at about 6.00–8.00 a.m. and it decreases during the rest of the day. However, the urinary excretion of 17-hydroxy corticosteroid lags behind the plasma concentration by about 3–4 hours. This endogenous circadian cycle of the adrenal cortex is considered as one of the mechanisms, which contribute to change in susceptibility to some drugs and chemical agents. For example, the acrophase of skin susceptibility to histamine, to antigens (home dust extracts, penicillin) and in allergic patients, differ roughly by 12 hours (180°) from the acrophase of adrenal cortex secretion. It has been shown that the cutaneous response to histamine and other agents is highest when the blood corticosteroid levels are lowest. For example, when the patients were challenged with 0.1 ml of freshly prepared house

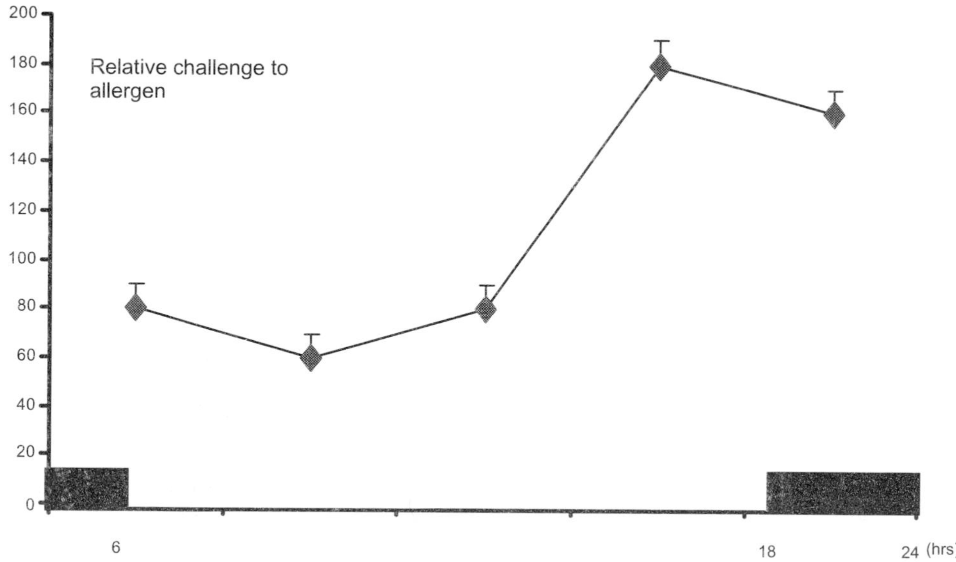

Fig 5. Relative challenge to allergen and the chronobiology.

dust extract, the maximum allergic reactions occur between 18.00–24.00 hr (Fig. 5) when cortisol levels are lowest. This has great relevance to the drug therapy of allergic diseases (such as asthma) and their management.

For example, the drug optimization by changing the physiological circadian rhythms is the treatment of asthma. The acrophase of the circadian rhythm in peak expiratory rate is shifted by the timed administration of prednisone to asthmatics.

Another good example is in the pathophysiology of heart diseases. Melatonin serves as an endogenous timer and is secreted by human pineal gland throughout the night but not during the day. In one group of patients with coronary heart disease, serum melatonin levels at night was more than five times lower than in control. The reason is unknown but melatonin reduces the sympathetic activity, which is higher during the day. In the morning, melatonin decreases and sympathetic activity rises. Such events may be important in heart disease patients who could not, therefore, reduce noradrenaline turnover in the heart. Therefore, more of the heart attacks are usually in the morning times as there are high levels of sympathetic discharges. Fig. 6 shows the melatonin levels in patients suffering from coronary heart diseases as compared with the control group.

CHRONOTOXICOLOGY

Halberg *et al.* (1959) were the first to report on circadian rhythms in drug toxicity by studying the mortality rate in groups of mice given the same lethal-range dose of ouabain but at different times of day. They found that the highest mortality to ouabain was at 08.00 hr when 60% of injected mice died, and the lowest susceptibility was at 20.00 hr when only 15% of an equivalent group of mice were killed due to the drug. Similar descriptive studies were repeated many times over the years with different drugs and xenobiotics. Table 2 lists the times of maximal and minimal mortality rates of

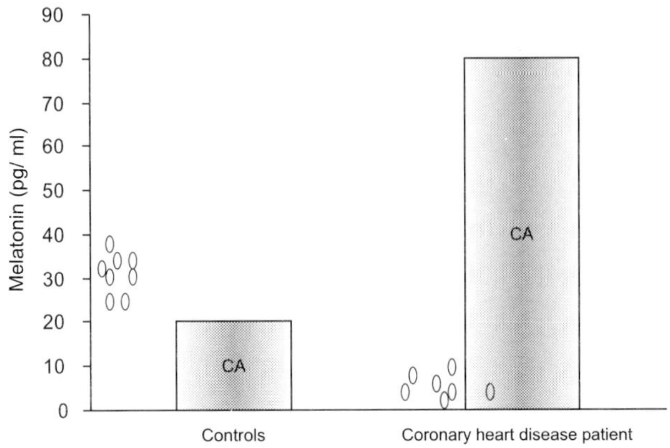

Fig. 6. Melatonin concentration at night (0200 hr) in patients with coronary heart disease and in controls.

selected drugs in mice, which are widely used in therapeutics. Two conclusions are deduced from these studies. First, each drug has its own particular rhythm of lethal toxicity and second, the temporal variations in mortality rate are prominent. For instance, the time-dependent variation in the LD_{50} value of propranolol was investigated in mice. A maximal LD_{50} value of about 130 mg/kg propranolol was obtained at 11.00 hr as compared to 95 mg/kg at 19.00 hr. These results demonstrate the unreliability of such a fundamental parameter as LD_{50} in drug research when the time of day of the experimentation is not considered.

Moreover, seasonal and circannual variations in the lethal toxicity of many compounds have also been reported further stressing the temporal variability of LD_{50} values. Few studies investigated the mechanisms of the chronotoxicity involved. Cal *et al.* (1985) correlated the chronotoxicity of cadmium, mercury and cisplatin and four aminoglycoside antibiotics reported by different authors. The data showed a similarity of rhythmic patterns among each series of chemically related compounds suggesting the involvement of one common mechanism of toxicity. They also studied temporal variations in the subacute nephrotoxicity of these compounds in rats and showed that the—peak and trough toxicity occurred at 20.00 hr and 02.00 hr, respectively, which corresponded to the times of lowest and highest rates of glomerular filtration. Thus, an increase in intraluminal concentration of the toxic agent consequent to a reduction in the glomerular filtration rate at 20.00 hr, and inversely, a decrease in the nephrotoxin concentration as a result of a higher filtration rate at 02.00 hr, could explain the chronotoxicity of these agents. The ototoxicity of gentamycin and kanamycin was also shown to vary depending on time of day in rats. Hearing loss was greater when treatment was given during the resting period of rats; i.e. 06.00–18.00 hr, than when given during the period of nocturnal activity. The reduced form of the tripeptide glutathione is the major nonprotein sulfhydryl compound present in the cell. One major function of the peptide is to form adducts with electrophilic chemicals or metabolites produced by the cytochrome P450 monooxygenase system thereby neutralizing the toxic effects of these intermediates as shown in Fig. 7. Examples of these toxic agents are chlorinated solvents, allyl alcohol, styrene, and drugs such as acetaminophen and anaesthetic agents.

One good example of the chronotoxicology is the circadian variations in the acetaminophen (paracetamol) lethality in mice. The lethality was inversely related to the temporal variations in the

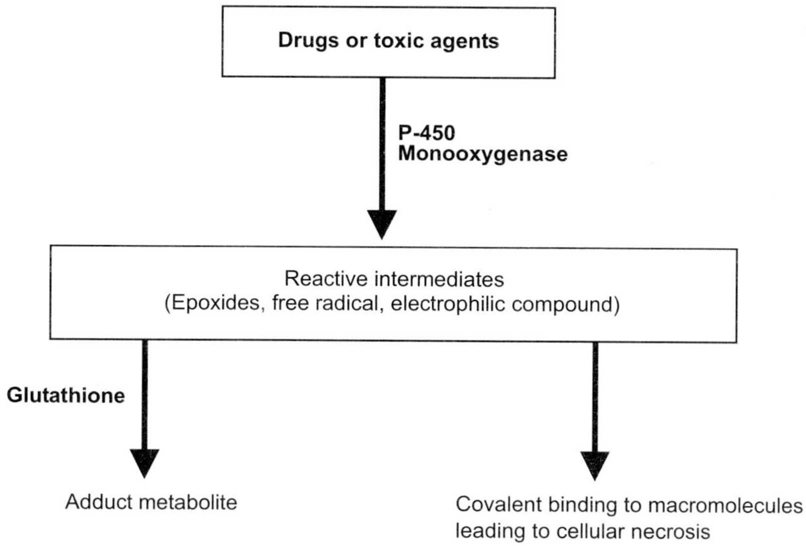

Fig. 7. Role of reduced glutathione in detoxification.

hepatic concentration of nonprotein sulfhydryl compounds as shown in Fig. 8. The time of maximum lethality occurred when concentration of nonprotein sulfhydryl compounds was at its lowest value and conversely, nadir mortality occurred when the nonprotein sulfhydryl concentrations were at the maximum (Table 2).

Table 2. Chronotoxicity of some drugs in mice

Drug	Time(s) of maximal and minimal toxicity		Ratio of maximal to minimal values
Atropine	18.00	06.00	1.2
Chlordiazepoxide	00.00	16.00	2.8
Cisplatin	14.00	06.00	2.0
Cyclosporine	15.00–21.00	09.00	50/0
Daunorubicin	20.00	06.00	1.25
Diazepam	21.00	09.00	1.4
Ethanol	20.00	00.00–04.00	7.00
Fluothane	00.00	08.00	4.0
Gentamicin	14.00	20.00	7.0
Hexobarbital	12.00	06.00	1.3
Indomethacin	13.00	21.00	2.0
Neostigmine	06.00	00.00	1.4
Phenobarbital	12.00	06.00	1.2

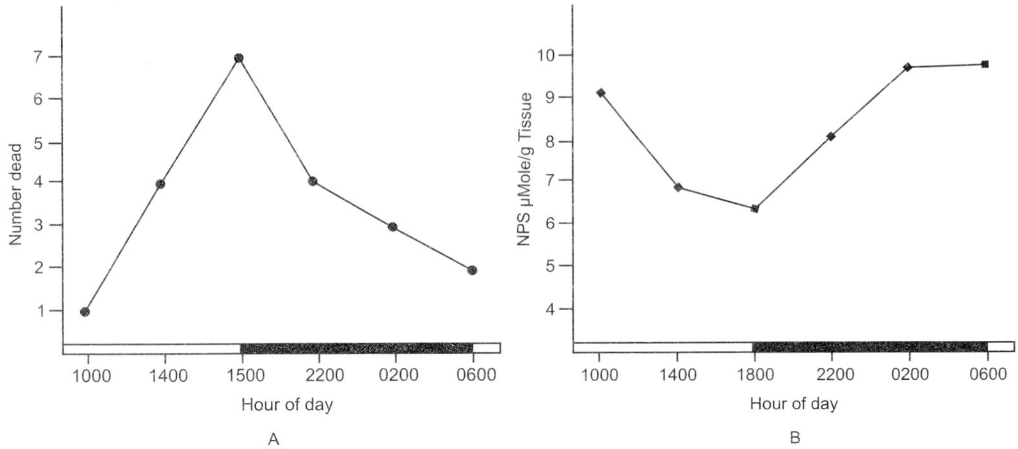

Fig. 8. Circadian variations in acetaminophen mortality (A) following intraperitoneal injection of 600 mg/kg, and in the hepatic nonprotein sulfhydryl (NPS) concentrations in mice (B).

Likewise, the chronohepatotoxicity of 1,1-dichloroethylene, allyl alcohol, chloroform and styrene has been found to be inversely related to the time-dependent variations in the hepatic concentrations of glutathione.

A circadian susceptibility to toxins has been reported in experimental animals. A lethal dose of *E. coli* endotoxin in one group of mice at one time was non-lethal to another group of animals at different circadian time. A calculation of biological potency of the endotoxin on the basis of these data would reveal two statistically significantly different potencies for endotoxin tested from the same batch but at two different circadian times. Such findings of dramatic differences in the susceptibility as a function of circadian rhythm of the organism will help in clinical medicine where *E. coli* septicaemia and endotoxin shock are still frequent causes of death in patients. Fig. 9 shows

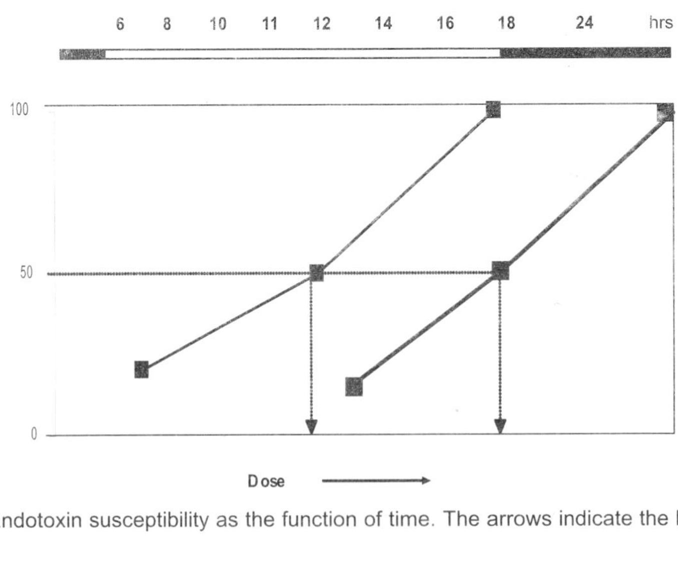

Fig. 9. Endotoxin susceptibility as the function of time. The arrows indicate the LD_{50} values.

the toxicity study done at different times of the day and gives different LD_{50} values for the same toxin/drug.

Similarly, animal experiments have demonstrated that one can achieve at least a reduction, if not elimination, of side effects of potent chemotherapeutic drugs or X-irradiation, 3 if rhythms are taken into account. An area of considerable interest is the application of chonopharmacologic concepts to chemotherapy and radiotherapy of cancer. One of the most common side effects with anti-neoplastic agents is the bone marrow suppression. It has been shown that bone marrow in rodents as well as in man is a circadian periodic organ. Careful controlled studies in animals have shown that animals irradiated during the middle of their dark cycle had high mortality rate as compared to animals irradiated during the middle of their light cycle (Fig. 10).

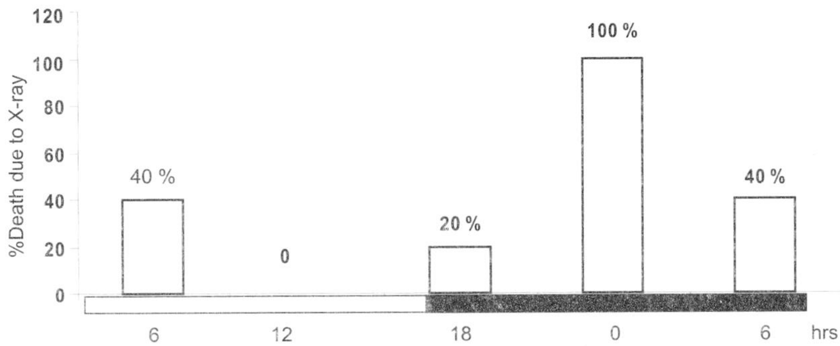

Fig 10. Showing the effect of radiation as function of time.

MELATONIN: A CHRONOBIOTIC DRUG

Melatonin (5-methoxy-N-acetyltryptamine) is a hormone which was first isolated from the bovine pineal gland in 1958. Pineal gland is a rudimentary gland (pine-shaped), a body attached to the roof of 3rd ventricle. Pineal gland acts as 'rhythm generator' or 'pacemaker' in birds, reptiles, amphibians, fishes and mammals. Melatonin, the principal component secreted by pineal gland is 'photoperiodic' in nature and influences the central nervous system (CNS) (melanosome-pigment granules). The principal innervations of the pineal gland emanates from the peripheral sympathetic nervous system. Light-dark information detected by the retina is relayed by the retino-hypothalamic pathway to the suprachiasmatic nuclei (SCN) of the hypothalamus, which acts as biological clock in mammals.

In mammals, the circadian clock is located in the hypothalamus within two paired, midline structure called the suprachiasmatic nuclei (SCN). Cells from the SCN project to the paraventricular nuclei (PVN) of the hypothalamus, fibres from this nucleus descend to synapse in the intermediolateral column of the spinal cord. The preganglionic sympathetic neurons from this region then project to the superior cervical ganglia (SCG) from which postganglionic neurons ascend along the internal carotid artery to enter the pineal gland where the synthesis of melatonin occurs. The circadian release of noradrenaline (NE) at night then determines the N-acetyltransferase activity and melatonin rhythms in the pineal gland

Melatonin synthesis is influenced by light and dark periods. Dark stimulates and light suppresses its synthesis. There are marked levels in nocturnal and low levels in daytime (undetectable). Retina

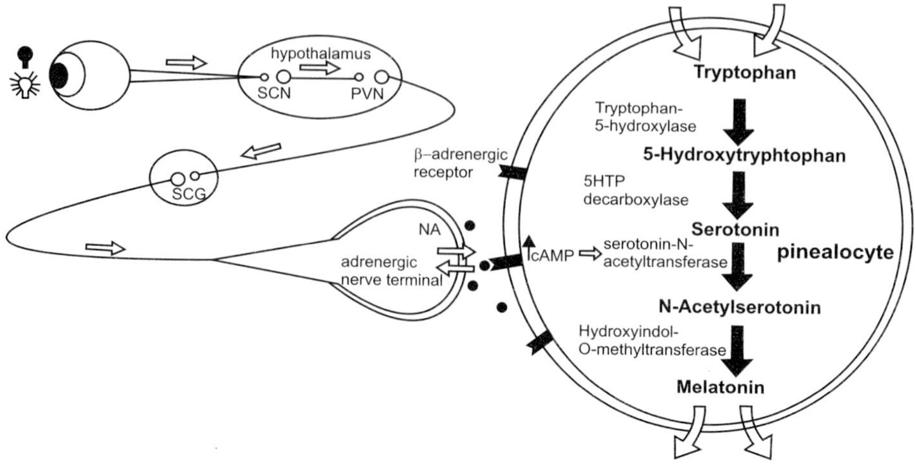

Fig. 11. Schematic view of melatonin synthesis in pineal gland.

has receptors and influences synthesis from pineal gland during dark in birds, reptiles, fishes, amphibians and mammals. It is synthesized from tryptophan with the help of enzyme tryptophan-5-hydroxylase. The synthesis of melatonin takes place primarily in the pineal gland, *via* a two-step process, acetylating of serotonin by arylalkylamine-acetyltransferase to give acetylserotonin, followed by methylation of the 5-hydroxy group by hydroxyindole-O-methyltransferase to yield melatonin. The whole process is summarized in Fig. 11.

Melatonin produces its action by acting on the three receptor subtypes i.e. MT_1, MT_2 and MT_3. The different characterizations of these receptors are shown in Table 3.

Table 3. Melatonin receptor subtypes

Receptor Subtypes (Other nomenclature)	**MT1 (ML1A, Mel1a)**	**MT2 (ML1B, Mel1b)**	**MT3 (ML2)**
Transduction Mechanism	Adenylyl cyclase (G_i)	Adenylyl cyclase (G_i)	Phosphoionositides turnover (Gq/11)
Distribution	SCN, pars tuberalis, hypothalamus, kidney	Retina, hippocampus	Hamster/mouse brain, kidney, testes
Selective Agonists	2-iodomelatonin> melatonin>> N-acetyl serotonin	Melatonin= N-acetyl serotonin	None
Selective Antagonists	GR128107 (Partial)	DH 97 (1218) 4-P-PDOT (1034) Luzindole (0877)	Prazosin (Non-selective)
Non-selective Ligands	6-chloromelatonin (0443) (agonist) 2-Iodomelatonin (0737) (agonist)	6-chloromelatonin (0443) (agonist) 2-Iodomelatonin (0737) (agonist)	6-chloromelatonin (0443) (agonist) 2-Iodomelatonin (0737) (agonist)
Physiological Functions			
Resynchronization of sleep and circadian rhythms disturbances (jet lag, sleep phase syndrome, hypnosis, mood disorder etc)			–

These melatonin receptors are known to interact with different receptor systems of the body and produces the physiological response. For example, melatonin interacts with GABA-benzodiazepine system resulting in hypothermia, sedation, hypolocomotion, anxiolysis, anticonvulsant, analgesic activity. Similarly, melatonin is known to interact with dopaminergic, serotonergic, cholinergic or opioidergic system of the body and influence a number of processes as shown in Fig. 12.

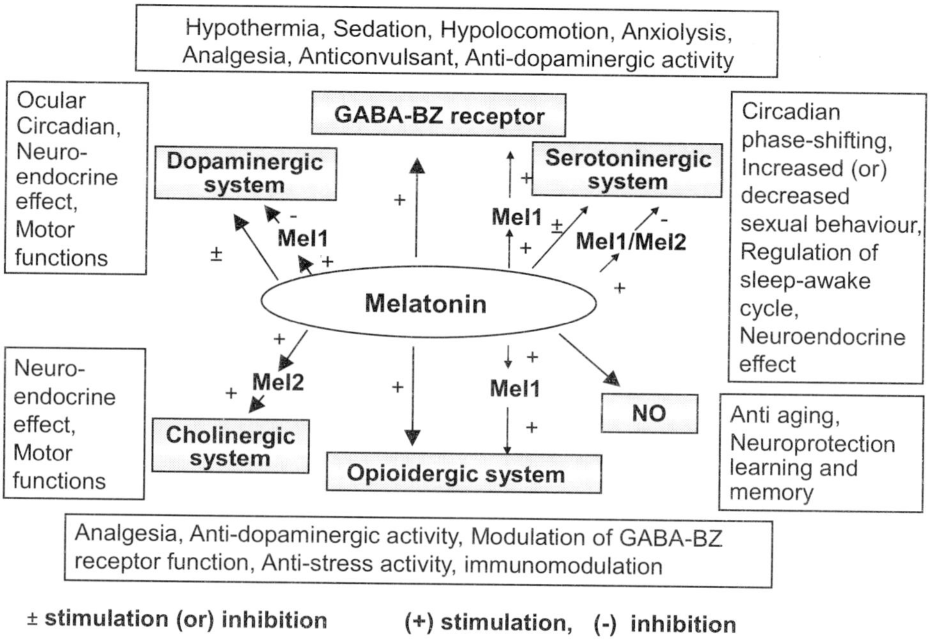

Fig 12. Neuromodulatory actions of melatonin.

OTHER CONCEPTS IN CHRONOBIOLOGY

Chronopharmacokinetics of a drug (or chronokinetics)

Several parameters can be used to characterize the pharmacokinetics of a drug. When the timing of administration is manipulated (e.g. single daily dose; 4 fixed times in the 24 hr scale each one being explored one week apart) statistically significant circadian rhythms can be demonstrated in the pharmacokinetics of substances such as ethanol (blood) and sodium salicylate (urine). This means that in the human organism (among other animal species), the metabolic fate of a pharmacologic agent as well as that of a nutrient is not constant, as it is a function of time. The metabolic pathways are neither open permanently nor open with a constant patency in the 24 hr scale among other bioperiodic domains. The chronokinetics of a drug includes both rhythmic (circadian) changes in the drug bioavailability (or pharmacokinetic) and in its excretion, urinary among others.

Chronesthesy of a biosystem

Chronesthesy of a biosystem (rhythmic changes in its susceptibility) includes both molecular and membrane phenomena, and related metabolic processes; it involves cells, tissues, organ, organ-systems as well as parasites, bacteria, etc.

Chronergy

It corresponds to rhythmic (e.g. circadian) change(s) either in any drug effect(s) or in biosystem response(s) to it.

Concepts in temporal pharmacology are summarized in Fig.13. In any pharmacological study, in addition to conventional items such as species, sex, age, dose, route and mode of the drug administration, one must take into consideration (a) environmental factors, mainly subject's synchronization and timing in administration of any drug (or food); (b) organismic bioperiodicity which involves rhythmic (e.g. circadian) changes in metabolic processes, biosystem chronesthesy, chronokinetics and chronergy of any drug (or food).

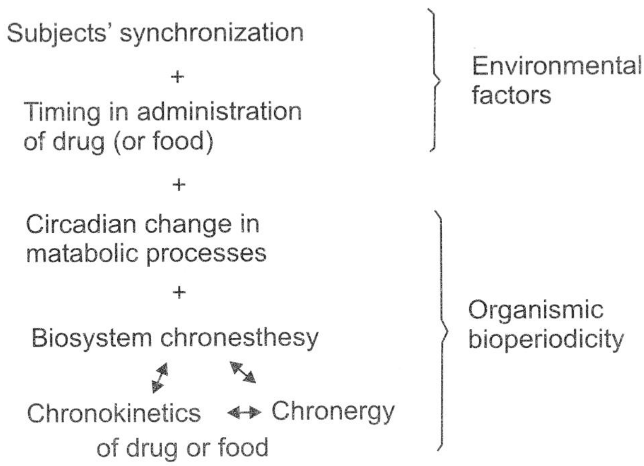

Fig 13. Concepts in temporal chronopharmacology.

CONCLUSION

One of the aims of the chronopharmacology and chronotoxicology is the better control and quantification of drug effects in a better way. The timing of experiments or administration of drug according to the inherent biological rhythms of organisms (synchronization) will improve the efficacy of testing methods in experimental pharmacology and the clinical effects of drugs and can prevent the toxicological outcome or undesirable effect of a therapeutic agent.

SUGGESTED READINGS

1. Belanger P.M, Chronopharmacology in drug research and therapy. Advances in Drug Research, 24, 1–71, 1993.
2. Borjigin J, Li X, Snyder S.H, The pineal gland and melatonin: Molecular and pharmacologic regulation. Annu. Rev. Pharmacol. Toxicol, 39, 53–65, 1999.
3. Kulkarni S.K, Chronopharmacology: A changing concept in drug administration. The Eastern Pharmacist. 25–27, 1979.
4. Kulkarni S.K, Shaji A.V, Behavioral pharmacology of melatonin, Methods Find. Exp. Clin. Pharmacol, 20, 237–247, 1998.
5. Kulkarni SK, Raghavendra V. Melatonin: A rhythm regulator. J Pun. Acad. Sci., 1, 199–203, 1999.
6. Sack R.B, Lewy A.J, Hughes R.J, McArthur A.J, Blood M.L, Melatonin as a chronobiotic drug. Drug News & Prespectives, 9, 325–332, 1996.

3

Bone Physiology in Health and Disease

Raj Bahadur, Harkiranpreet Kaur and D.V. Rai

INTRODUCTION

Bone is a rigid connective tissue having highest power of differentiation among the supporting tissues. It constitutes most of the skeleton of higher vertebrates. Bone consists of matrix made up of collagen and non-collagenous proteins (osteoid). Specialized cells called osteoblasts initiate bone formation, osteocytes monitor bone mechanical stresses, and osteoclasts resorb bone. Bones are continually being altered, reformed, remodeled, and refurbished. Bone is a magnificent combination of structural and physical properties—high tensile and compressive strength. Bone also possesses a little property of elasticity due to its collagen content. When there is an imbalance between bone resorption and bone formation i.e., either bone resorption is excessive, or bone formation is diminished, it leads to bone disease known as osteoporosis. It is a pressing health problem that is increasing tremendously. In United States, it causes more than 1.5 million fractures every year, most involving the spine, hip or wrist. Although often considered a women's disease, it also affects many men. About 8 million American women and 2 million American men have osteoporosis, and nearly 18 million more Americans may have low bone density. Even children are not immune. The announcement of new drugs to strengthen bones made fragile by osteoporosis is welcome news, but raising awareness about the disease and how to prevent it is equally urgent. At present, too many remain unaware of osteoporosis, fail to assess their risks or get the simple tests that can detect it. The objective of the present document is to explain the normal bone physiological structure and changes in bone physiology during diseased state i.e., osteoporosis. This document also deals with various risk factors and prevention and management of osteoporosis.

BONE PHYSIOLOGY

Bone is a dynamic tissue, which is composed of compact and spongy bone tissues that differ in their packing density.

Compact Bone (Cortical or Dense Bone)

It is the dense bone tissue forming the hard outer layer of bones with minimum gaps or spaces. It accounts for 80% of the total bone mass of an adult skeleton. It primarily consists of orderly osteons or haversian systems as shown in Fig. 1. Haversian systems are bundled together in parallel to form a compact bone. The osteon consists of osteonic (haversian) canal in the centre, surrounded by (concentric rings) lamellae interspread with osteocytes in lacunae. Haversian canals are the route for the blood vessels, lymphatic vessels and nerves fibers serving the osteons. Canaliculi (small canals) radiate from the haversian canal to the lacunae to provide nutrients and extract waste materials from an individual osteocyte. Haversian canals of adjacent osteons are interconnected by lateral canals. Interstitial systems, the remnants of old osteons are the irregularly shaped areas outside the circular osteons. The proportion of osteons to interstitial systems is fairly predictable with increasing age, making it possible to age a piece of bone fairly accurately. The osteons of the compact bone provide mechanical strength to manage compressive and bending forces.

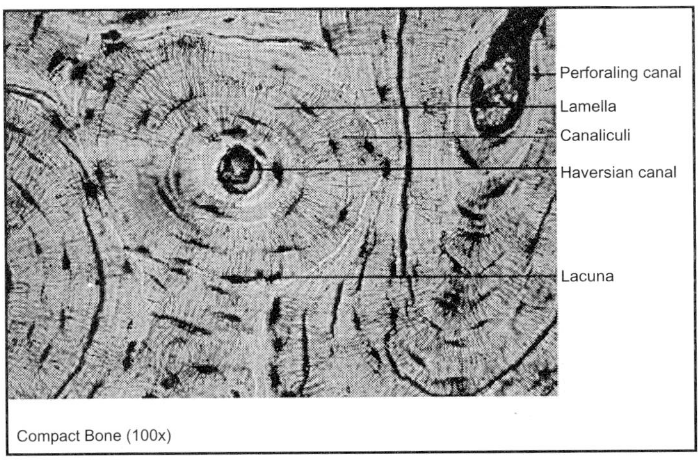

Fig. 1. Osteon—A basic structural unit of compact bone

Trabecular Bone (Spongy or Cancellous Bone)

It is a lighter and less dense bone as it consists of a network of rod (spicules) and plate (trabeculae)-like elements adjacent to marrow cavities having blood vessels as shown in Fig. 2. Trabecular bone accounts for 20% of total bone mass, but has nearly 10 times the surface area of compact bone. Regular arrangement of osteons and interstitial systems is absent but trabeculae are organized in a manner such as to provide maximum strength. They also realign in response to change in direction of stress. Trabeculae have osteocytes enclosed within lacunae. The osteocytes are nourished by diffusion from the bone marrow via the canaliculi, which connect to the adjacent cavities, instead of a central haversian canal.

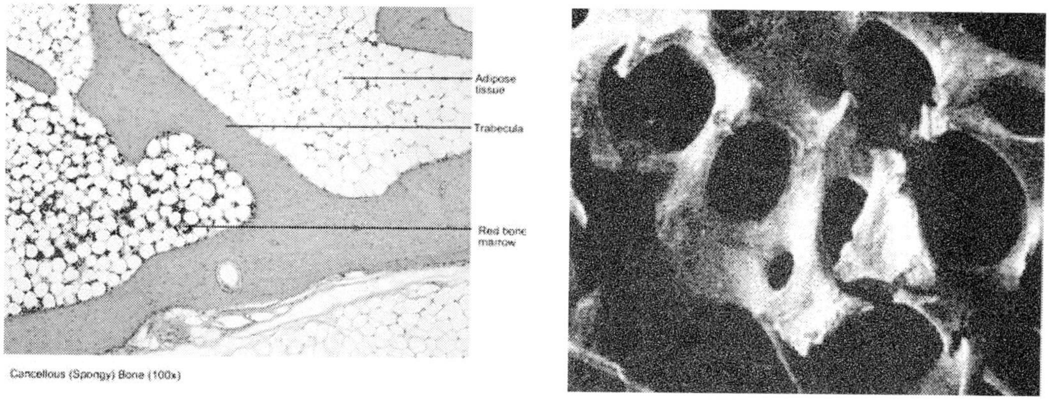

Fig. 2. Human cancellous bone.

TYPES OF BONE

There are five types of bones in the human body: long, short, flat, irregular and sesamoid.

Long Bones

They are longer than wide. Long bones consist of a shaft (the diaphysis) made up of compact bone, two articular surfaces (epiphyses) at each end, with metaphysis in between made up of spongy bone with a thin outer layer of compact bone. During growing age the epiphyseal phase separates epiphysis region from shaft and metaphysis. Elongation of the anatomic bone occurs at the epiphyses. The

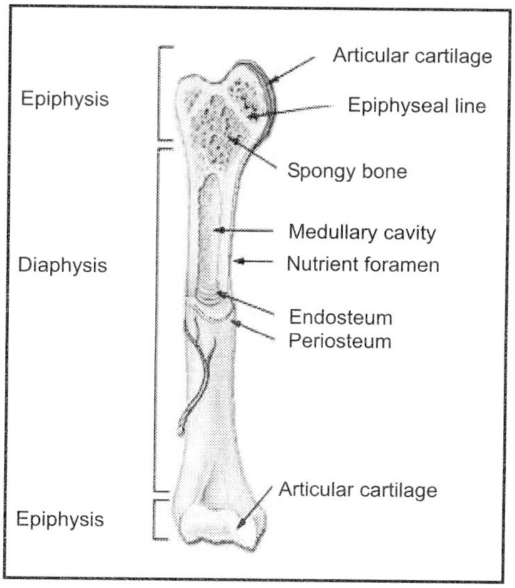

Fig. 3. Long bone.

outer surface of the bones (except where they interact with other bones through joints) is covered with the periosteum, which consists of a tough external collagenous layer and an internal osteogenic layer, as shown in Fig. 3.The periosteum is attached to the bone through sharpey's fibres and is richly supplied with blood, lymph and nerve vessels. In the centre of the long bones is medullary cavity, which cantains yellow marrow, whereas red marrow is present in the spongy bone of the epiphysis and metaphysis. Red marrow is present in the spongy bone at the epiphyseal ends of bone. Red marrow functions in the formation of red blood cells, white blood cells and blood platelets. With age, red marrow is largely replaced by yellow marrow for fat storage. The major bones of the arms and legs, femur, tibia, fibula, humerus, radius and ulna are long bones.

Short Bones

They have tubular shafts and articular surfaces at each end but are much smaller than long bones. They have only a thin layer of compact bone surrounding a spongy interior. These include metacarpals, metatarsals, phalanges and clavicle (paired).

Flat Bones

They are thin and generally curved, with spongy bone sandwiched between the two parallel layers of compact bones. These include innominate, scapula, ribs and sternum.

Irregular Bones

They are irregular and complicated in shape. These bones consist of thin layers of compact bone surrounding a spongy interior. These include the entire vertebral column including coccygeals carpal bones and tarsal bones as shown in Figs. 4 and 5.

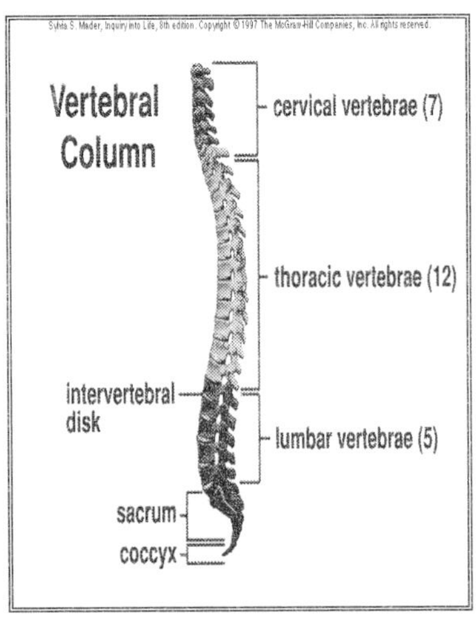

Fig. 4. Bones of the vertebral column.

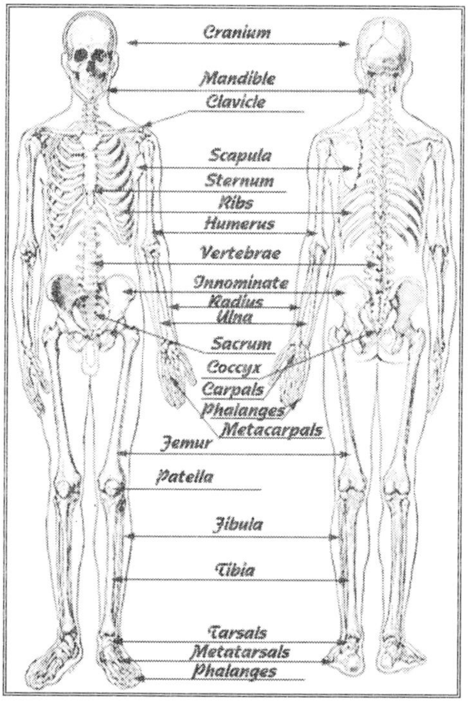

Fig. 5. The basic types of bones in human body.

Sesamoid Bones

They are embedded in tendons. They hold the tendon further away from the joint, the angle of the tendon is increased and thus the force of the muscle is increased. Examples of sesamoid bones are patella (knee cap) and pisiform, as shown in Fig. 5.

MOLECULAR STRUCTURE OF BONE

The bone matrix surrounding the cells is composed of two major components: inorganic-hydroxyapatite (65% of dry weight) and organic (25% of dry weight) 2–5% of which are cells.

Inorganic: The inorganic matter, which provides toughness and rigidity to bone, mainly consists of crystalline mineral salts, citrate ions, carbonate ions and calcium phosphate, called hydroxyapatite $[Ca_{10}(PO_4)_6OH_2]$. Osteoblasts form the initial matrix as unmineralized osteoid. During mineralization, alkaline phosphatase is secreted by osteoblast vesicles. This enzyme cleaves the phosphate groups and acts as the foci for calcium and phosphate deposition, rupturing the vesicles and consequently acts as a centre for crystals to grow on. The crystallization may involve the interaction of collagen fibres with the glycosaminoglycans (GAGs) of the matrix material.

Organic: The organic matrix consists of collagen fibres embedded in a ground substance rich in proteoglycans-sulfated GAGs (chondroitin sulfate, keratin sulfate) and non-collagenous proteins as

Table 1. Non-collagenous proteins in bone

Protein	Role
Osteonectin (SPARC)	Calcium, apatite and matrix protein binding Modulates cell attachment
α-2-HS-Glycoprotein	Chemotactic for monocytes Mineralization via matrix vesicles
Osteocalcin (Bone GLA protein)	Involved in stabilization of hydroxyapatite Binding of calcium Chemotactic for monocytes Regulation of bone formation
Matrix-GLA-protein	Inhibits matrix mineralization
Osteopontin (Bone Sialoprotein I)	Cell attachment (via RGD sequence) Calcium binding
Bone Sialoprotein II	Cell attachment (via RGD sequence) Calcium binding
24K Phosphoprotein (α-1(I) procollagen N-propeptide)	Residue from collagen processing
Biglycan (Proteoglycan I)	Regulation of collagen fibre growth Mineralization and bone formation Growth factor binding
Decorin (Proteoglycan II)	Collagen fibrillogenesis Growth factor binding
Thrombospondin and Fibronectin	Cell attachment (via RGD sequence) Growth factor binding Hydroxyapatite formation
Others (including proteolipids)	Mineralization
Growth factors:	
IGFI & IGFII	Differentiation, proliferation and activity of osteoblasts
TGF α	
Bone morphogenetic proteins (BMPs)	Induction of bone and cartilage in osteogenesis and fracture repair

shown in Table 1. Ninety percent of the organic matrix constitutes type I collagen. It has a triple helical structure, and specific points along the collagen fibres serve as nucleation sites for the bone mineral crystals. The collagen fibres are orderly in their arrangement. Collagen-I is made intracellularly as tropocollagen and then exported to form fibrils. Various growth factors are required in organic matrix formation. Elastic fibres are minimal or absent. Bone matrix is distinguishable from other cells in being hard. Fibrous component provides tensile strength and flexibility to bone.

CELLULAR STRUCTURE

Four cell types peculiar to an actively growing bone are:

Osteoprogenitor Cells

Osteoprogenitor cells are the stem cells derived from embryonic mesenchymal cells that give rise to osteoblasts under the influence of growth factors such as bone morphogenetic factors such (BMP's).These are spindle-shaped cells with elongated or oval-shaped nuclei found near bone surfaces, in the inner portion of the periosteum, in the endosteum and within the vascular canals of compact bone. Osteoprogenitor cells are active during bone growth and are reactivated during the repair of a fractured bone.

Osteoblasts

Osteoblasts are mononucleate bone-forming cells, which descend from osteoprogenitor cells as shown in Fig. 6. Osteoblasts are the immature bone cells. These are found on the margin of growing bone where osseous matrix is being deposited. These cells get arranged in epitheloid layer of cuboidal or columnar cells during deposition of the new protein matrix called osteoid. Matrix components secrete type I collagen, proteoglycans, osteocalcin, osteonectin, osteopontin and hormones such as prostaglandins to act on the bone itself. When osteoid mineralizes, it becomes bone. Osteoblasts secrete basophilic proteins and contain the enzyme alkaline phosphatase, which has a role in the mineralization of bone. In adults majority of the bone surfaces that are not being remodeled are covered by bone lining cells, inactive form of osteoblasts that act as a barrier for ions. In short, osteoblasts actively synthesize the bone matrix material and the collagen fibres.

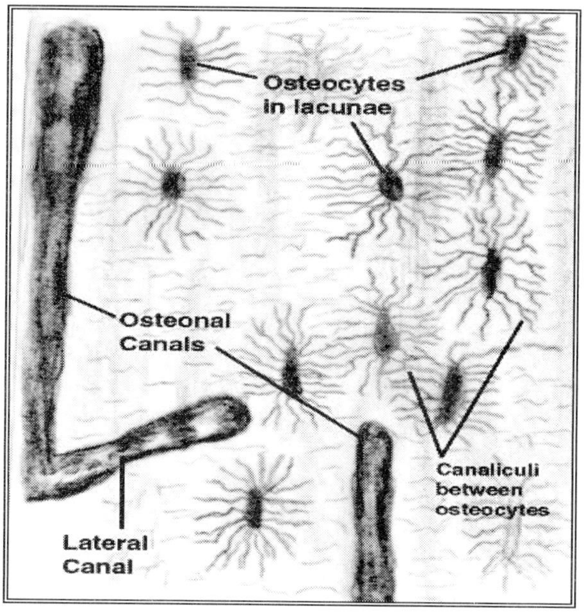

Fig. 6. Osteocytes.

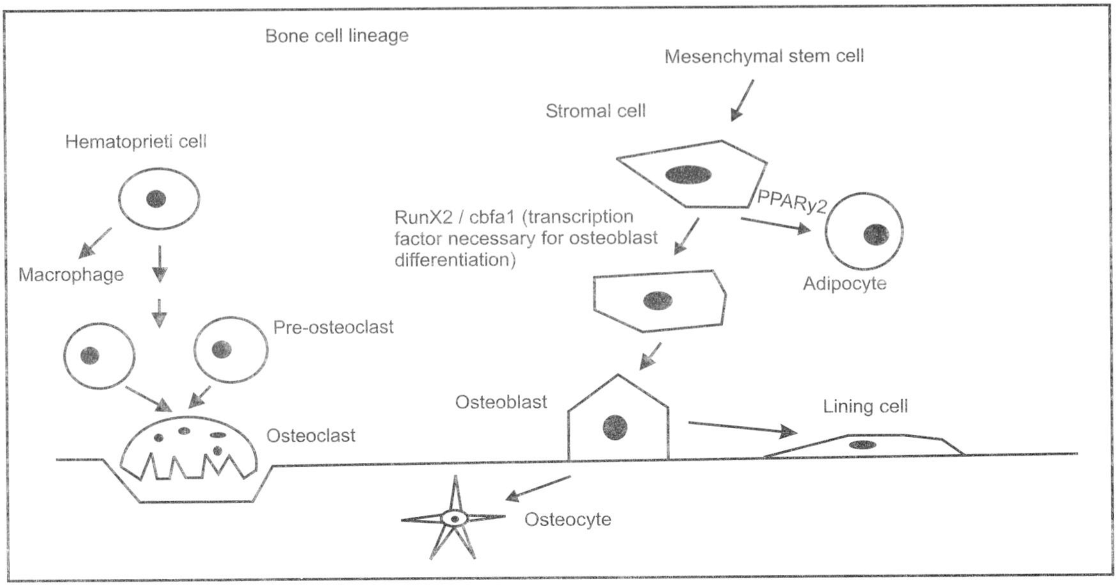

Fig. 7. Pathway of osteogenesis.

Osteocytes

Osteocytes are the principal cells of mature bone. These cells originate from osteoblasts, which become trapped in the lacunae within the calcified matrix. Osteocytes have numerous slender processes that extend for some distance into canaliculi, which radiate out from the lacunae as shown in Fig. 6 and are used for exchange of nutrients and wastes. Gap junctions are present at points of contact between processes and canaliculi. But in mature bone, processes are absent, therefore, the exchange of nutrients between osteocytes and blood stream is carried out by canaliculi. Formation of bone, matrix maintenance and calcium homeostasis through mechano-sensory mechanisms are the main functions of osteocytes.

Osteoclasts

Osteoclasts are the agents of vascularization of bone and bone resorption, by which they help in bone remodeling. Osteoclasts originate by fusion of macrophage-like mononuclear precursors as shown in Fig. 7. Osteoclasts are potential catabolic cells, equipped with engulfment strategies similar to circulating macrophages. These are large, multinucleated cells occupying shallow concavities called Howship's lacunae. These lacunae, or resorption pits, are left behind after the breakdown of bone. Osteoclast seals to bone matrix, acidifies extracelluler compartment, dissolves the mineral phase and secretes collagenase and other proteolytic enzymes, such as tartrate-resistant acid phosphatase on the bone matrix to degrade bone proteins. In a single day, one active osteoclast is capable of degrading the total amount of bone that can be produced by over hundred osteoblasts. The specific osteoclast differentiation factor RANKL, its receptor RANK, and its inactivating soluble decoy receptor osteoprotegerin (OPG), locally regulate the populations of osteoclasts in various regions of the skeleton.

BONE FORMATION

New bone arises by the replacement of some pre-existing tissue. The formation of bone occurs by two methods: intramembranous and endochondral ossification. The two methods differ in the replacing material for bone formation.

Intramembranous Ossification

Intramembranous ossification primarily occurs in flat bones of the skull (e.g., the frontal, parietal, occipital, and temporal). Here, the pre-existing material for bone formation is embryonic mesenchyme. In intramembranous ossification, bone formation takes place at numerous sites simultaneously, one characteristic that distinguishes it from endochondral ossification. The steps in intramembranous ossification are:

1. Development of ossification centre—Group of mesenchymal cells within a highly vascularized area of the embryonic connective tissue proliferates and cells differentiate directly into osteoblasts.
2. Collagen fibres are made and calcification of matrix occurs.
3. Osteocytes formation—Osteoblasts are trapped in their lacunae, causing their conversion to osteocytes.
4. Woven bone is formed and blood vessels are incorporated between the woven bone trabeculae and form the hematopoietic bone marrow.
5. Development of periosteum and this woven bone is replaced by mature lamellar bone during remodeling.

Endochondral Ossification

Endochondral ossification occurs in the long bones as shown in Fig. 8. Here the pre-existing tissue is hyaline cartilage, which is replaced to form bone. The steps in endochondral ossification are:

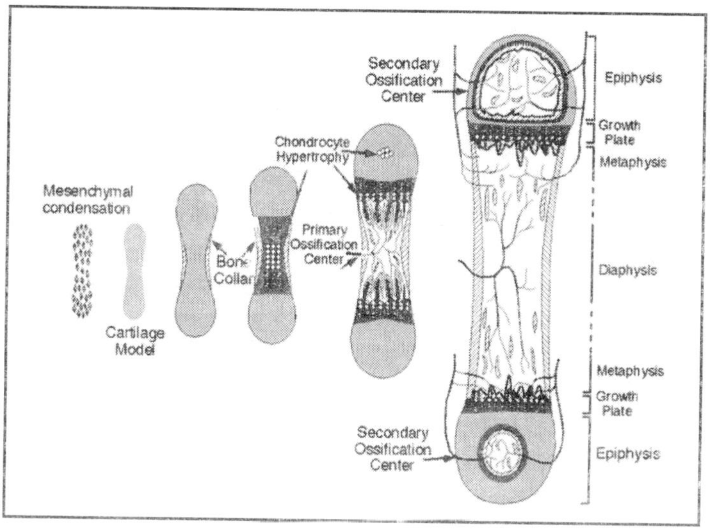

Fig. 8. Different stages of chondrocyte differentiation involved in endochondral bone formation.

1. Development of cartilage model—The mesenchymal cells proliferate to form prechondroblasts and then chondroblasts. Then these cells secrete the cartilaginous matrix and form chondrocytes when get trapped in the lacunae.
2. Growth of cartilage model—Growth is of two types: appositional growth, where mesenchymal cells at the periphery of cartilage continue to differentiate, and interstitial growth, where new matrix is formed in between chondrocytes and cells proliferate.
3. Development of the primary ossification center—It consists of five zones: zone of resting cartilage, zone of proliferation, zone of hypertrophy, zone of cartilage calcification and zone of ossification. The chondrocytes further proliferate and become hypertrophic, the mineral matrix is formed and calcification of cartilage occurs and osteoclasts act on. Then the woven bone is formed on the top of calcified cartilage by osteoblast differentiation and later lamellar bone is formed from woven bone.
4. Development of medullary cavity—As calcification of woven bone occurs, blood vessels and osteoclasts enter in the primary ossification centre and bone marrow is formed in the centre of the new formed bone. Osteoclasts continue to resorb calcified cartilage.
5. Development of the secondary ossification center—It is formed at the epiphyseal ends of the cartilaginous model. Here also trabeculae and marrow are formed.
6. Formation of articular cartilage and epiphyseal plate—Epiphyseal (growth) plate is formed between the two ossification centres.

Longitudinal bone growth occurs by endochondral ossification, where differentiation of chondrocytes, cartilage mineralization and remodeling occurs. Intramembranous ossification occurs on surface of the bone, proper shape and thickness of the bone is maintained. Primary ossification occurs during fetal development and is responsible for formation of diaphysis region of long bone, short bones and certain parts of irregular bones, while secondary ossification occurs after birth and forms the epiphysis region of a long bone and the extremities of irregular and flat bones, and these processes continue in the epiphyseal disk until the disk "closes". By 25 years of age, all the cartilage gets replaced by bone.

Remodeling of Bone

Bone remodeling maintains the mineral homeostasis, bone mass and the skeleton integrity throughout the lifetime, although it gets slowed down with age. Osteoclasts and osteoblasts act in a coupling manner i.e., breakdown and reconstruction occurs simultaneously. This process takes about 100 days and is under PTH control. The basic principle of cellular activity at the remodeling site is the Activation-Resorption-Formation (ARF) sequence. In remodeling process, first the preosteoclasts are stimulated and differentiate under the influence of cytokines and growth factors into mature active osteoclasts and then these osteoclasts digest mineral matrix (old bone). After the end of resorption osteoblasts come into action by synthesizing new bone matrix.

MAIN FUNCTIONS OF THE BONES

1. To provide structural support to the body.
2. To provide protection to internal organs, such as the skull protecting the brain or the ribs protecting the heart and lungs.

3. To produce blood cells in a process called haematopoiesis, within red marrow of bones.
4. To provide fat storage site (yellow marrow).
5. To act as a mineral reservoir most notably for calcium and phosphorous. It acts as a mobilizable source of calcium.
6. To provide attachment points for skeletal muscles, tendons, ligaments and joints to enable movement in body.
7. To provide acid-base balance; bone buffers the blood against excessive pH changes by absorbing or releasing alkaline salts.
8. To act as a detoxifying agent by storing heavy metals and other foreign elements, removing them from the blood and reducing their effects on other tissues.
9. To provide help in sound transduction (auditory ossicles for hearing in ear).

BONE DISEASE—OSTEOPOROSIS

Osteoporosis is a disease of bone in which the bone mineral density (BMD) is reduced, bone micro architecture is disrupted, and the amount and variety of non-collagenous proteins in bone is altered. Osteoporotic bones are more susceptible to fracture. Osteoporosis is defined by the World Health Organization (WHO) as either a bone mineral density 2.5 standard deviations below peak bone mass (20-year-old sex-matched healthy person average) as measured by DEXA, or any fragility fracture. Due to its hormonal component, more women, particularly after menopause, suffer from osteoporosis than men. In addition, it may be caused by various hormonal conditions, smoking and medications (specifically glucocorticoids). This disease can occur because bones are small and weak to start with (partly because of inadequate calcium intake) or through later bone loss. Whether it happens from an initial deficit or accelerated bone loss, it involves a reduction in bone density and quality. Bone is not static but constantly in flux, with a dynamic "remodeling" process that repairs damage and replaces brittle areas with new bone. Basically, osteoporosis stems from an imbalance in this bone-remodeling process, whereby bone loss or "resorption" (by the osteoclasts) outpaces its build-up (by the osteoblast cells). It results in loss of the normal density of bone, which becomes more porous and more compressible like a sponge. Normal bone is composed of protein collagen and calcium. Osteoporosis depletes both the calcium and the protein from the bone, resulting in either abnormal bone quality or decreased bone density. Bones that are affected by osteoporosis can fracture with only a minor fall or injury that normally would not cause a bone fracture. The fracture can be either in the form of cracking (as in a hip fracture) or collapsing (as in a compression fracture of the vertebrae of the spine). The spine, hips and wrists are common areas of osteoporosis-related bone fractures, although fractures can also occur in other skeletal areas such as the ribs. Photographs of normal and osteoporotic bone are shown below in Fig. 9 and 10. Collapse of a vertebra ("compression fracture") can cause one or a combination of the following: acute onset of back pain; a hunched forward or bent stature; loss of height; limited mobility and possibly disability.

Fractures of the long bones acutely impair mobility and may require surgery. Hip fracture, in particular, usually requires prompt surgery, as there are serious risks associated with a hip fracture, such as deep vein thrombosis and a pulmonary embolism. While treatment modelities are becoming

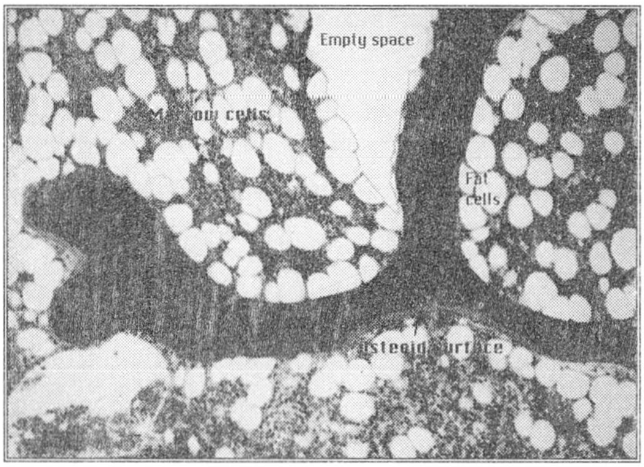

Fig. 9. Normal bone.

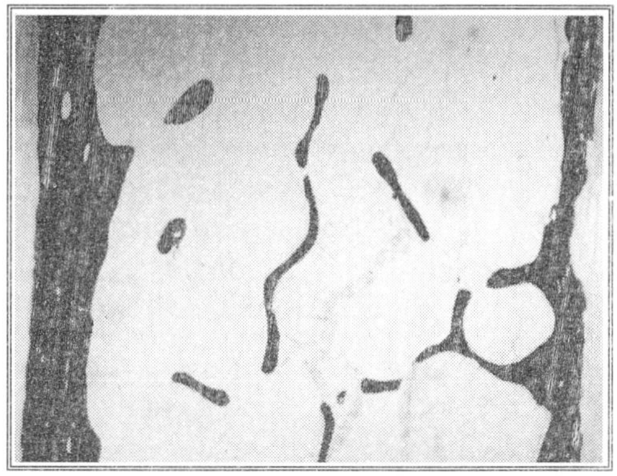

Fig. 10. Osteoporotic bone.

available (such as the bisphosphonates) prevention is still considered the most important way to reduce fracture.

Types of Osteoporosis

Osteoporosis can be classified as either primary or secondary osteoporosis. Primary osteoporosis can be further divided into two types:

Type 1 osteoporosis is also known as postmenopausal osteoporosis, as it occur mostly in women of age between 50–75 years. In this, estrogen levels decrease, which lead to bone resorption and calcium depletion making the bones weak and brittle and increasing the chances of fractures related to vertebrae, hip, wrist and forearm due to minor falls and accidents.

Type 2 osteoporosis is also known as senile or age-related osteoporosis, as it occurs with increase in age after 70. Although it affects both genders, it is more prevalent in women. During this the remodeling process of bone gets disrupted as bone resorption starts to overcome bone formation. Fractures occur in femoral neck, proximal humerus, proximal tibia and pelvis.

Secondary osteoporosis occurs mainly due to genetic factors, certain diseases, medications or changes in lifestyle.

Risk Factors

Overall risk factors that predispose to the development of osteoporosis, either primary or secondary, can be split between modifiable and non-modifiable.

Non modifiable:

1. Advanced Age and Gender: During growth, maturation and ageing, bone undergoes remarkable changes both at the structural level and the tissue morphological level. Bone loss occurs as part of the normal ageing process. The skeleton is being renewed continuously in the remodeling process, which prevents accumulation of damage and adapts the bone architecture to external loads. But at approximate age of 35, in both sexes, after peak bone mass has been attained, a gradual decrease in bone volume occurs, due to deficit of osteoblast formation relative to osteoclast resorption. This bone loss and the concomitant changes in bone architecture decrease the strength and stiffness of the bone. With ageing, there is continuous decrease in both compact and trabecular bone as revealed by histological and radiogram metric studies. The decrease in the bone density of postmenopausal women is a much more serious problem due to decrease in estrogen levels, as compared to men. However, by age 65 or 70, the gap closes where both genders become equally prone to bone loss which leads to osteoporosis in future.

2. Ethnicity and Heredity: Racial and ethnic characteristics largely influence the risk of osteoporosis. Caucasian and Asian or white women have lower peak bone mass than African, African-American, American Indian, east Indian, and Latin women, which is due to higher levels of bone turnover. A family history of weak bones, thin body frame, particularly in females increases the risk of bone loss and in future osteoporosis. Each bone cell population is under the direction of various endocrine factors and is influenced by various local biochemical and bioelectrical factors. Besides this cells themselves are endowed with genetic instructions that determine their ability to form, resorb or maintain bone. The heritability of peak bone mass has been studied extensively, but the genetic contribution to bone loss is largely unknown. Additional work to locate genes affecting bone loss may be fruitful. A flowchart of the non-modifiable risk factors of the osteoporosis is shown in Fig. 11.

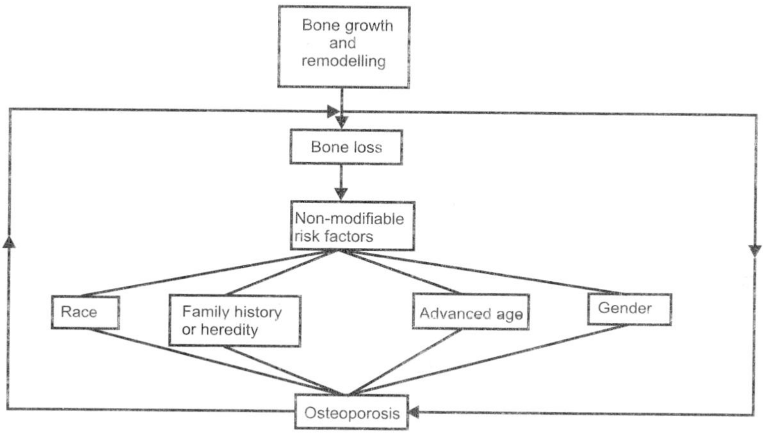

Fig. 11. Flowchart of non-modifiable risk factors of osteoporosis.

Potentially Modifiable

1. Nutritional Factors

Insufficient nutrient intake is an important modifiable factor for bone health, as it is associated with low bone mass and increased risk of osteoporosis. Lack of vitamins and minerals in the diet results in overt symptoms of deficiency.The process of bone formation requires an adequate and constant supply of nutrients, such as calcium, protein, magnesium, phosphorus, vitamin D, potassium, and fluoride. However, there are several other vitamins and minerals needed for metabolic processes related to bone, including manganese, copper, boron, iron, zinc, vitamin A, vitamin K, vitamin C, and vitamin B. Women over 70 do not get sufficient amount of these nutrients in their diet and they are not able to absorb these nutrients properly. This is the reason they are more prone to osteoporosis.

(a) Low calcium intake: Calcium, a major structural component of bone tissue, is deposited in bone in the form of a mineral complex called hydroxyapatite. Ninety-nine per cent calcium is present in bone and teeth. Calcium also plays a role in nerve and muscle functioning. Deficiency of calcium leads to lowering of calcium in blood level, stimulating parathyroid gland to produce parathyroid hormone (PTH) which in turn causes calcium to be released from the bone 'reservoir' into the blood stream to maintain the level of calcium in body. This process makes bone thin and brittle. Calcium absorption efficiency is high during teenage years, the time of rapid growth of the skeleton, and it decreases with increase in age. Therefore, low calcium intake during childhood and premenopausal years is linked to lower bone mass.

(b) Low vitamin D intake: Vitamin D is a fat-soluble vitamin that is essential for maintaining normal mineral metabolism, maintaining intracellular levels of calcium, mineralization of bone tissue, cell differentiation, muscle growth and strength. Low intake of vitamin D leads to decrease in calcium absorption and the consequent increase in PTH levels causes increased bone resorption. In children, severe vitamin D deficiency results in inadequate mineralization of the bone matrix, leading to growth retardation and bone deformities known as rickets. In adults, the same condition is known as osteomalacia. There are some evidence, which show that mothers deficient in 25-hydroxy vitamin D in pregnancy give birth to children with reduced bone mass, which could in turn be a risk factor for bone loss and osteoporosis later in life.

(c) Low vitamin K intake: Vitamin K is a fat-soluble vitamin, essential for functioning of several proteins involved in blood clotting and bone health and also in reducing calcium loss. Three known bone matrix vitamin K dependent proteins important for bone formation are—osteocalcin, matrix Gla protein (MGP) and anticoagulant protein S. Osteocalcin is the second most abundant protein in bone after collagen, synthesized by osteoblasts. Vitamin K is a cofactor of γ-carboxylase, which mediates the conversion of undercarboxylated osteocalcin to carboxylated osteocalcin by transforming three glutamic acid residues of osteocalcin into carboxyglutamic acid residues, which have a high affinity for calcium ions in hydroxyapatite crystal formation, thus help in bone mineralization. There are experimental reports, which suggest that low dietary intake of vitamin K could contribute to low BMD and increased risk of fragility fractures in the elderly, due to reduced functioning of osteocalcin, MGP and anticoagulant protein S. Although role of anticoagulant protein S in bone metabolism is unclear, but evidence suggests that children with inherited protein S deficiency suffer complications related to increased blood clotting and decreased bone density. The literature suggests that less vitamin K intake affects the bone metabolism, although the possible mechanisms of this affect need to be further explored.

(d) High homocysteine levels and B vitamins: Vitamin B_6, Vitamin B_{12} and folic acid are important for homocysteine metabolism, and their deficiency is associated with an increase of homocysteine levels in the blood. Homocysteine is a metabolite of amino acid L-methionine. It is linked with

increased risk of cardiovascular disease and increased hip fracture rate by interfering with the formation of collagen protein. It has been reported that in older women, low serum levels of vitamin B_{12} (below 280 pg/ml) are associated with increased hip bone loss. Both vitamin B_{12} and folate have been reported to decrease with age. Therefore, it is still not clear whether it is vitamin B_{12} or folate levels that are related to bone mineralization. Even the adequate mechanism of the interrelationship between B vitamins and bone loss is not known.

(e) High vitamin A levels: The role of vitamin A in bone loss is very controversial. Vitamin A is present in two forms—carotenoids as a precursor of vitamin A in foods of plant origin and retinol in foods of animal origin. It is already known that consumption of very high amounts of vitamin A has adverse effects on bone, liver and skin. An observational study among postmenopausal American women found an association between high levels of vitamin A intake and hip fracture risk. Similarly, a study found an association between blood retinol levels (but not blood carotenoid level) and fracture risk in Swedish men. In concordance with the above given results, it can be said that beta carotene does not carry the same risk to bones as that of retinol. However, in another study of elderly British women no indication of increased risk of hip or other fractures with higher vitamin A intake was found, either from the diet or fish oil supplements. Therefore, further research is clearly needed in this area.

2. Lifestyle

(a) Low body weight and weight loss diets: Being thin or underweight or bone mass index (BMI) < 22 and weight below 58 kilograms is associated with more risk of osteoporosis. Body fat increases certain types of estrogen, so thin women tend to have lower levels of estrogen. As a result, age- and estrogen-related declines in bone mass will have a greater effect on them. Some diets can also cause bone loss. Loosing a lot of weight, putting it back on, and losing it again can lead to poor nutrition and shifts in hormone levels. These ups and downs can cause significant bone loss. Moreover, the overweight people who restrict calories to lose weight experience greater loss of bone mineral density than those who exercise for weight loss.

(b) Lack of exercise: Prolonged bed rest or inactivity (lack of bone loading or exercise) also contributes to bone loss as bones are not working very hard and they lose strength more easily. A sedentary lifestyle may lead to a decrease in osteoblast activity, resulting in lower bone mass and an increased risk of osteoporosis.

(c) High alcohol intake: Heavy alcohol consumption also leads to bone loss or osteoporosis (more than two standard units of alcoholic drinks per day, and binge drinking may be especially harmful). Excessive alcohol intake is known to have direct detrimental effects on bone-forming cells and on the hormones, which regulate calcium metabolism directly or indirectly (e.g., steroid reproductive hormones and growth hormone). In addition, chronic, heavy alcohol consumption is associated with reduced food intake (including low calcium, vitamin D and protein intakes) and overall poor nutritional status, which will in turn have adverse effects on skeletal health. Moderate alcohol intake is not thought to be harmful to bone.

(d) Smoking: Regular smoking is associated with reduced bone density. Smoking results in increased estrogen degradation, earlier menopause, poor nutrition, chronic ill health and less activity. Smokers tend to be thinner, and those with low body mass index tend to have greater rates of bone loss. Further, lower serum 25-hydroxy vitamin D levels have been reported in older smokers, but it is unclear if they are related to changes in PTH production. Reduced serum osteocalcin levels have also been reported in women smokers, post menopause. In some studies, it has been demonstrated that rates of bone loss are faster in older smokers in proximal femur and ulna. Biochemical markers of bone resorption are reported to be higher in older smokers. Therefore, it can be concluded that bone loss in smokers occurs due to increase in bone resorption.

(e) High cadmium levels: Cadmium is also considered to be a risk factor for bone loss and osteoporosis. It can accelerate net loss of bone mineral and increase the incidence of osteoporosis in elderly. It has been found that women who smoke cigarettes, which contain cadmium, are more prone to bone loss, especially after menopause. Bone loss response begins within hours of first exposure to cadmium, increase in expression of enzymes of bone resorption is seen and increase in the formation and activity of osteoclasts is observed. Various cadmium exposure levels are given in Table 2.

Table 2. Cadmium Exposure Levels in Perspective

Group	Blood Cadmium Concentration
Start of bone response, dogs	2–5 μg/L
Non-smokers	0.26 + 0.03 μg/L
Smokers (> 20 cigarettes/day)	2.5 + 1.5 μg/L
Battery plant workers	22 μg/L
OSHA action level	5 μg/L

(f) High caffeine intake: Caffeine is often implicated in the development of bone loss, but without any convincing evidence. Caffeine does produce a small increase in urinary calcium excretion and a very small decrease in calcium absorption, but the body appears to balance this out by reducing calcium excretion later on, therefore, the net effect is negligible. Studies examining the effects of caffeine on rates of bone loss in postmenopausal women showed that as long as calcium intake was sufficient (above about 800 mg/day), caffeine intake had no detrimental effects. However, if calcium intake was low, caffeine intake equivalent to about 3 cups of brewed coffee per day was associated with increased risk of hip fracture.

(g) Carbonated beverages: The evidences that show the association between high carbonated beverage consumption and either decreased BMD or increased fracture rates in teenagers, are not convincing. The negative effect of carbonated beverage on calcium metabolism can be either due to phosphorus or caffeine content, but this has not been demonstrated in experimental studies. No evidence is known, which shows the detrimental effect of high phosphorus intake on bone. It is being considered that negative effects of carbonated beverages are more likely to be due to the fact that these drinks displace milk in the diet, and hence cause negative impact on calcium intake. Further, many commercial

mineral waters are carbonated but they have not shown any negative impact on bone health. Therefore, carbonated beverages are no longer regarded as a risk factor for osteoporosis.

3. Pathological factors

(a) Endocrine Diseases: Hypogonadism, prolactinoma, hypopituitarism, hyperparathyroidism, hyperthyroidism, cushing syndrome, acidosis, diabetes type I and type II, androgen insensitivity, gaucher's disease, hemochromatosis are the endocrine and metabolic diseases associated as a risk factor for osteoporosis. A few of these are discussed below:

Hyperthyroidism: The strength of our bones depends on the amount of calcium and other minerals which they contain. Hyperthyroidism is a condition in which thyroid produces too much thyroid hormone, which interferes with our body's ability to incorporate calcium into bones. Even mild hyperthyroidism can have adverse effects on the bone.

Hyperparathyroidism: In hyperparathyroidism, over-activity of the parathyroid glands is seen. These glands produce parathyroid hormone, which helps to regulate calcium concentrations in the body. The increased secretion of parathyroid hormone increases the removal of calcium from bone, raising blood calcium levels (hypercalcemia) and potentially leading to osteoporosis.

Hypogonadism: Hypogonadism is a state in which the activity of the ovaries or testes decreases, resulting in low amounts of estrogen or testosterone. This may occur as a result of aging, but it can also occur in younger men and women due to medications that cause hypogonadism (e.g., chemotherapy agents), block estrogen synthesis (aromatase inhibitors), or induce testosterone/estrogen deficiency (GnRH agonists). It may also occur as a result of low body weight, excessive exercise, or pituitary abnormalities. Men who have low levels of testosterone and women who have a low level of estrogen are at increased risk of bone loss and hence osteoporosis.

Cushing's syndrome: Cushing's syndrome is a condition that results from an excess of cortisol, a hormone produced by the adrenal glands. Cortisol or hydrocortisone is classified as a glucocorticoid. Cushing's syndrome affects women three times more than men. Evidences suggest that Cushing's syndrome can lead to bone loss or thinning of the bones, but the underlying mechanisms still need to be elucidated.

Diabetes mellitus: Diabetes mellitus is an emerging lifestyle disease, which is on the verge of becoming epidemic. In patients with diabetes, metabolic effects of poor glycemic control lead to increased bone resorption and bone loss in young adults, and low bone turnover retards age-related bone loss. The mechanisms underlying diabetes-mediated bone loss are not well defined. It has been reported that the advanced glycation end-products (AGEs) and receptor for AGEs (RAGEs) are involved in diabetic complications. Type 1 diabetes people are often associated with bone loss. People with type 2 diabetes have been shown to have low osteocalcin levels. Adequate mechanisms still need to be explored further.

(b) Hormonal Changes due to Menopause: Estrogen plays an important role in skeletal growth and bone homeostasis. The primary risk factor for osteoporosis is the decrease in estrogen levels that occurs with menopause or other estrogen deficient conditions. Rate of bone loss increases soon after

the menopause, particularly in trabecular bone. This increased rate of loss lasts for approximately 10 years. At this point, the rate of bone loss slows to near the premenopausal rate, but the premenopausal rate of bone formation is absent. Therefore, the condition of bone deteriorates further, if hormone therapy is not taken. In postmenopausal women diagnosed with breast cancer, sudden removal of hormone replacement therapy (HRT) can cause significant decrease in levels of estrogen due to sudden withdrawal of hormones. Surgical removal of ovaries also leads to early menopause and decrease in estrogen level. Osteoclast apoptosis also appears to be regulated by estrogens. Hence, decrease in estrogen is related to bone resorption. Estrogens also retard the bone resorbing effects of PTH. Experiments on mice have shown underlying mechanisms involving the TGFb and over-expansion of T cells in the immune system in associating the estrogen level and bone loss. A recent study showed that over expression of TGFb *in vivo* prevents bone loss caused by removal of the ovaries. Lack of estrogen decreases the levels of TGFb in the bone marrow, accelerating bone loss. Various factors interconnect to explain the mechanism of bone loss due to estrogen deficiency.

(c) Psychiatric diseases: BMD has been reported to decrease in psychiatric diseases such as schizophrenia, anorexia nervosa or major depression. The decrease in BMD is considered to be due to hypercortisolemia, which is common in psychiatric disorders and especially in depression, which leads to bone loss and in future can lead to osteoporosis, if left untreated.

Anorexia nervosa: Anorexia Nervosa (AN), with a prevalence of 0.5–1% among college-age women, is a psychiatric disorder complicated by severe bone loss; cognitive impairment; and a high prevalence of mood and anxiety disorders, including major depressive disorder. Osteopenia is present in 92% and osteoporosis in 38% of young women with AN. This eating disorder is a chronic psychiatric illness with an onset usually during adolescence—the time of life when maximal bone mass accrual occurs, thereby putting these patients at very high risk for compromising their peak bone mass. The extreme body thinness in female anorexia patients leads to estrogen deficiency and amenorrhea. Estrogen deficiency in younger women contributes to bone loss in much the same way that estrogen deficiency after menopause does. The low body weight and specific nutrient deficiencies are also risk factors for low bone mass, as the multiple hormonal and metabolic disturbances are seen in anorexia patients. Anorexia patients with average illness duration of about 6 years are found to have an annual fracture rate 7 times greater than that of healthy women of the same age. Even recovery from anorexia nervosa does not confer full establishment of normal bone density, and fracture risk is increased throughout life. Particular attention needs to be paid to the skeletal health of anorexic patients. In order to prevent or treat osteoporosis, they need to be identified early and given appropriate support.

Depression and bone loss: Depressive disorders affect the person's body, mood and cognition and take different forms, including major (or unipolar) depression, dysthymia and bipolar disorder, as described in the fourth edition of the American Psychiatric Association's Diagnostic and Statistical Manual of Mental Disorders (DSM-IV-TR).The pathogenesis of bone loss in major depressive disorder is a matter of debate. Studies of bone loss in nonpsychiatric medical disorders have found an association between the activation of osteoclastic cells and an imbalance of pro- and anti-inflammatory cytokines. Major depressive disorder is also associated with alterations in serum cytokine concentrations. Several endocrine changes have been described in depressed male patients. Some of them have negative influence on bone metabolism as discussed below. Depression leads to increased central corticotropin releasing hormone (CRH) release, provoking higher adrenocorticotropic hormone (ACTH) and cortisol secretion, while lowered growth hormone (GHRH) and gonadotrophin-releasing (GnRH) hormone release from the hypothalamus suppress the secretion of GH and gonadal steroids, respectively as shown in Fig. 12. CRH also stimulates catecholamine release, inducing interleukin-6 production. The hyperactivity of hypothalamo-pituitary-adrenal and sympatho-adrenal axes, together with the suppression of somatotrophic and gonadal axes, results in decreased bone formation, ultimately leading to net bone loss.

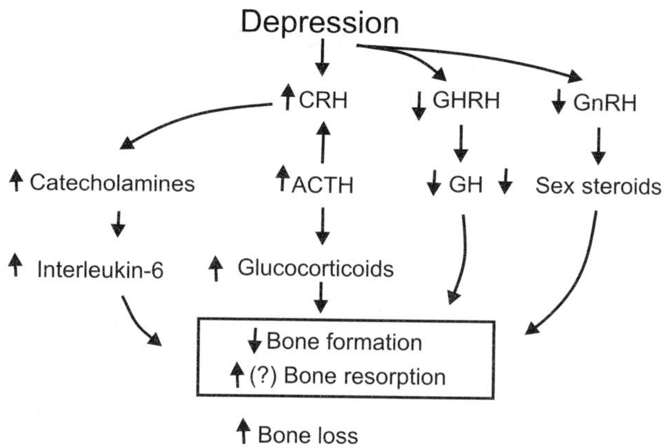

Fig. 12. Proposed endocrine model of bone loss in depression.

(d) Other Diseases as a risk factor for osteoporosis:

Gastrointestinal diseases: Celiac sprue, malabsorption, malnutrition, inflammatory bowel disease and chronic hepatic disease.

Hereditary skeletal /connective tissue diseases: Osteogenesis imperfecta, rickets, hypophosphatasia, homocystinuria, marfan's syndrome, pseudoglioma and Ehlers-Danlose syndrome.

Marrow diseases: Myeloma, mastocytosis, thalassemia and leukemia.

Neurologic diseases: Spinal cord injury, stroke, Parkinson's disease and history of polio.

Pulmonary diseases: Chronic obstructive pulmonary disease and cystic fibrosis.

Rheumatologic diseases:Systemic lupus erythematosus, ankylosing spondylitis and rheumatoid arthritis.

Renal diseases: Chronic kidney failure, renal tubular acidosis and hypercalciuria.

Other diseases: Organ transplantation, lactation, AIDS, transient regional osteoporosis, any disease that causes serious weight loss or prolonged bed rest, lactose maldigestion and intolerance.

4. Treatment Induced Bone Loss (TIBL)

Treatment-induced bone loss or risk of fracture can result from systemic glucocorticoid use in chronic conditions as well as from immunosuppressant therapy.

(a) Glucocorticoid Therapy: Glucocorticoids (a class of steroidal immunosuppressive medications) are used widely by transplant patients to prevent organ transplantation rejection and by patients affected with chronic diseases such as rheumatoid arthritis, lupus and asthma. Commonly prescribed glucocorticoids are: cortisone, hydrocortisone, prednisone and dexamethasone. But these drugs have some shortcoming too, as they can induce substantial bone loss early in the course of treatment, if given in high doses and it subsequently puts the body at risk for fractures and disability. An average loss of 8% of trabecular bone density and 2% of cortical bone density has been seen in the lumbar spine over a 20-week period in people treated with a mean prednisone dose of 7.5 mg/day. The process of programmed cell death, also called apoptosis, occurs at increased rates in bone cells (osteoblasts) during glucocorticoid treatment i.e., decreased bone formation and other evidences imply that bone resorption increases. Studies confirm that glucocorticoids promote osteoclastogenesis by inhibiting OPG and concurrently stimulating RANKL production by osteoblastic lineage cells, thereby enhancing bone resorption. These drugs also interfere with calcium metabolism and also affect the levels of sex hormones. Even the primary diseases (rheumatoid arthrios, gastrointestinal diseases), predispose to bone loss because of nutritional deficiency or due to increased production of inflammatory cytokines which enhance bone loss. Before the start of treatment with glucocorticoids, patients should be counseled for preventive measures that should be taken like good nutrition, adequate calcium and vitamin D intake, supplements and weight-bearing exercise.

(b) Non-glucocorticoid Immunosuppressive Drugs: Contribution of immunosuppressive agents to osteoporosis is still conflicting, but such agents appear to have an analogous effect when used independently or in combination with glucocorticoids. Two nonglucocorticoid drugs commonly used in transplant patients are cyclosporine A and tacrolimus, which act by inhibiting osteoclastic bone resorption and suppressing T-cell activation that in turn releases cytokinins.

(c) Schizophrenia treatment induced bone loss: Bone mineral loss is severe in some antipsychotic-treated patients with schizophrenia. Prolactin elevation is associated with reduced bone mineral density in both men and women. Reduced oestrogen levels affect bone mineral density directly, but low oestrogen levels also decrease synthesis of 1,25-dihydroxy vitamin D_3, and thereby reduce calcium absorption from the intestine as shown in Fig. 13. But the association between alterations in calcium metabolism and bone loss in antipsychotic-treated patients is still not clear. A study of ten antipsychotic-treated women found that 4 had a bone density less than 95% of that of matched controls. All 10 had reduced menstrual function, 3 were amenorrhoeic, and 9 had galactorrhoea. Therefore, it can be concluded that anti-psychotic drugs contribute to osteoporotic fracture.

(d) Other medications: Other medications which cause osteoporosis include corticosteroids (e.g. prednisone, inhaled steroids for asthma); anti-epileptic drugs—valproic acid, phenobarbitol, carmazepine, phenytoin, clonazepam, dilantin; some diuretic-thiazides; a class of antidiabetic drugs—

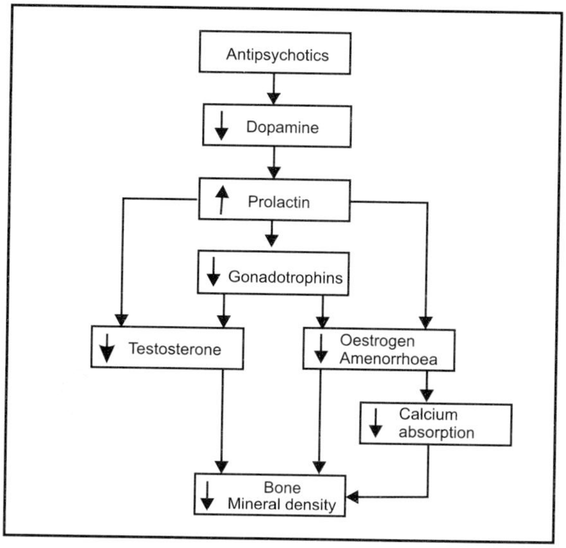

Fig. 13. Possible mechanisms for antipsychotic-induced bone loss.

TZDs; antidepressants (SSRI), chronic use of barbiturates, gonadotropin releasing hormone agonists, methotrexate, excess thyroid, heparin, depo-medroxyprogesterone acetate, anti-neoplastic agents cyclosporin, proton-pump inhibitors, selective serotonin reuptake inhibitors and aromatase inhibitors.

5. Cancer Treatment Induced Bone Loss

(a) Hormone ablative therapy: Hormone ablative therapy is usually used for the treatment of breast cancer and prostrate cancer. In this treatment, the production of certain hormones is suppressed. In women with breast cancer, treatment involves antiestrogens e.g., tamoxifen and aromatase inhibitors which may lead to bone loss due to deficiency of estrogen, and through the OPG/RANK/RANKL pathway. It has been shown in a study that women treated with aromatase inhibitors have 1.5- to 2-fold higher risk of experiencing vertebral fracture, compared with patients receiving tamoxifen alone.

(b) Androgen deprivation therapy: In androgen deprivation therapy (ADT), gonadotropin-releasing hormone (GnRH) agonist is used to treat advanced or metastatic prostate cancer. GnRH agonist inhibits the production of testosterone, growth factor of prostrate cancer cells, which in turn leads to decrease in bone mass and subsequently increase in bone turnover within a span of 2 years through OPG/RANK/RANKL pathway. Hormone ablative therapies lead to excess RANKL activity, which results in osteoclast-driven bone resorption and subsequent bone loss.

(c) Chemotherapy: Chemotherapy also leads to bone loss by inducing early menopause, which leads to decrease in estrogen level significantly.

6. Space and Bone Loss

Astronauts experience bone loss during their space flights as a result of the zero gravity environment. However, the exact mechanism of this bone loss i.e. whether it is due to conversion of mechanical signal to chemical (a hormone), which then regulates bone growth or it is due to the decrease in formation rate, is still not clear. The bone loss in space affects load-bearing bones, which include ankles, tibia, femoral neck, greater trochanter, pelvis and the lower vertebrae. In zero gravity conditions, these bones experience rapid depletion of calcium leading to the bone loss at the rate of 1–2% per month (6–24% per year), which is 6% greater than the bone loss in women with severe osteoporosis. The zero g conditions cause a great decrease in blood pressure to the legs, therefore, the blood gets pooled in upper body regions as shown in Fig. 14(a) and (b). Due to decreased blood flow in legs build up of bones or healing become extremely difficult. Poor lighting in the aircraft and high CO_2 concentration, also have adverse impact on the bones.

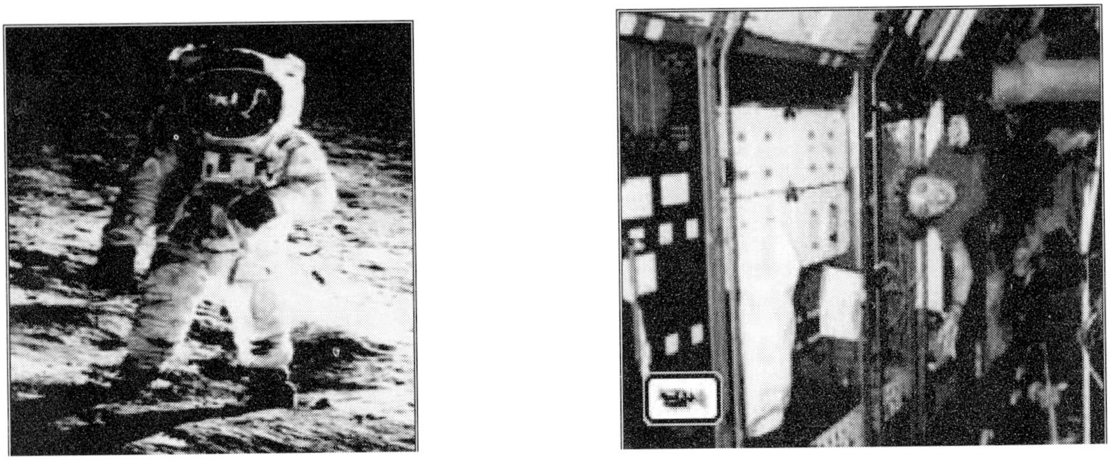

(a) (b)

Fig. 14. (a) Lack of exercise and weightlessness on muscle and loss during space flight. (b) The lack of gravity, in which bone allows a shift of body fluids toward head.

The extent of bone loss also differs from person to person as studies have shown. A cosmonaut who spent longest time in space showed no significant bone loss while one astronaut David Wolf, who spent 4½ months on Mir, showed 12% bone loss and 40% muscle loss. Even recovery time varies from person to person. Two astronauts who stayed on Mir did not recover bone loss after 2 years of return, while other members recovered within 6 months to 3 years. To help the astronauts to combat with bone loss combination of drug therapy, physical exercise and vibration treatments are being used. A drug known as osteoprotegerin (OPG) is being tested for its use for preventing bone loss.

A flowchart of the modifiable risk factors of the osteoporosis is shown in Fig. 15.

PATHOGENESIS

The underlying mechanism in all cases of osteoporosis is an imbalance between bone resorption and bone formation. Either bone resorption is excessive, and/or bone formation is diminished. Bone matrix is manufactured by the osteoblast cells, whereas bone resorption is accomplished by osteoclast cells. The mechanisms influencing the formation of the disease are complex. Most cases do not result from inadequate calcium intake, but include other factors like cigarette smoking, sedentary

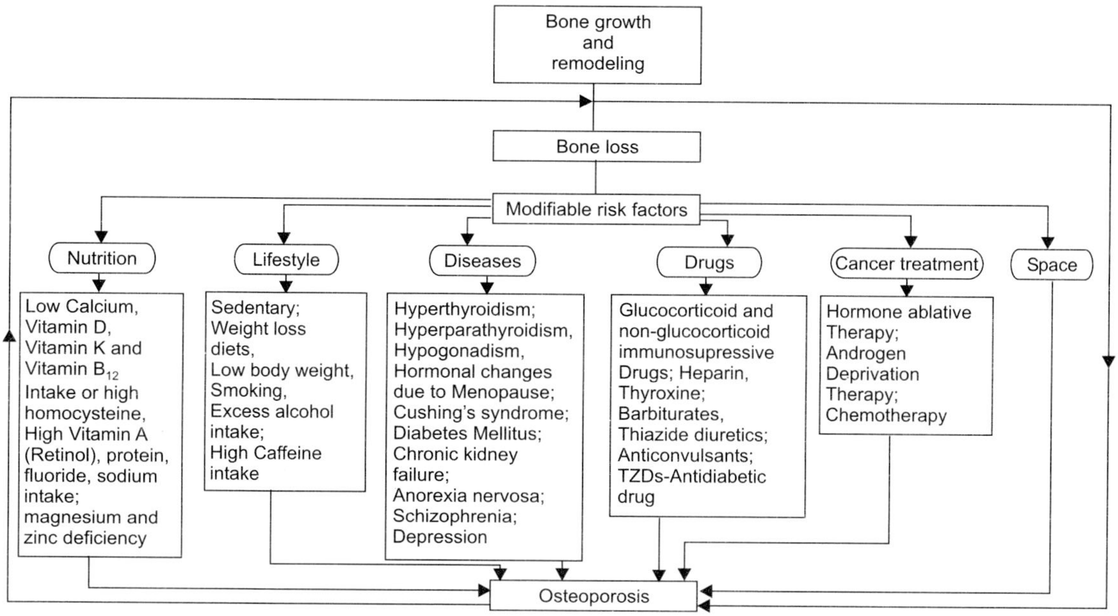

Fig. 15. Flowchart of modifiable risk factors of osteoporosis.

lifestyle affecting bone matrix formation and resorption. Trabecular bone is the sponge-like bone in the center of long bones and vertebrae. Cortical bone is the hard outer shell of bones. Since osteoblasts and osteoclasts inhabit the surface of bones, trabecular bone is more active, and more subject to bone turnover and to remodeling. Long before any overt fractures occur, the small spicules of trabecular bone break and are reformed in the process known as remodeling. Bone will grow and change shape in response to physical stress. The bony prominences and attachments in runners are different in shape and size than those in weightlifters. It is an accumulation of fractures in trabecular bone that are incompletely repaired and leads to the manifestation of osteoporosis. Common osteoporotic fracture sites, the wrist, the hip and the spine, have a relatively high trabecular bone to cortical bone ratio. These areas rely on trabecular bone for strength. Low peak bone mass is important in the development of osteoporosis. Achieving a higher peak bone mass through exercise and proper nutrition during adolescence is important for the prevention of osteoporosis. Achieving a higher peak bone mass through exercise and proper nutrition during adolescence is important for the prevention of osteoporosis. Bone remodeling is heavily influenced by nutritional and hormonal factors. Calcium and vitamin D are nutrients required for normal bone growth. Parathyroid hormone regulates the mineral composition of bone, with higher levels causing resorption of calcium and bone. Glucocorticoid hormones cause osteoclast activity to increase, causing bone resorption. Calcitonin, estrogen and testosterone increase osteoblast activity, causing bone growth. The loss of estrogen following menopause causes a phase of rapid bone loss. Similarly, testosterone levels in men diminish with advancing age and are related to male osteoporosis. In addition to estrogen, follicle-stimulating hormone (FSH) affects BMD. In mice, lower levels of FSH mean less resorption by osteoclasts. Role of different hormones and vitamins in bone metabolism is shown in Table 3.

Table 3. Mode of action of hormones and vitamins on bone metabolism

Hormones	**Mode of Action**
Calcitonin	Inhibits osteoclast function
Cortisol	Osteoblast apoptosis
Estrogens	Inhibit osteoclasts, osteoclast precursors, decrease IL-6, may enhance osteoblast survival
Growth hormone/IGF-1	Increases osteoblast function
Parathyroid hormone (PTH)	Stimulates osteoblast activity, increases expression of RANK-L
Thyroid hormone	Stimulates osteoclast
Vitamins	**Mode of Action**
Vitamin A	Directly stimulates osteoclast
Vitamin D-1,25(OH), and	Promote differentiation of osteoblast, osteoclast precursor

Physical activity has its greatest impact during adolescence, affecting peak bone mass most. However, excessive exercise can lead to constant damage to the bones. There are numerous examples of marathon runners who developed severe osteoporosis later in life. Age-related sarcopenia, or loss of muscle mass, loss of balance and dementia contribute greatly to the increased fracture risk in patients with osteoporosis. Physical fitness in later life is associated more with a decreased risk of falling than with an increased bone mineral density.

DIAGNOSIS

Bone strength constitutes the two main features—bone density and bone quality. Bone mineral density (BMD) contributes for 70% of bone strength. Bone density is expressed as grammes of mineral per area or volume; it is determined by the peak bone mass and amount of bone loss. Bone quality refers to the micro architecture turnover and damage accumulation. Accurate measurement cannot be done for bone strength but dual energy X-ray absorptiometry (DXA, formerly DEXA) is used to measure bone density and quantitative ultrasound (QUS) to measure bone quality.

The DEXA technique basically consists of an X-ray tube and scintillation detector (made up of $CdWO_4$ or NaI (Tl) scintillator coupled to a photomultiplier tube) mounted on a C-arm as shown in Fig. 16, so that the patient is exposed to a pencil X-ray beam scanning in a rectilinear fashion. The pencil beam is used to reduce the detection of scattered radiation. The filter assembly is used to switch filters and calibration standards into and out of the pencil beam at appropriate intervals. By rotating the C-arm as shown in Fig. 17 around the patient during the scan in the second generation devices, a CT image can be formed. The output of the scintillation detector is fed to a computer for dual-energy data processing and image display (from where data about bone mineral density is derived).

DXA is considered the gold standard for diagnosis of osteoporosis. Diagnosis is made when the bone mineral density is less than or equal to 2.5 standard deviations below that of a young adult reference population. This is translated as a T-score.

The World Health Organization has established diagnostic guidelines as T-score –1.0 or greater is "normal", T-score between –1.0 and –2.5 is "low bone mass" (or "osteopenia"), –2.5 or below as

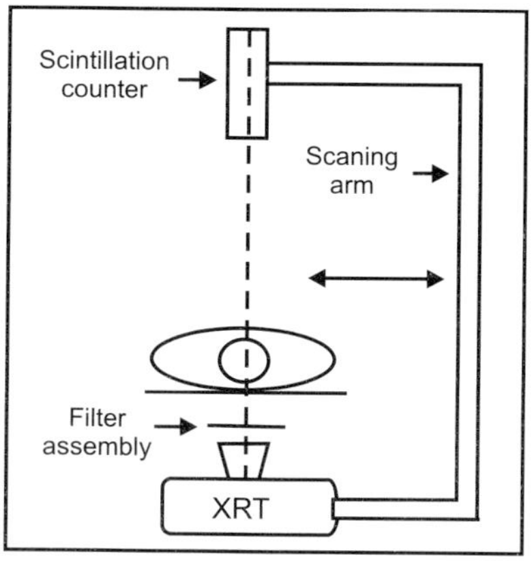

Fig. 16. Diagram of a DEXA scanning apparatus.

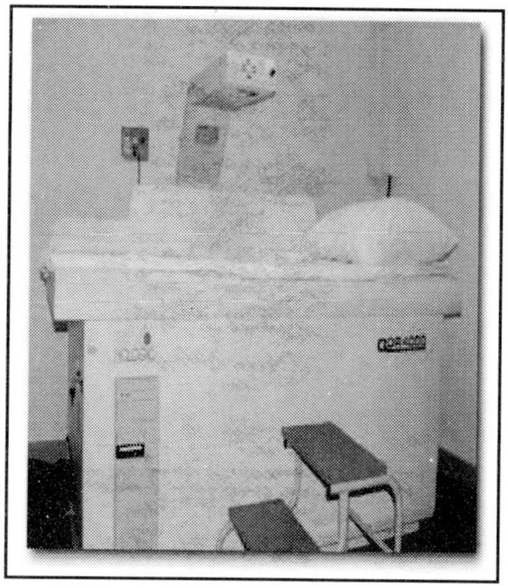

Fig. 17. DEXA scanning apparatus with an under-couch XRT and pencil beam detector.

osteoporosis and established osteoporosis includes the presence of traumatic fractures. A low trauma or osteoporotic fracture, which occurs as a result of a fall from a standing height, is also diagnostic of osteoporosis regardless of the T-score. In order to differentiate between "primary" (postmenopausal, regardless of age, or senile-related to age) and "secondary" osteoporosis, blood tests and X-rays are usually done to rule out cancer with metastasis to the bone, multiple myeloma, cushing's disease and other causes mentioned above.

QUS is a simple and portable technique that takes an ultrasound scan of a person's heel. A QUS machine is shown in Fig. 18. Heel is an easily accessible area that contains trabecular bone—the type of bone that changes with age like the lumbar spine does. This technique doesn't involve exposure to radiation. It depicts the bone strength and elasticity, which determines the risk of fracture.

Fig. 18. Quantitative Ultrasound Machine.

Biochemical Markers of Bone Metabolism

Bone remodeling occurs continuously to repair minor fissures and to maintain its structural integrity. In adults, about 10% of the skeleton is renewed each year. This process occurs as localized episodes of bone resorption by osteoclasts followed by replacement of the bone by osteoblasts. The rate of total bone turnover can be indirectly assessed by measuring the serum and urine biochemical markers of bone formation and bone resorption. Currently available biochemical markers of bone are given in Table 4.

Table 4. Currently available biochemical markers

Marker	Type
Total alkaline phosphatase (AP, TAP)	Bone formation
Bone specific AP (BAP)	Bone formation
Osteocalcin (OC, BGP)	Bone formation
Carboxyterminal propeptide of type1 procollagen (PICP)	Bone formation
Aminoterminal propeptide of type1 procollagen (PINP)	Bone formation
Hydroxyproline (OH-Pro, OHP)	Bone resorption
Pyridinoline (PYD, Pyr)	Bone resorption
Deoxypyridinoline (DPD, d-Pyr)	Bone resorption
Carboxyterminal cross-linked telopeptide of type1 collagen (ICTP)	Bone resorption
Aminoterminal cross-linked telopeptide of type1 collagen (INTP, NTX)	Bone resorption
Hydroxylysine-glycosides	Bone resorption
Tartrate-resistant acid phosphatase (TRAP)	Bone resorption

TREATMENT

Patients at risk for osteoporosis (e.g., steroid use) are generally treated with vitamin D and calcium supplements. In renal disease, a different form of vitamin D (1,25-dihydroxycholecalciferol or calcitriol, which is the main biologically active form of vitamin D) is used, as the kidney cannot adequately generate calcitriol from calcidiol (25-hydroxycholecalciferol), which is the storage form of vitamin D.

In osteoporosis (or a very high risk), bisphosphonate drugs are prescribed. The most often prescribed bisphosphonates presently are sodium alendronate 10 mg a day or 70 mg once a week, risedronate 5 mg a day or 35 mg once a week or and ibandronate once a month. Other medicines prescribed for prevention of osteoporosis include raloxifene, a selective estrogen receptor modulator (SERM). Estrogen replacement remains a good treatment for prevention of osteoporosis but, at this time, is not recommended unless there are other indications for its use as well.

Recently, teriparatide, recombinant parathyroid hormone has been shown to be effective in osteoporosis. It is used mostly for patients who have already fractured, have particularly low BMD or several risk factors for fracture or cannot tolerate the oral bisphosphonates. It is given as a daily injection with the use of a pen-type injection device. Teriparatide is only licensed for treatment, if bisphosphonates have failed or are contraindicated.

Oral Strontium ranelate is the first in a new class of drugs called a Dual Action Bone Agents (DABA's), and has proven efficacy in the prevention of vertebral and non-vertebral fractures (including hip fracture). Strontium ranelate works by stimulating the proliferation of osteoblast (bone building) cells and inhibiting the proliferation of osteoclast (bone absorbing) cells. This means that strontium ranelate increases BMD by forming new bone, rather than just preserving existing bone. In comparison to bisphosphonates, which only act on one aspect of bone remodeling, strontium ranelate also preserves bone turnover, allowing the micro architecture of the bone to be continuously repaired as it would in healthy bone. Strontium ranelate is taken as a 2 g oral suspension daily and is licensed for the treatment of osteoporosis to prevent vertebral and hip fracture. Strontium ranelate has shown significant efficacy at reducing both vertebral and non-vertebral fractures in patients over the age of 80, who are most at risk. This is unique to strontium ranelate as bisphosphonates have only proven efficacy in vertebral fracture reduction, not non-vertebral (hip, wrist) fracture reduction. Strontium ranelate also has less side effect as compared to the bisphosphonates, as it does not cause any form of upper GI side effect, which is the most common cause for medication withdrawal in osteoporosis. Changes in lifestyle factors and diet are also recommended; the "at-risk" patient should include 1500 mg of calcium daily either through dietary means (for instance, an 8 oz glass of milk contains approximately 300 mg of calcium) or supplementation. The body will absorb only about 500 mg of calcium at one time and so intake should be spread throughout the day. However, the benefit of supplementation of calcium alone remains, to a degree, controversial since several nations with high calcium intakes through milk-products (e.g. the USA, Sweden) have some of the highest rates of osteoporosis worldwide. A few studies even suggested an adverse effect of calcium excess on bone density and blamed the milk industry for misleading customers. Some nutritionists assert that excess consumption of dairy products causes calcification, which leeches calcium from the system, and argue that vegetables and nuts are a better source of calcium and that in fact milk products should be avoided. In a recent study that examined the relationship between calcium supplementation and clinical fracture risk in an elderly population, there was a significant decrease in fracture risk in patients that received calcium supplements versus those that received placebo. However, this benefit was only applied to patients who were compliant to their treatment regimen. According to recent studies, increased vitamin D intake also has been shown to reduce fractures by up to 25 percent in older people. There is some evidence to suggest bone density benefits from taking the folowing supplements (in addition to calcium and vitamin D): boron, magnesium, zinc, copper, manganese, silicon, strontium, folic acid, and vitamins B_6, C and K.

Thirty minutes of weight-bearing exercise such as walking of jogging, three times a week, has been shown to increase bone mineral density and reduce the risk of falls by strengthening the major muscle groups in the legs and back. There is no controversy on the beneficial Affect of physical activity in preventing and treating osteoporosis.

Reconstruction of osteoporotic bone is a challenging problem, but recently two methods, vertebroplasty and kyphoplasty, are being used for the management of acute vertebral fractures, which involve the injection of polymethylmethacrylate bone cement into the fractured vertebra.

These are used in rare cases which do not involve neurological deficits or unstable spine. Long term effects of the procedures are unknown.

Management: Prevention

1. Regular weight bearing Exercise
    (a) Best Exercises
        (i) Jogging
        (ii) Stair climbing
        (iii) Walking
        (iv) Hiking
    (b) Not as helpful to strengthen bone
        (i) Cycling
        (ii) Swimming
    (c) See fall prevention the elderly
    (d) Hip protector (underwear with trochanter pads)
2. Smoking cessation
3. Moderate alcohol use
4. Dietary supplementation
    (a) Calcium supplementation (1200 to 1500 mg per day)
    (b) Vitamin D supplementation
        (i) Daily: 800 IU per day or
        (ii) Weekly: Fosamax with Vitamin D (Fosamax plus D)
            - Standard weekly Fosamax dose with Vitamin D
            - Includes 2800/IU week of Vitamin D
            - Additional Vitamin D 400 IU needed daily
        (iii) Quarterly: 100,000 IU per 4 months
5. Medication
    (a) Postmenopause
        (i) Estrogen replacement
        (ii) Selective Estrogen Receptor Modulator (SERM)
        (iii) Bisphosphonate (e.g. Alendronate)

Management: Treatment

1. Calcium supplementation (1500 mg per day)
    (a) Does not increase bone density (but slows loss)
    (b) Overdosage weakens bone
2. Vitamin D supplementation (800 IU per day)
    (a) Increases bone density 1% per year
3. Bisphosphonates (Alendronate, Risedronate)
    (a) Increases bone density 5–6% per year
    (b) Alendronate
    (c) Risedronate

4. Estrogen Replacement Therapy (ERT or HRT)
    (a) Increases bone density 3–4% per year
    (b) Fracture protection lost 5 years after stopping ERT
    (c) Standard dosing
        (i) Minimum preventive plasma estradiol level: 60 pg/ml
        (ii) Maximal effect requires Premarin 0.625
    (d) Alternative Estrogen dosing
        (i) Some effect seen at 0.3 mg or Transdermal Estrogen
        (ii) 17-beta-estradiol 0.25 mg increases BMD
5. Selective Estrogen Receptor Modulator (e.g. Raloxifene)
    (a) Similar benefit to estrogen replacement
    (b) Indicated if unable to take estrogen replacement
6. Calcitonin
    (a) Increases bone density 1–2% per year
    (b) Not indicated for prevention
7. Fluoride supplementation
    (a) Increases bone density 10% per year
    (b) Unclear whether bone strength is increased

Management: Treatment (Investigational)

1. Teriparatide (Forteo)
    (a) Recombinant Parathyroid Hormone
    (b) Dose: 20 mcg SC daily
    (c) Reduced risk for osteoporotic fractures
    (d) Recently approved for high risk men and women
    (e) Do not use with bisphosphonate
    (f) Use limited to 2 year maximum
2. Thiazide Diuretics (e.g. Hydrochlorothiazide)
    (a) Decreases urinary calcium loss
    (b) Reduction in hip fracture if used >10 years
    (c) Consider in hypertensive patients
3. Dietary magnesium 600 mg/day or more
4. Dietary soya protein 40 g/day or more

Potential New Medications in Osteoporosis

1. HMG-CoA Reductase Inhibitor (Statin drugs)

CONCLUSION

Bone is a hard connective tissue that is being remodeled throughout life. However, when disruption in the process occurs due to nutritional deficiency, lifestyle, pathological factors and due to use of some medications, it leads to bone loss and if left untreated it leads to bone disease—osteoporosis. It is estimated that 1 in 3 women and 1 in 5 men over the age of 50 worldwide have osteoporosis. It is responsible for millions of fractures annually, mostly involving the lumbar vertebrae, hip and wrist. Today, most cases of osteoporosis are diagnosed before symptoms develop. This is due to widespread screening for osteoporosis using the DEXA scan. With treatment, bone mineral density increases and fracture risk decreases. Some of the newer drugs such as alendronate, calcitonin and slow-release fluoride, along with calcium supplements, promise to restore some bone loss, but preventing loss is better yet. The main strategies for preventing osteoporosis are to ensure adequate calcium intake throughout life, to do regular weight-bearing exercise, avoid smoking and excess alcohol consumption. Menopause women can consider taking estrogen supplements to retard bone loss. People should be counseled about the changes in lifestyle and diets that are beneficial for their health. Those suffering from diseases that cause osteoporosis should be given guidelines for their screening and treatment of osteoporosis.

SUGGESTED READINGS

1. Behari J, Jha SC, Rai DV, Prim A. Bioelectric Behavior and Ultrasonic Electromagnetic Field Coupling in Bone. Medical and Life Science Engg. Vol.9, 33-42, 1987.
2. Behari J, Rai DV, Saha S, Mathur M. Bone metabolism in calcium and phosphorus deficient rats. Med. And Life Sci. Engg. 14-23, 1991.
3. Bourne Geoffrey H, Ed. The Biochemistry and physiology of bone, 1972.
4. Lindsay R. Prevention and treatment of osteoporosis. Lancet. Vol. 50:833-846, 1993.
5. Notelovitz M. Identifying women at risk for osteoporosis. In: Osteoporosis: Prevention, Diagnosis and Management. Oklahoma City, Okla: Professionals Communications Inc; 55-66, 1994.
6. Rai DV, Behari J. Biophysical Characterization of Osteoporotic Bone. Environmental Research, Vol. 40, 68-83, 1986.
7. Rai DV, Behari J, Saha S. The effect of Mineral Deficient Diet on the Structural and Mechanical properties of Long Bones. Biomedical Engineering V: Recent Development, edited by Saha S, Published by Pergmon Press, New York, 1986, pp. 456-460 (Abstract Published in Intl. J. of Biomaterials, Medicals Devices and Artificial Organs, Vol.14, 1 &2,1986.
8. Rai DV, Darbari R, Agarwal LM. Age-related changes in the elemental constituents and molecular behavior of bone. I.J.B.B. Vol42, 127-130, 2005.
9. Riggs BL. Overview of osteoporosis. West J Med. 63-77. 1991.
10. Seibel MJ, Pols HAP. Clinical application of biochemical markers of bone metabolism. In: Bilezikian JP, Raisz LG, Rodan GA, Eds. Principles of Bone Biology, San Diego, Calif: Academic Press; 1293-1311, 1996.

4

Targetting Carcinogen Biotransformation: An Emerging Cancer Control Strategy

S.C. Gangar, D.V. Rai and A. Koul

INTRODUCTION

Cancer is perhaps the most dreadful disease that causes huge mortality in human beings around the world. Approximately 80–90% of cancers are considered to be of environmental origin, exposure to chemical carcinogens being one of the most prominent factors contributing to the induction of carcinogenesis. Much has been explored as far as the causation/development of cancer is concerned; however, restless efforts are in progress worldwide for combating this disease. Several treatments/therapies have been discovered/used for controlling cancer, but these are not efficient in their actual task. Rather these often result in undesired side effects, induction of second primary malignancies and other compromises leading to poor quality of life. Prevention of cancer is proving itself as an effective as well as practical approach. Recent progress in molecular biology and pharmacology enhances the likelihood that cancer prevention will increasingly rely on chemoprevention. The latter is the use of natural or synthetic agents for blocking and/or reversing the process of carcinogenesis. The multi-step nature of the process of chemical carcinogenesis offers a number of targets that could be hit by chemoprotectants, targeting biotransformation of carcinogens with agents of natural origin for preventing the initiation of carcinogenesis appears to be the most pledging approach for cancer chemoprevention.

CANCER

The body is made up of many types of cells, the basic building blocks that grow and divide to produce more cells as they are needed to keep the body healthy. Sometimes, this orderly process gets altered so that new cells get generated when the body does not need them, and/or old cells do not die when they should. The extra cells form a mass of tissue called a neoplasm or tumor that can be benign or malignant as far as its broad classification is concerned. Benign tumors are not cancers and they can often be removed with minor surgery after which they do not regenerate. Cells in benign tumors do not spread to other parts of the body. They rarely pose a threat to life. Malignant tumors are called cancer. Cells in malignant tumors are abnormal and divide without control or order. Cancer cells invade and destroy the tissue around them. They can break away from a malignant tumor and enter the bloodstream or lymphatic system. By moving through the bloodstream or lymphatic system, cancer can spread from the primary cancer site to form new tumors in other organs. This spread of cancer cells is called metastasis that often proves to be fatal for the affected animal/humans.

Cancer has afflicted humans throughout recorded history. Some of the earliest evidence of cancer is found among fossilized bone tumors, human mummies in ancient Egypt, and ancient manuscripts. Bone remains of mummies have revealed growths suggestive of the bone cancer, osteosarcoma. In other cases, bony skull destruction as seen in cancer of the head and neck has been found. The oldest description of cancer (although the term *cancer* was not used) was discovered in Egypt and dates back to approximately 1600 B.C. The origin of the word "cancer" is credited to the Greek physician Hippocrates (460–370 B.C.), considered the "Father of Medicine." Hippocrates used the terms *carcinos* and *carcinoma* to describe non-ulcer forming and ulcer-forming tumors. In Greek these words refer to a crab, most likely applied to the disease because the finger-like spreading projections from a cancer.

TYPES OF CANCER

Different cancers are classified by the type of cells that resemble the tumor and, therefore, the tissue, presumed to be the origin of the tumor. The following general categories are usually accepted:

- Carcinomas are the malignant tumors derived from epithelial cells and this represents the most common cancers, including the common forms of breast, prostate, lung and colon cancer, squamous or epidermoid carcinoma of skin, lip, and tongue, cervix cancer.
- Sarcoma refers to the malignant tumors derived from connective tissue, or mesenchymal cells. e.g., cancers of bone, cartilage, skeletal muscle, fibrous and neurongenic connective tissue etc.
- Mesotheliomas are the tumors derived from the mesothelial cells lining the peritoneum and the pleura.
- Glioma are the tumors derived from glia, the most common type of brain cells.
- Germinoma refers to the tumors derived from germ cells.
- Choriocarcinoma is a name given to the malignant tumors derived from the placenta.
- Myeloma refers to the malignancy of plasma cell series, the cells that produce some of the proteins found in the blood. These cells are generally found in the bone marrow.
- Leukemia is the malignancy of the blood-forming elements of the bone marrow.
- Lymphoma represents the cancer of cells of the lymph nodes or of similar cells that may occur elsewhere.
- In addition to above described forms, cancer composed of different tissue types are called mixed types. The type of components may be within one category or from different categories. Examples include *malignant mullerian mixed* (composed of adenocarcinoma and sarcoma cells), *mixed adenosquamous carcinoma* (malignant glandular and squamous epithelial cells) etc.

CAUSES BEHIND INDUCTION OF CANCER—CARCINOGENS

Cancer is caused by alterations in genes that normally control the division, differentiation as well as death of cells. The altered forms of genes allow the growth and metastasis of cancer. Alterations in genes result from exposure to carcinogens, the cancer-causing agents, in the environment and/or workplace. Approximately 80–90% of cancers worldwide are considered to be of environmental origin and chemical carcinogens play a vital role in the induction of carcinogenesis. Some gene alterations are inherited; however, having an inherited gene alteration does not always mean that the person will develop cancer; it only means that the chance of getting cancer is increased. When compared to environmental factors, the hereditary factors contribute much less to cancer induction as shown in Fig. 1.

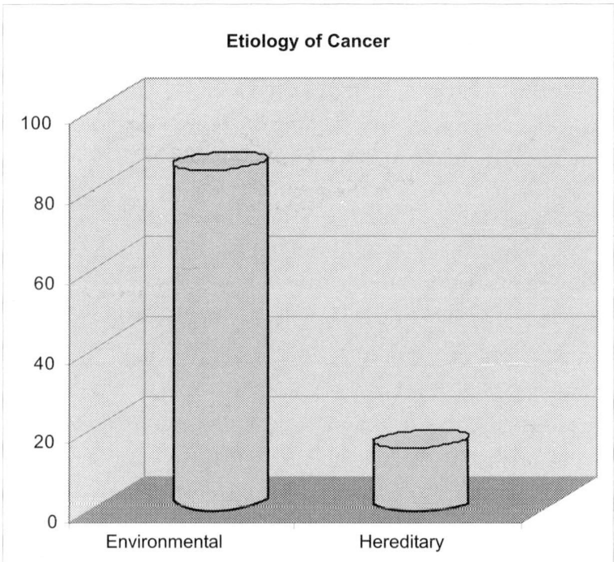

Fig. 1. Impact of environmental and hereditary factors on induction of cancer.

Many definitions have been suggested to define carcinogen but most of them seem to be inefficient in covering various associated aspects. However, considering current standards, the definition cited here seems to be quite appropriate. Carcinogen is an agent whose administration to previously untreated animals leads to statistically significant increased incidence of neoplasms of one or more histogenetic types as compared with the incidence in appropriate untreated animals. Carcinogens have broadly been divided into types as exemplified in Table 1.

The most studied class of carcinogens is perhaps polycyclic aromatic hydrocarbons (PAHs) that are ubiquitous environmental contaminants that are formed by the incomplete combustion of organic materials. PAH molecules are made up of three or more benzene rings, at least two of which are fused with two neighboring rings sharing two adjacent carbon atoms. An increase in the incidence of cancer (lung, skin, and urinary cancers) in humans upon exposure to mixtures of PAHs has been well

Table 1. Different types of carcinogens

ORGANIC:
Polycyclic Aromatic Hydrocarbons, Benz[a]anthracene, Benzo[a]pyrene

Epoxides: Ethylene oxide, Propylene oxide

Organohalogen Compounds: Vinyl chloride, Carbon tetrachloride, Chloroform, Hexachlorobenzene, Trichloroethylene

Hydrazines: Hydrazine (and salts), 1,2-Dimethylhydrazine

N-Nitroso Compounds: N-Nitrosodimethylamine

Aromatic Amines: Benzidine, Aniline, o-Anisidine, o-Toluidine

Aflatoxin B1: produced by *Aspergillus flavus*

MISC. ORGANIC COMPOUNDS: Formaldehyde, Acetaldehyde, 1,4-Dioxane, Ethyl carbamate, 2-Nitropropane, Styrene, Thiourea, Thioacetamide

INORGANIC: Arsenic, Chromium, Beryllium, Cadmium, Lead, Nickel etc and their respective compounds, Thorium dioxide and Selenium sulfide

PHYSICAL CARCINOGENS: UV exposure, Ionizing radiation

BIOLOGICAL CARCINOGENS: Viruses - HIV, HPV, EBV **Bacteria** - *Helicobacter pylori*; **Parasites** - *Schistosoma* species

PARTICLES : Asbestos

MEDICAL DRUGS: Anticancer drugs; Hormones; Contraceptive Drugs

established by now. The primary source of PAHs is from the incomplete burning of carbon-containing compounds, wood and fuel. They are also contained in gasoline and diesel exhaust, soot, coke, cigar and cigarette smoke, and charcoal-broiled foods. Items that usually contain considerable amounts of PAHs include smoked, barbecued, or charcoal-broiled foods, roasted coffees, and sausages. PAHs have been associated with cancers of the lung, mouth, bladder, colon, kidney, throat, nasal cavity, voice box, esophagus, lip, stomach, cervix, liver, and pancreas, and with leukemia.

Vinyl chloride, a colorless gas, is a human carcinogen associated with lung cancers as well as angiosarcomas of the liver and brain. It is used in the plastics industry for manufacturing many consumer products including containers, wrapping film, electrical insulation, water and drain pipes, hosing, flooring, windows, and credit cards. Human exposure occurs primarily in workers in the plastic industry and people living near a plastics plant through inhalation of contaminated air. Benzidine, recognized as being associated with increased bladder cancer risk in humans, is used in the production of more than 250 benzidine-based dyes for textiles, paper, and leather products.

Aflatoxins are cancer-causing substances produced by certain types of fungi growing on food, especially over the grains and peanuts. Meat, eggs, and milk from animals that eat aflatoxin contaminated feed are other sources of exposure. Agricultural workers are potentially at risk if they inhale contaminated air-borne grain dust. Exposure to high levels of aflatoxins increases the risk of liver cancer. The incidence of aflatoxin exposure is higher in developing countries where there is no screening for the fungus before processing of the peanuts.

Several solvents, such as benzene, carbon tetrachloride, chloroform, dichloromethane (methylene chloride), tetrachloroethylene, and trichloroethylene used in paint thinners, paint and grease removers, and in the dry-cleaning industry are known or suspected of being carcinogenic. Workers employed in the petrochemical industry, pharmaceutical industry, leather industry, rubber industry, gas stations, and in the transportation industry are exposed to most of these chemicals.

Some pesticides have been banned or their use has been restricted. These include ethylene oxide, amitrole, chlorophenoxy herbicides, DDT, dimethylhydrazine, hexachlorobenzene, hexamethylphosphoramide, chlordecone, lead acetate, lindane, mirex, nitrofen, and toxaphene. Studies of people with high exposure to pesticides (farmers, pesticide applicators, crop duster pilots, and manufacturers) have found high rates of blood and lymphatic system cancers, cancers of the lip, stomach, lung, brain, and prostate, as well as melanoma and other skin cancers.

Arsenic compounds are associated with skin, lung, bladder, kidney and liver cancers, particularly when high levels are consumed in drinking water. In addition, occupational exposure to inhaled arsenic (especially in mining and copper smelting) has been consistently associated with an increased risk of lung cancer. Arsenic is also used in wood preservatives, glass, herbicides, insecticides and pesticides, and it is a general environmental contaminant of air, food, and water.

Beryllium compounds are known to cause lung cancer in humans. These compounds are used as metals for aerospace and defence industries; for electrical components, X-ray tubes, nuclear weapons, aircraft brakes, rocket fuel additives, light aircraft construction, and the manufacture of ceramics; and as an additive to glass and plastics, dental applications, fiber optics and cellular network communication systems. Workers can be exposed through jobs related to the above activities, whereas the general population can be exposed to trace amounts of beryllium by inhaling air and consuming food contaminated with beryllium residues.

Cadmium metal and cadmium compounds are associated with an increased risk of lung cancer. Cadmium metal is primarily used to coat metals to prevent corrosion. Other uses are in plastic and synthetic products, in batteries, as stabilizers for polyvinyl chloride, and in fungicides. The industrial processes involved in making these products release cadmium into the air, surface water, ground water, and topsoil where it can be taken up by both land and water plants and, in turn, transferred to animals and human beings.

Some chromium compounds are known to cause lung cancer. In steel industry it is used for protection against corrosion of metal accessories, including automotive parts, as well as for electroplating, layering one metal over another. Other uses include nuclear and high-temperature research; the textile and leather-tanning industry; pigments for floor covering products, paper, cement, and asphalt roofing; and creating an emerald color in colored glass. The highest exposure being in occupations related to stainless steel production, welding, chrome plating, and leather tanning; the entire population is exposed to chromium compounds to some degree.

Lead acetate and lead phosphate are suggested to be linked to kidney and brain tumors. Lead acetate is used in cotton dyes; as a coating for metals; as a drier in paints, varnishes, and pigment inks; as a colorant in certain permanent hair dyes, in explosives, and in washes to treat poison. Lead phosphate is used as a stabilizer in certain plastics and specialty glass. Primary exposures occur through dermal contact, ingestion and inhalation.

Nickel and nickel compounds are associated with cancers of the nasal cavity, lung and larynx in human populations. Nickel is used in steel, dental fillings, copper and brass, permanent magnets, storage batteries, and glazes. Because nickel is present in the air, water, soil, food, and consumer products, humans get exposed through dermal contact, ingestion and inhalation

Infectious agents such as viruses and bacteria clearly contribute to the development of several types of cancer. A sexually transmitted virus called human papilloma virus (HPV) is the primary cause of cervical and anal cancer. Hepatitis B and hepatitis C viral infections are major causes of liver cancer. Risk factors for hepatitis B virus include occupational exposure to blood products, injection drug use, and high-risk sexual behavior (unprotected sex with multiple partners). Almost all adults are infected with Epstein-Barr virus (EBV), which is linked to some types of lymphoma. The risk of cancer for people infected with EBV is low, except for those whose immune systems are weakened, such as people infected with the human immunodeficiency virus (HIV). Infection with *Helicobacter pylori*, a bacterium, is widespread and is the primary cause of peptic ulcers and chronic gastritis (inflammation of the stomach). *H. pylori* contributes to the development of stomach cancer.

Ionizing radiation is invisible, high frequency radiation that can damage the DNA/genes. Everyone is exposed to very small doses of ionizing radiation from cosmic rays (rays that enter the earth's atmosphere from outer space). Elevated levels of radon, a naturally occurring radioactive gas (produced by the breakdown of uranium) found at low levels in most soil, could be observed at certain residential areas. The health effects of radon were first seen in the elevated levels of lung cancer found in underground uranium miners around the world. Ionizing radiation may also be the radioactive substances released by atomic bombs or nuclear weapons known as "fallout." The doses of ionizing radiation received by the atomic bomb survivors in Japan resulted in increased risks of leukemia and cancers of the breast, thyroid, lung, stomach, and other organs. People exposed, especially as children, to one radioactive form of iodine (Iodine^{-131}), which collects in the thyroid gland, may have an increased risk of thyroid disease, including thyroid cancer.

Human beings who receive radiation to treat cancer or other conditions may be at increased cancer risk. Persons treated with radiation in childhood to treat acne, ringworm, and other head and neck conditions have been reported to be at increased risk for thyroid cancer and other tumors of the head and neck. X-rays used to diagnose or screen for a disease are also forms of ionizing radiation. Most studies on the long-term effects of exposure to radiation used to diagnose or screen for cancers or other diseases have not shown an elevated cancer risk, but it is possible that there is a small risk associated with this exposure. Children, whose mothers received diagnostic X-rays during pregnancy, were found to have increased risks of childhood leukemia and other types of cancer, which led to the current ban on diagnostic X-rays in pregnant women. Studies of women who received small weekly X-ray doses to the chest over extended periods to monitor treatment for tuberculosis showed a radiation-related increased risk of breast cancer. Ultraviolet radiation from the sun, sunlamps, or tanning beds causes premature aging of the skin and DNA damage that can lead to melanoma and other forms of skin cancer. The incidence of skin cancers is rapidly increasing.

Certain medical drugs used to treat cancer (e.g., cyclophosphamide, chlorambucil, melphalan) have been shown to increase the occurrence of second primary malignancies, including leukemia. Others that are used as immuno-suppressants, such as cyclosporin and azathioprine for patients having organ transplants, also are associated with increased cancer risks, especially lymphoma. Estrogens used to treat symptoms of menopause and other gynecological conditions have been shown to increase the incidence of endometrial cancer. In addition, some studies have shown an increased risk of breast cancer with estrogen use, but a reduced risk of colon cancer. Progesterone, another hormone now taken in combination with estrogen for hormone replacement therapy in older women, helps to protect against the increased endometrial cancer risk with estrogen alone. However, increased risks of breast cancer, heart disease, stroke, and blood clots have recently been shown to be associated with the use of estrogen plus progestin, a synthetic form of progesterone. Long-term users of combination oral contraceptives experience increases in early-onset breast cancers and liver cancer. Tamoxifen, a synthetic hormone used to prevent cancer, may increase the risks of endometrial cancer.

Exposure to various fibers, fine particles, and dust are associated with increased cancer risks. Asbestos fibers and all commercial forms of asbestos are human carcinogens. Increased rates of mesothelioma, a rare cancer of the lining of thc lung and abdominal cavity, and cancer of the lung have been consistently observed in a variety of occupations involving asbestos exposure. Asbestos fibers are released into the environment from the use and deterioration of more than 5,000 asbestos products, including roofing, thermal, and electrical insulation; cement pipe and sheet; flooring; gaskets; plastics; and textile and paper products. Workers in asbestos insulation, brake maintenance and repair, and building demolition jobs being highly exposed, the entire population may have been exposed to some degree of asbestos.

Ceramic fibers are now used as insulation materials and are a replacement for asbestos. Because they can withstand high temperatures, they are used to line furnaces and kilns. These fibers cause lung cancer in experimental animals. Silica dusts are associated with an excess risk of lung cancer in humans and are found in industrial and occupational settings such as coalmines, mills, granite quarrying and processing, crushed stone and related industries, and sandblasting operations. Wood dust, associated with cancers of the nasal cavities and sinuses, is a known carcinogen for unprotected workers who are exposed regularly from sanding operations and furniture manufacturing.

Dioxins are surplus byproducts of chemical processes that contain chlorine and hydrocarbons. There are at least 100 different kinds of dioxins. They are unintentionally produced by paper and pulp bleaching; incineration of municipal, toxic, and hospital wastes; certain electrical fires; and smelters (plants where metal is extracted from ores). Dioxins are widespread environmental contaminants and accumulate in fats and break down slowly. A particular dioxin that is likely to be carcinogenic to humans is called 2,3,7,8-tetrachlorodibenzo-p-dioxin (TCDD) is highly carcinogenic in animals, and, in highly exposed workers, increased overall cancer death rates have been reported. The general population is exposed to low levels of TCDD primarily from eating dairy products, fish, and meat, including poultry.

It is evident from the above facts that environmental carcinogens especially chemical substances constitute the largest group causing cancer in human beings.

HISTORICAL OUTLOOK OF CHEMICAL CARCINOGENESIS

John Hill (1571) observed an increased incidence of nasal cancer among snuff users. However the first causal relationship between environmental mixtures and induction of carcinogenesis was reported by an English physician/surgeon Percivall Pott in 1575. He observed an increased incidence of scrotal cancer in patients with the history of employment in chimney sweeps. Yamigiwa and Itchikawa (1918) substantiated Pott's observations by demonstrating that multiple topical applications of coal tar to rabbit skin produced skin carcinomas. It was the first demonstration that a chemical could produce cancer in animals and these two scientists were later considered the Fathers of Experimental Chemical Carcinogenesis. In 1932, Kennaway and his coworkers reported the production of carcinogenic tars by means of pyrolysis of organic compounds consisting only of carbon and hydrogen.

In 1930s, Kennaway isolated the first chemical carcinogen from coal tar that was then identified as benzo[a]pyrene, a prototypical PAH (Fig. 2) and perhaps the most studied chemical carcinogen.

In 1935, Sasaki and Yoshiada demonstrated that feeding of azo dye o-aminoazotoluene to rats resulted in development of liver cancer. Kinosita (1936) demonstrated that the administration of 4-dimethylazobenzene in the diet also causes neoplasia in the liver. Acetylaminoflourene was demonstrated to induce neoplasm of the mammary gland, ear duct and liver in rats; and of bladder in mice. 2-Napthytlamine and several other aromatic amines have been known to be carcinogenic for the urinary bladder in humans (Fig. 3). These observations, in contrast to those that were obtained for PAHs (Fig. 2) suggested the induction of carcinogenesis at sites distant from the site of application of the carcinogen.

The major question that came out from above observations was: How a carcinogen can induce carcinogenesis sites distant from the site of their application? In their attempts to solve this mystery, Elizabeth and James Miller in 1950s for the first time demonstrated that azo dyes became bound to

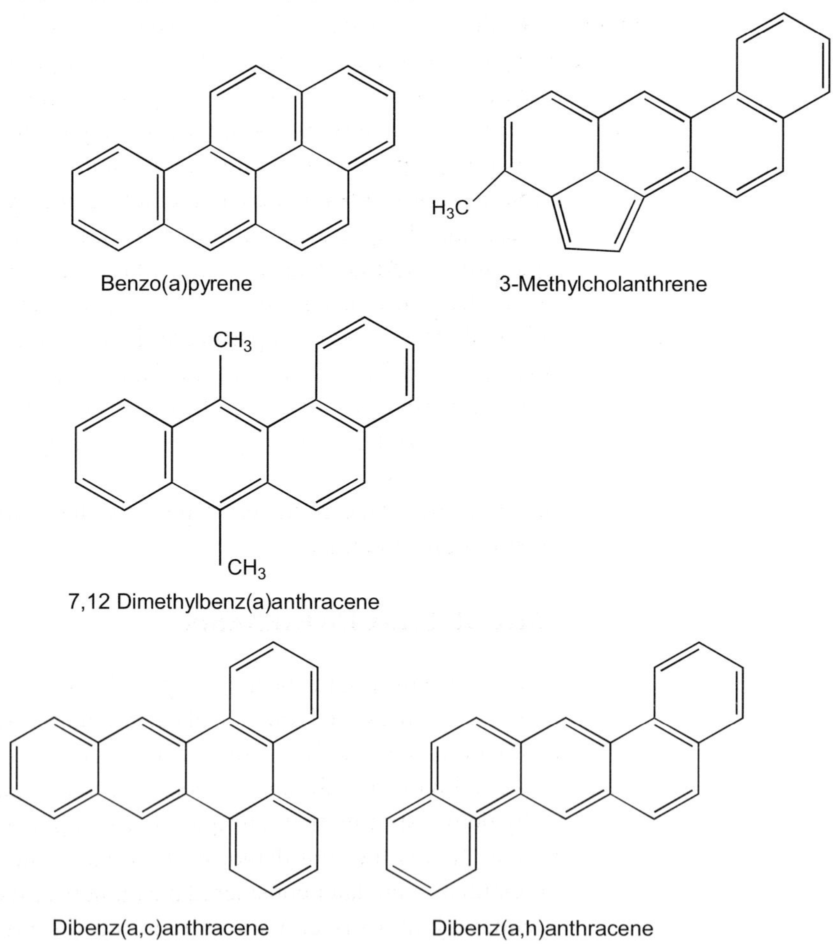

Fig. 2. Structures of some carcinogenic PAHs.

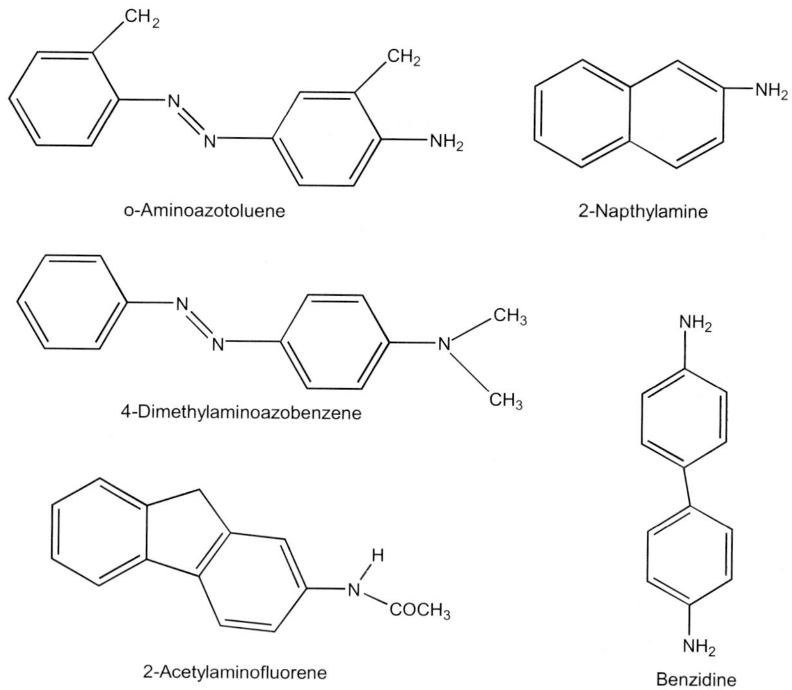

Fig. 3. Some carcinogenic aromatic amines.

proteins of the liver but not to the proteins of the resulting neoplasms. Number of studies done later on by Millers and other researchers strongly suggested that a critical step in the induction of cancer by chemicals is the covalent interaction of some form of the chemical with cellular macromolecules. They also observed that the parent compound was incapable of covalent binding directly with the macromolecules hence concluding that interaction of the chemical with the macromolecules was the result of metabolic alteration of the parent compound. Their later findings on azo dyes and PAHs paved the way to further investigations of the activation of carcinogens by means of their metabolic activation.

GENOTOXIC VS. EPIGENETIC CARCINOGENS

Genotoxic carcinogens are electrophilic (attract electrons) directly or converted by enzymes to positively charged electron deficient chemical species. These reactive electrophiles interact with negatively charged electrons on biological molecules such as nucleic acids and proteins forming covalent adducts. Covalent adducts on DNA bases if not ezymatically repaired lead to fixation of nucleotide substitution or frame-shift mutations. Cells with un-repairable DNA damage usually undergo programmed cell death (apoptosis). Un-repaired DNA damage can be passed on to daughter cells. Dysfunction of gene products leads to development of a cancer cell. Epigenetic carcinogens are the carcinogens that do not interact with DNA. These increase rates of DNA synthesis, mitosis and cell replication. These promote tumour growth (tumour promoters). Generally, do not require metabolism to exert a biological effect.

PROTO-ONCOGENES AND TUMOR SUPPRESSOR GENES

There are approximately 100 known human proto-oncogenes and tumour suppressor genes.

Proto-oncogenes first identified in viruses, these are normal cellular genes that control cell growth (proliferation), specialization (differentiation), death (apoptosis). Almost all proto-oncogenes encode signal transduction proteins. Upon activation proto-oncogenes get converted into oncogenes i.e. oncogenes are activated, unregulated versions of proto-oncogenes. Single copies of oncogenes are sufficient to result in malignant transformation. 20% of human tumours involve activation of proto-oncogenses. *ras* oncogene is a typical example of proto-oncogenes. It is an intermediate on/off switch in cell signal transduction to nucleus activating point mutations found in animal carcinogenesis models and in human cancer. Model carcinogens include methylnitrosourea and dimethylbenz[a]anthracene. It has been observed to be activated in certain human cancers including colon, lung, pancreatic malignancies.

Tumour suppressor genes are functional opposites of oncogenes, hence originally named anti-oncogenes. These are normal cellular genes that limit tissue growth and contribute to the destruction of cells with damaged genomes (i.e. they are involved in cell cycle regulation). Almost all tumour suppressor genes encode signal transduction proteins. They cause carcinogenesis upon deactivation only. Mutational inactivation results in loss of function. Examples include p53 tumor suppressor gene. This is a transcription factor that up-regulates factors (e.g. p21) that inhibit cell cycle progression. It is inactivated by point mutations that results in loss of function. p53 is found mutated in many human cancers.

DIRECT AND INDIRECT CHEMICAL CARCINOGENS

The activity of chemical carcinogens is directly related to the ability to covalently modify DNA. For most of the carcinogens, this process requires metabolism (indirect) whereas, for others metabolism is not required (direct). Various states of indirect carcinogens could be described as follows:

Pro-carcinogen: These are the parent compounds which are upstream of the ultimate mutagen.

Ultimate-carcinogen: These are the reactive metabolites that covalently bind to DNA and modify it.

Proximate-carcinogen: It is an intermediate metabolite between pro- and ultimate-carcinogen.

METABOLISM OF CARCINOGENS

Direct chemical carcinogens do not require metabolic activation. Indirect chemical carcinogens require metabolic activation to generate the ultimate carcinogens. Carcinogen metabolism may be divided into Phase I and Phase II reactions.

Phase I Reactions

The Phase I monooxygenase system contains three main components: Cytochrome P450, NADPH-cytochrome P450 Reductase, and Lipid. Cytochrome P-450 (CYP450) acts as the terminal oxidase in this system and catalyses the insertion of one atom of molecular oxygen into a substrate to convert the substrate to a more polar compound while converting the other atom of oxygen into water, using electrons donated by NADPH via NADPH P450 reductase. In the general reaction of CYP450, the oxidized ferric form of the P450 first interacts with a specific substrate. Subsequently an electron is transferred from NADPH via the reductase to produce the ferrous P450 substrate complex, which then binds molecular oxygen. After the oxidation of ferrous cytochrome back to the ferric form the oxygen becomes activated. Another electron is then transferred from NADPH via the reductase and this results in stereo-specifc oxidation of the substrate, formation of water, and regeneration of the ferric form of CYP450. The substrate free oxidized CYP450 can then re-enter the catalytic cycle (Fig. 4).

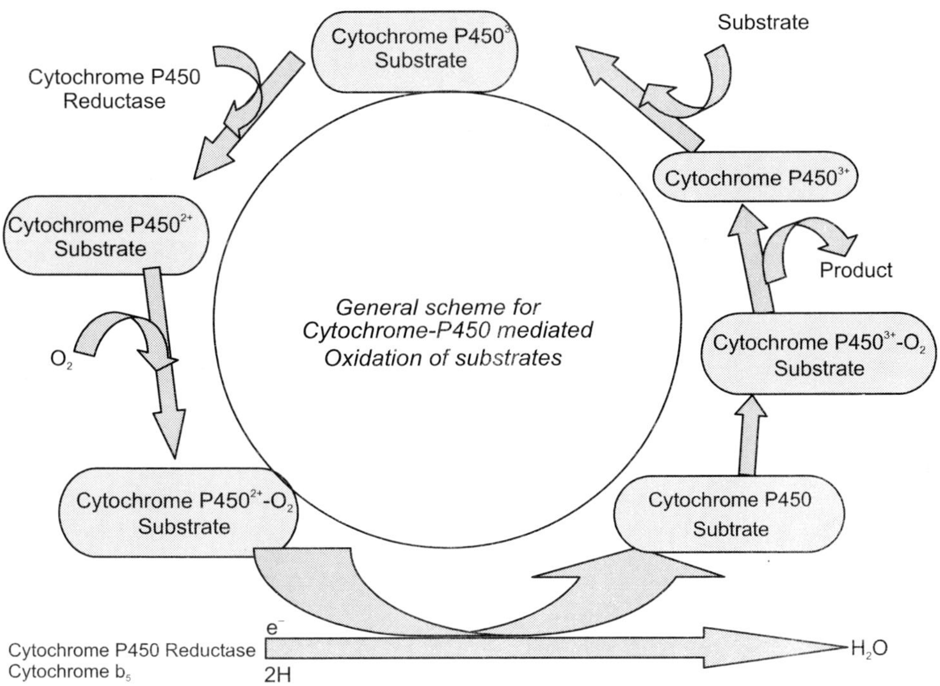

Fig. 4. Cytochrome P450 mediated monooxygenase reactions.

Although a wide variety of oxidative reactions are mediated by the CYP450-dependent monooxygenase system, majority of these reactions can be visualized essentially as hydroxylations.

C-Hydroxylation

Aromatic groups can be hydroxylated or epoxidized, sometimes resulting in the production of metabolites with toxic potential. Examples include metabolites of benzene and PAHs. Aliphatic compounds can also be hydroxylated e.g. n-hexane is metabolized to the ultimate neurotoxin, 2,5-hexanedione through this process. The C-hydroxylation of N-nitrosodialkylamines produces unstable intermediates that rearrange to release reactive carbonium ions. The latter act as the ultimate carcinogenic metabolites by resulting in the methylation of oxygen or nitrogen atoms in the purines and pyrimidines of DNA (Fig. 5).

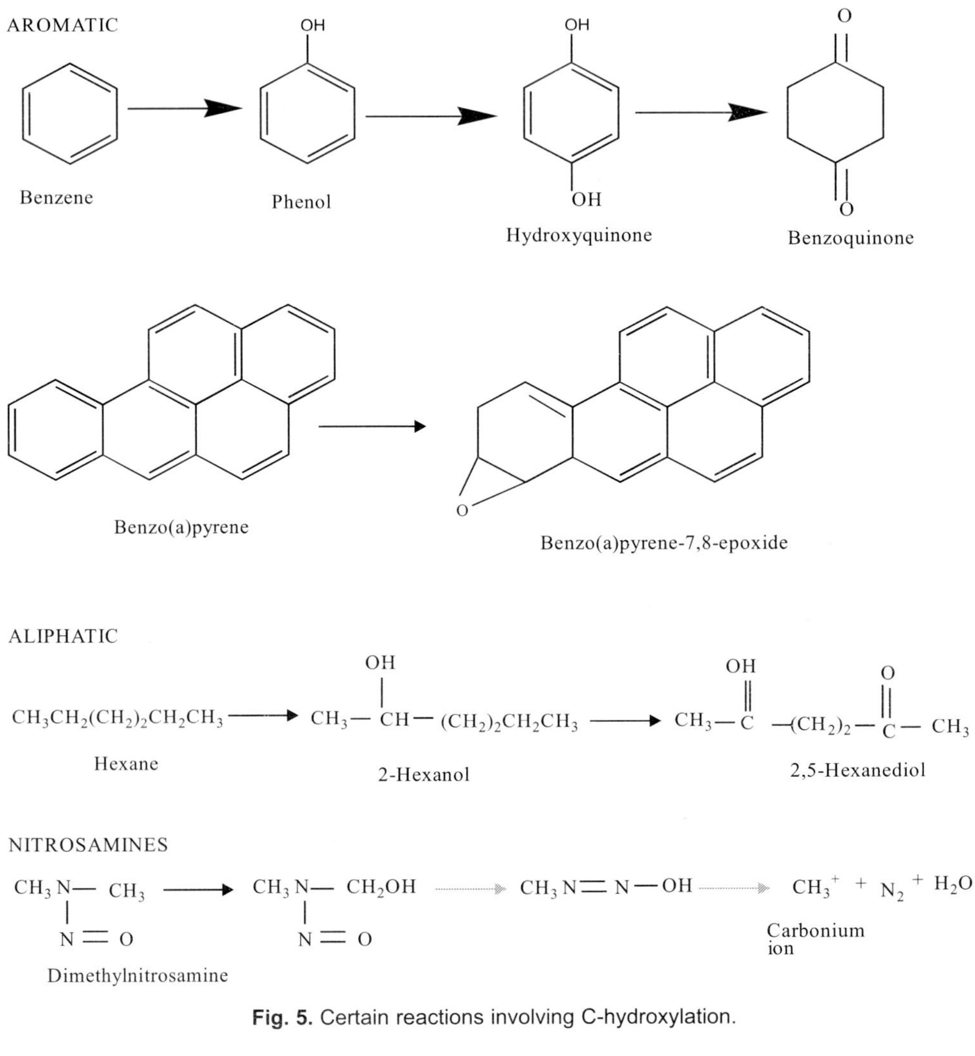

Fig. 5. Certain reactions involving C-hydroxylation.

N-Hydroxylation

The N-hydroxylation of arylamines and acetylamines (such as benzidine, 4-aminobipheynyl, 2-acetyleaminofluorene and hetrocyclic arylamines formed by food pyrolysis) is a crucial initial step in the production of the ultimate carcinogens. The N-hydroxylamines are considered proximate carcinogens because they are generally more mutagenic and carcinogenic than the parent compounds (Fig. 6).

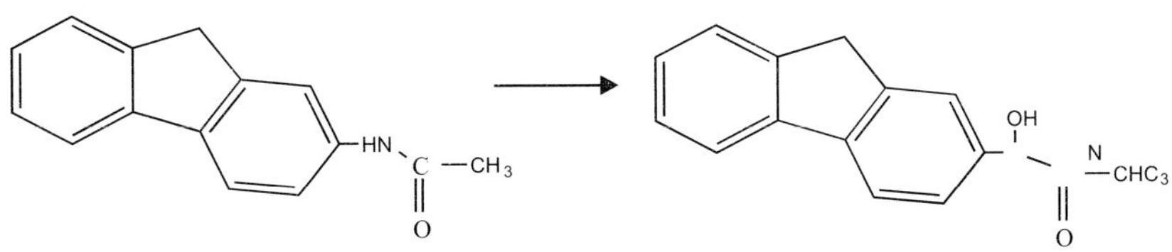

2-Acetylaminofluorene N-Hydroxy-acetylaminofluorene

Fig. 6. N-hydroxylation of 2-acetylaminoflourene.

S-Oxidation

As an example of S-oxidation of thionosulfur compounds, the insecticide phorate is oxidized to the S-sulfoxide, which can be further metabolized to the sulfone and the phorateoxon sulfone. Many of these sulfoxide metabolites are cytotoxic (Fig. 7).

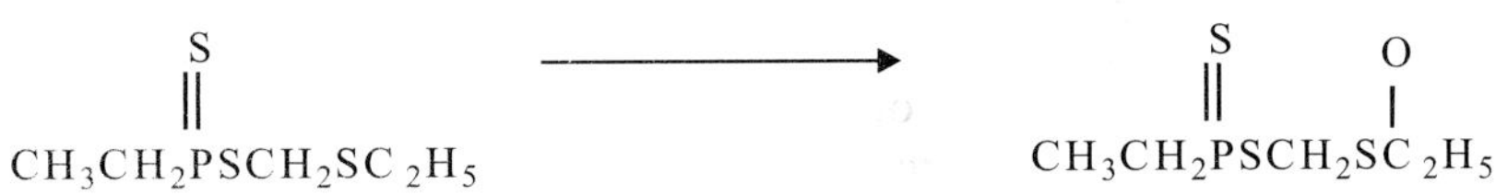

Phorate Phorate sulfoxide

Fig. 7. S-Oxidation of phorate.

Dealkylation

Dealkylation reactions at nitrogen, oxygen, or sulfur are also catalysed by the cytochrome P-450 enzymes. Examples of these reactions include the O-dealkylation of various drugs such as pheacetin to acetaminophen, the N-demethylation of aminopyrine, and the S-dealkylation of the 6-methylthiopurine to 6-mercaptopurine (Fig. 8).

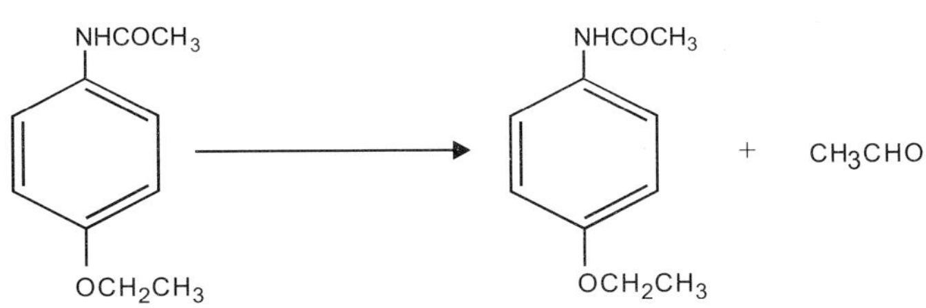

Phenacetin Acetaminophen

Fig. 8. Dealkylation of phenacetin.

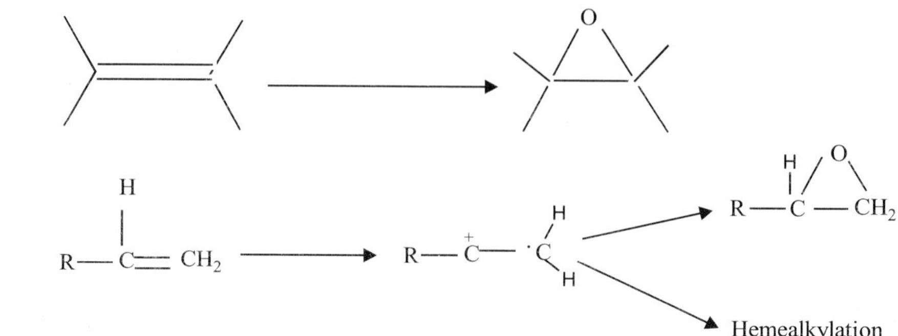

Fig. 9. Epoxidation of olefinic centres.

Epoxidation of Olefinic Centres

Olefinic compounds such as styrene, haloalkenes such as vinylidene chloride and thrichloroethylene, and dihydrofurans such as aflatoxin are activated at the carbon-carbon π-bond by expoxidation (Fig. 9). Many of these expoxides can react with cellular macromolecules and are genotoxic. These reactions can also result in the alkylation and suicide-based inactivation of the heme group of cytochrome.

Oxidative and Reductive Dehalogenation

A variety of halogenated alkanes, alkenes and alkynes are oxidized at the carbon-hydrogen bond to produce unstable intermediates that dehalogenate to form aldehydes or ketones (Fig. 10). Examples are the oxidation of ethylene dibromide to 2-bromacetaldehyde and the oxidation of chloroform to phosgene. The metabolites to the mono- and di-halomethanes are only weakly cytotoxic. However, trihalomethate metabolites such as phosgene produce hepatic and renal toxicity. The tetra-halogenated methanes produce toxicity by a different mechanism. These compounds undergo reductive dehalogenation to produce a free radical. Carbon tetrachloride produces both hepatic and renal damage via this mechanism.

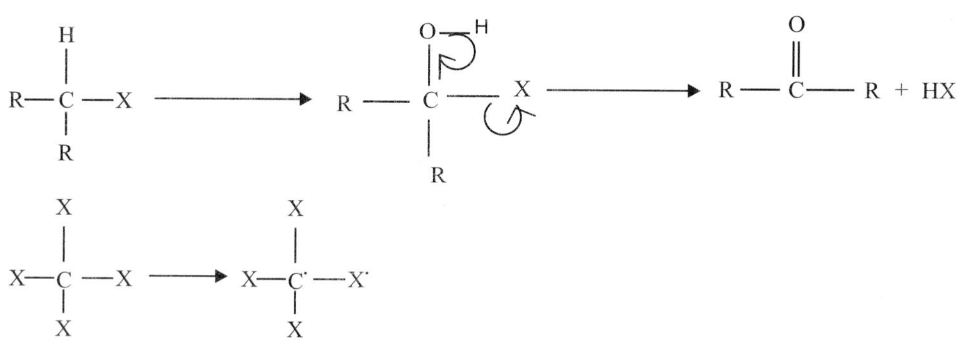

Fig. 10. Oxidative and reductive dehalogenation of halogenated hydrocarbons.

Phase II reactions

Epoxide Hydrolases

Epoxides are reactive metabolites that are generated principally by the CYP450 dependent pathways. The epoxide hydrolases are considered to be protective enzymes that hydrate the epoxides to form

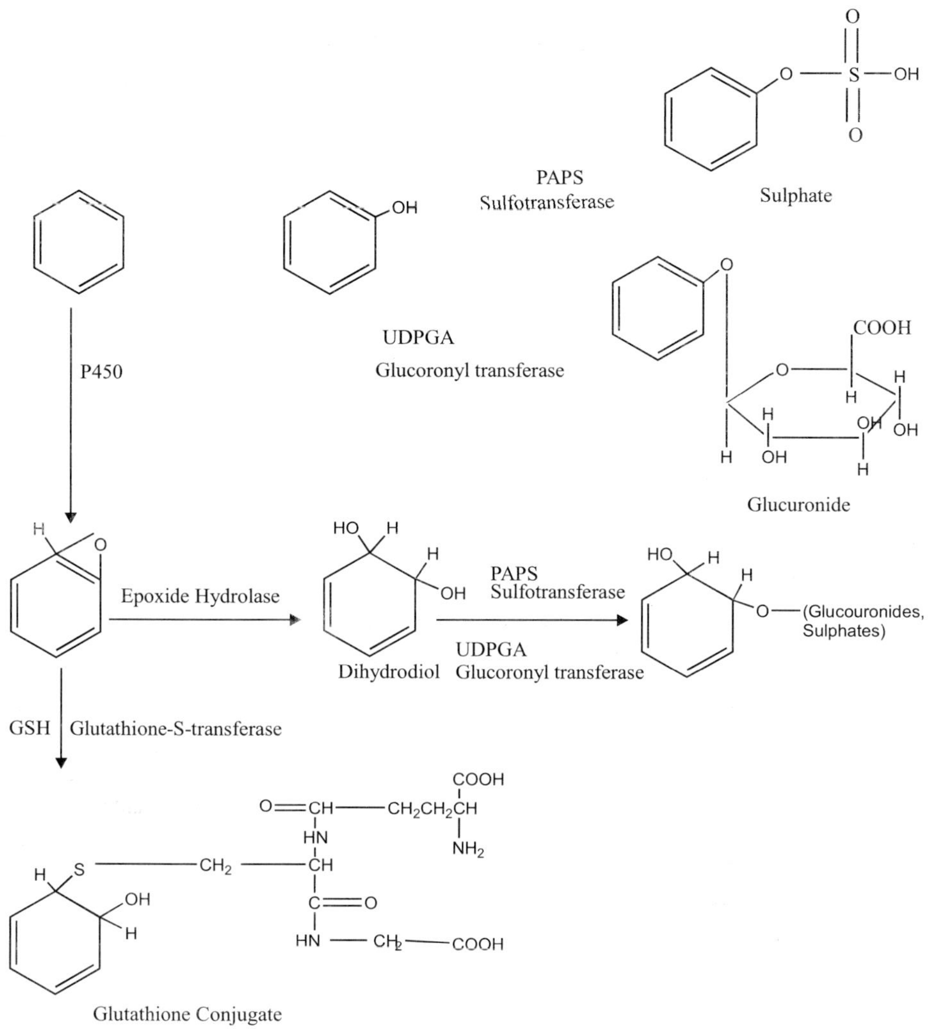

Fig. 11. Some phase II biotransformation reactions.

trans 1,2-dihydrodiols. These less reactive dihydrodiols can then be conjugated with glucuronic acid, sulfate, or glutathione and excreted. However, the activity of epoxide hydrolase can also result in the production of more reactive species, as in the hydrolysis of benzo(a)pyrene-7,8-oxide to its dihydrodiol, which is then metabolised by CYP450 to the ultimate carcinogen, benzo(a)pyrene-7,8-dihydrodiol-9,10-epoxide. There are multiple forms (isoforms) of epoxide hydrolase. The two forms that metabolise xenobiotics are microsomal exoxide hydrolase (mEH) and cytosolic exoxide hydrolase (cEH). The mEH metabolises a wide variety of substrates, including both K-region and non-K-region arene oxides of polycyclic hydrocarbons, butadiene, vinyl chloride, exoxides derived from herbicides and insecticides, and epoxides derived from therapeutic drugs such as phenytoin. Bay region diol-epoxides of polycyclic hydrocarbons, on the other hand, are poor substrates. Microsomal epoxide hydrolase also has a small role in the detoxification of aflatoxin B_1 8,9-oxide, the major species responsible for genotoxicity of aflatoxin B_1. These compounds are detoxified primarily by glutathione conjugation. Cytosolic epoxide hydrolase is found primarily in the cytosol and peroxisomes. It specifically catalyses the hydrolysis of mono- and di-substituted oxiranes, but does not metabolise arene oxides appreciably. Substrates include trans-substituted styrene oxides, stilbene oxides, and lipid epoxides (e.g., epoxide derivatives of arachidonic acid) (Fig. 11).

Glutathione S-Transferases

There are multiple forms of the soluble glutathione S-transferases that can be divided into four gene families based on their electropheretic mobilities. The group consists of basic proteins, is a neutral class, and the enzyme is acidic. These enzymes transfer glutathione (the tripeptide-glutamyl cysteinyl glycine) to compounds that the electrophilic per se as well as those generated by Phase-I reactions. Thus, glutathione transferases can play a protective role by covalently binding and removing reactive electrophiles. Examples are the detoxification of benzene epoxide, aflatoxin B_1-8,9-oxide, and benzo[a]pyrene-7,8-dihydrodiol-9,10-oxide. However, some pro-carcinogens are activated by conjugation with glutathione to produce unstable reactive metabolites. Examples of such genotoxic metabolites are the glutathione conjugates of ethylene dibromide and methylene chloride (Fig. 12).

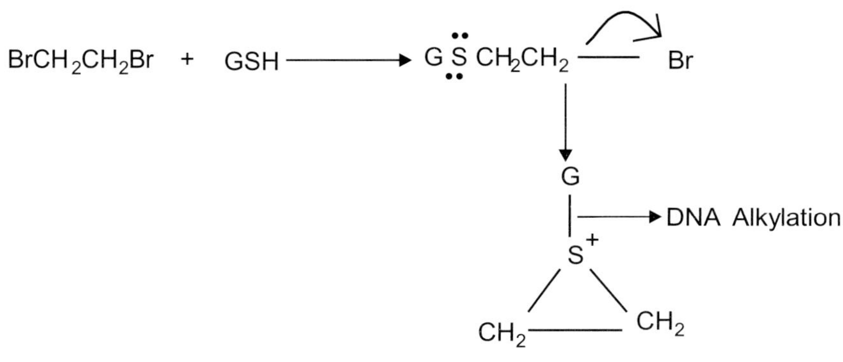

Fig. 12. GST mediated biotransformation of dibromoethene.

UDP – Glucuronosyltransferases

Uridine diphosphate glucuronosyltransferases represent a major phase II family of endoplasmic reticulum-bound enzymes that utilize the cofactor uridine diphosphoglucuronic acid (UDPGA) to conjugate glucuronic acid to a large number of potentially reactive compounds including phenols, dihydrodiols, quinines, and quinols. The addition of glucuronic acid renders these compounds more water-soluble and enables the body to excrete them more readily. Examples of substrates that are conjugated with glucuronic acid by this system include the phenols, dihydrodiols, quinines, and quinols of numerous compounds such as benzene, benzo[a]pyrene, and the chlorinated benzenes. Conjugates can include ether O-glucuronides, ester O-glucuronides, N-glucuronides, S-glucuronides, and C-glucuronides. Although glucuronidation is a major route of detoxification for many compounds, glucuronidation is involved in the etiology of human colorectal cancer from heterocyclic amines. These compounds are first N-oxidated and then N-glucuronidated. The N-hydroxy-metabolites can be transported to the colon where they are hydrolysed by glucuronidases. The N-hydroxy derivative

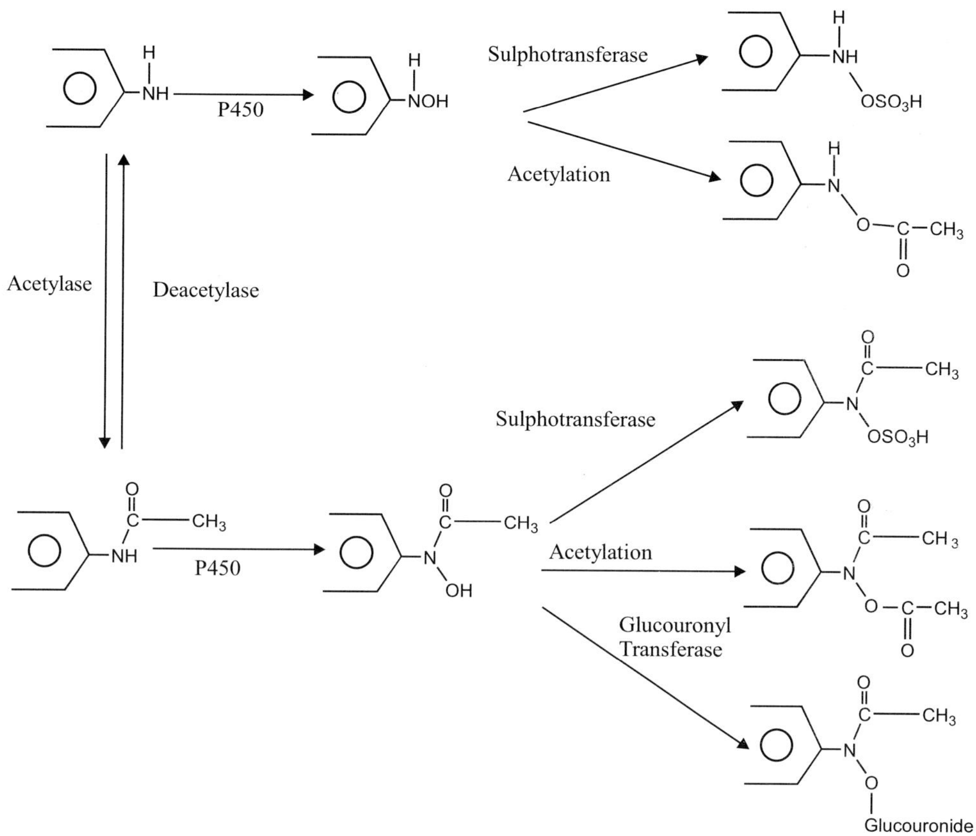

Fig. 13. Acetylation and sulfation of acetyl amines.

can then be O-acetylated in the colon mucosa, which results in formation of N-acetoxyarylamines, which can react with DNA to form covalent adducts.

Acetylation

Acetylation involves the enzymatic or non-enzymatic transfer of the acetyl group from acetyl coenzyme A to molecules containing a primary amine, hydroxyl, or sulfhydryl group. These enzymes are distributed in many organs, but the liver is a major site of N-acetylation. Acetylation of the primary amine group of arylamines and hydrazines is a major route of bio-transformation of these compounds. Arylamines of primary arylamines are inter-converted by N-acetyltransferases and deacetylases. They are initially activated by the cytochrome P-450s and flavin monooxygenase enzymes to form N-hydroxyarylamides and N-hydroxyarylamines that can be inter-converted by enzymatic N-deacetylation/N-acetylation. These compounds are considered proximate carcinogens because they are more mutagenic and carcinogenic than the parent compounds. N-hydroxyarylamides are converted to the ultimate carcinogens through conjugation with sulphuric, acetic or glucuronic acids. N-acetoxyarylamines are highly reactive and appear to have an important role as ultimate carcinogens. Therefore, O-acetylation appears to play an important role in the activation of aromatic amines and food pyrolysates classified as human carcinogens, including 2-aminofluorene, benzidine, 4-aminobiphenyl, and 2-naphthylamine. Acetylation may affect the carcinogenicity of aromatic amines in two ways. In bladder cancer, acetylation appears to be a detoxification mechanism for aromatic amines such as 2-naphthylamine and benzidine. Epidemiological studies in various populations including workers in the dye industry have found that bladder cancer is more common in the slow acetylator phenotype than the rapid acerylator phenotype. On the other hand, epidemiological studies indicate that slow acetylators have a decreased risk for colorectal cancer, presumably because the glucuronides of aromatic amines are hydrolysed in the colon, and then acetylated to reactive N-acetoxyarylamines in colon mucosa as discussed earlier. Studies of the aromatic amine 2-aminofluorene indicate that acetylation and sulfation yield more reactive electrophiles which react with nucleic acids to produce adducts Fig. 13.

Sulfation

Sulfate conjugation improves water solubility and excretion. The sulfotransferases, are a family of enzymes that transfer the sulfuryl group from adenosine 3-phosphate 5-phosphosulphate to hydroxyl or amino groups. Conjugation with sulfate is generally considered a detoxification mechanism. However, the addition of sulfate also introduces a good learning group on unstable compounds such as hydroxylamines. The N-sulfonyloxy derivatives are considered to be among the predominant metabolites of 2-acetylaminofluorene responsible for its carcinogenicity via the formation of DNA adducts.

MULTISTAGE NATURE OF CHEMICAL CARCINOGENESIS

Initiation

It refers to heritable change in a cell resulting from the action of an initiating agent as a result of a direct alteration in the molecular structure of the genome of the cell. Chemical carcinogen reacts with DNA to produce a mutation. Initiation involves carcinogen metabolism, DNA repair and cell proliferation. Initiation is irreversible but an initiated cell is not a cancer cell and does not necessarily become one (Fig. 14).

Fig. 14. Initiation of carcinogenesis.

Promotion (and Malignant Conversion)

The stage of neoplastic development characterized by a change in the expression of genetic information, resulting from the epigenetic action of a promoting agent. Promoting agents influence proliferation of initiated cells to form benign lesions, which may regress or acquire additional mutations to become malignant neoplasms. Malignant conversion is another sub-phase of promotion phase of carcinogenesis that involves transformation of initiated cell to transformed cell and then the clonal expansion of the latter (Fig. 15).

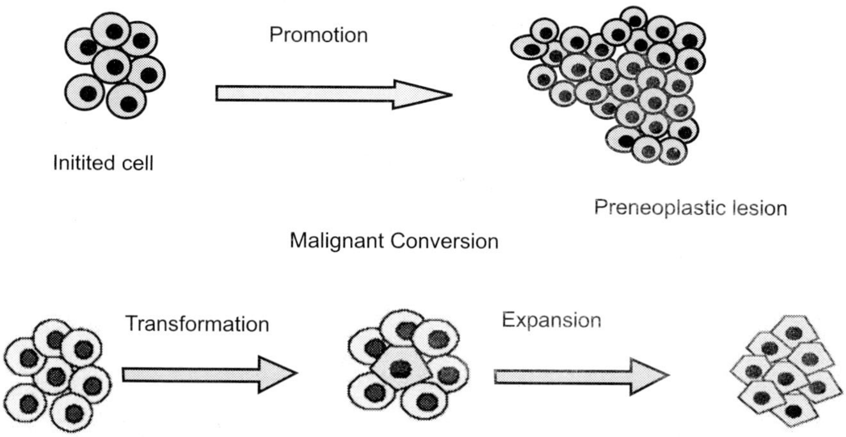

Fig. 15. Promotion of carcinogenesis

Progression

It is characterized either by demonstrable changes in the number and/or arrangement of genes (as evidenced by nucleic acid sequence and/or hybridization studies) or by visible karyotypic alterations (as evidenced by light microscopic techniques) within a majority of the neoplastic cells that make up the tumor. These alterations in turn are associated with increased growth rate, increased invasiveness, metastases, and alterations in biochemical and morphologic characteristics of the neoplasm. Tumors acquire ability to invade and establish distant metastases (Fig. 16).

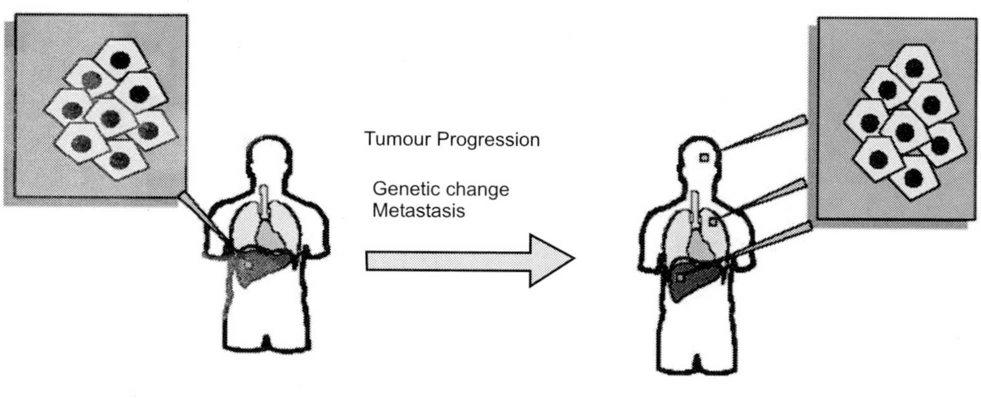

Fig. 16. Progression of carcinogenesis

CONTROL MEASURES FOR CANCER

The number of treatment choices depends on the type of cancer, the stage of the cancer, and other individual factors such as age, health status, and personal preferences. Cancer treatment may include surgery, radiation therapy, chemotherapy, hormone therapy, and biological therapy. Sometimes combination therapy is used depending on the type and location of the cancer, status of the metastasis of the disease, the patient's age and general health etc. Because treatment for cancer can also damage healthy cells and tissues, it often causes undesirable side effects and sometimes the side effects of treatment prove to be worse than the disease itself. Various treatment options available are described below:

Surgery

Surgery deals with operation to remove cancer. In theory, cancers can be cured if entirely removed by surgery, but this is not always possible, especially when the cancer has metastasized to other sites in the body prior to surgery. The goal of the surgery can be either the removal of only the tumor, or the entire organ. Examples of surgical procedures for cancer include mastectomy for breast cancer and prostatectomy for prostate cancer. A single cancer cell is invisible to the naked eye but can regrow into a new tumor, a process called recurrence. In addition to removal of the primary tumor, surgery is often necessary for staging, e.g. determining the extent of the disease and whether it has metastasized to regional lymph nodes. Staging is a major determinant of prognosis and of the need for adjuvant therapy. The side effects of surgery depend on many factors, including the size and location of the tumor, the type of operation, and the general health. One of the major side effects of surgery is the spread of the disease if not handled with utmost care.

Radiotherapy

Radiation therapy (also called radiotherapy, X-ray therapy, or irradiation) is one of the most common treatments for cancer. It uses beams of high-energy waves or particles (for example, x-rays, gamma rays, or alpha and beta particles) to kill or damage cancer cells. The powerful stream of energy, which is thousands of times more intense than the rays used for a routine chest x-ray, damages the DNA of cancer cells, rendering them unable to reproduce and grow. Radiation can be given externally by a machine that aims radiation at the tumor area. It can also be given internally: needles, seeds, wires, or catheters containing a radioactive substance are placed directly in or near the tumor. Radiation therapy is the primary treatment for many types of cancer, including certain cancers of the lung, breast, cervix, prostate, testicles, bladder, thyroid, larynx and brain, as well as early-stage Hodgkin's disease and non-Hodgkin's lymphoma. In some cases, radiation therapy is the only treatment needed; in others, it is used in combination with surgery or chemotherapy. Radiation can be used before surgery to shrink a tumor so that it is easier to remove. After surgery, it is used to destroy microscopic extensions of cancerous tissue around a tumor that surgery might have missed. Like all cancer therapies, radiation therapy can have side effects. These include: skin irritation, temporary change in skin color in the treated area, temporary or permanent loss of hair in the area being treated, and fatigue. Other side effects depend on the area of the body being treated. Radiation therapy may also cause a decrease in the number of white blood cells, the body protection against infection.

Chemotherapy

While surgery and radiation therapy are used to treat localized cancers, chemotherapy is used to treat cancer cells that have metastasized (spread) to other parts of the body. Chemotherapy is the treatment of cancer with drugs ("anticancer drugs") that can destroy cancer cells, however, healthy cells can also be harmed, especially those that divide quickly. Chemotherapy drugs are designed to kill or disable rapidly dividing cells. Many of these drugs act on human cells much like some antibiotics act on bacteria: They prevent the replication of DNA, a critical step in cell division, thereby preventing the cells from multiplying. Other anticancer drugs interfere with enzymes produced by tumor cells that help them to grow. The most effective treatment involves two or more anticancer agents to tackle unrestrained cell division from several angles. This is called combination chemotherapy. Adjuvant means "in addition to," and in cancer treatment it refers to treatments used in addition to the primary treatments of surgery and radiation. Chemotherapy is usually an adjuvant therapy. Sometimes, however, it is used to shrink a tumor before surgery or radiation therapy and makes it easier to remove or destroy the cancerous mass. This is called neo-adjuvant chemotherapy. Chemotherapy can also be used to shrink a tumor in order to alleviate symptoms it may be causing, such as back pain. Though some anticancer drugs can be taken orally or through an injection, most require intravenous (IV, through a vein) administration. Chemotherapy is often given in cycles; for example, two weeks of therapy followed by two weeks without therapy, over a period of several months. Cancer cells do not all divide at the same time and staggering treatments can capture different populations of cells in the midst of growing and dividing, when they are most vulnerable to the chemotherapy drugs. The disadvantage of chemotherapy is that healthy cells can also be harmed by anticancer drugs, especially those cells that normally divide quickly. These include cells in the hair follicles, bone marrow, and lining of the gastrointestinal tract. The result can be hair loss; depressed red and white blood cell counts, causing anemia and an inability to fight off infections, respectively; and nausea, vomiting and mouth sores. Chemotherapy can also have several neurological side effects, such as fuzzy thinking and difficulty concentrating. The side effects of chemotherapy depend mainly on the drug(s) and the dose(s) the patient receives.

Hormone therapy

It is used to treat certain cancers that depend on hormones for their growth. It works by keeping cancer cells from getting or using the hormones they need to grow. This treatment may include the use of drugs that stop the production of certain hormones or that change the way hormones work. Another type of hormone therapy is surgery to remove organs that make hormones. For example, the ovaries may be removed to treat breast cancer, or the testicles may be removed to treat prostate cancer. Hormone therapy can cause a number of side effects. Patients may feel tired, or have fluid retention, weight gain, hot flashes, nausea and vomiting, changes in appetite, and, in some cases, blood clots. Hormone therapy may also cause bone loss in pre-menopausal women. Depending on the type of hormone therapy used, these side effects may be temporary, long lasting, or permanent.

Biological therapy

There is a lot of evidence that suggests that the immune system, the body's natural defence mechanism, plays a major role in the body's response to cancer. At least some forms of cancer occur when the immune system fails to destroy cancer cells or to prevent their growth. Biologic therapy is an effective treatment for certain cancers. It is sometimes called immunotherapy, biotherapy, or biological response modifier therapy. Biologic therapies use the body's immune system to fight cancer or to lessen the side effects of some cancer treatments. Biologic therapies can act in several ways in cancer treatment. These include interfering with cancer cell growth, acting indirectly to help healthy immune cells control cancer, and helping to repair normal cells damaged by other forms of cancer treatment. There are several kinds of biologic therapy now in use. More than one kind of biologic therapy may be used, or biologic therapy may be combined with chemotherapy or radiation therapy to treat cancer. Monoclonal antibodies, interferon, interleukin-2, and colony-stimulating factors are some types of biological therapy. The side effects caused by biological therapy vary with the specific treatment. In general, these treatments tend to cause flu-like symptoms, such as chills, fever, muscle aches, weakness, loss of appetite, nausea, vomiting, and diarrhea. Patients also may bleed or bruise easily, get a skin rash, or have swelling. These problems can be severe, but they go away after the treatment stops.

CANCER PREVENTION

Although there is no 'magic bullet' that can completely conquer cancer, but the risk of development of cancer may be avoided. Epidemiological studies indicate that approximately 80% of human cancer is caused by exposure to chemical carcinogens in tobacco smoke, in the diet, and in the workplace. Given these observations, at least three approaches to the prevention of cancer can be envisioned. First, reduce human exposure to environmental carcinogens through careful monitoring of the workplace and through educational approaches to encourage changes in lifestyle.

Since we have the names of many of the chemicals that are known to cause cancer, it only makes sense that eliminating carcinogenic chemicals from regular use would be the first way of dealing with cancer—preventing it. Cancer risk can be reduced by eliminating the identified carcinogens or at least minimizing exposure to them but, without complete identification of the corresponding risk factors, such primary prevention might be difficult to implement. Furthermore, the avoidance of some risk factors could require large lifestyle changes, which are not easy to implement. Second, identify individuals at high risk for cancer development through predisposing genetic or biochemical factors, followed by appropriate clinical follow-up. Third, provide chemoprevention by dietary or synthetic means. For several reasons, chemoprevention has received growing consideration as a means of cancer control. Primary cancer prevention requires removal of exposure to etiologic agents. Although this is an important approach to cancer prevention, it is not always effective, as evidenced by the marginal success of tobacco cessation programs. Moreover, numerous populations at high risk for certain types of cancer may already have received considerable exposure to etiologic agents, and many human cancers cannot be ascribed to specific agents. Thus, preventive strategies that do not require prior knowledge of specific etiological factors have great appeal. Additionally, the success obtained in chemoprevention of cancer in animal models provides a strong mandate for this approach to cancer prevention in humans.

NEED FOR CHEMOPREVENTION

Treatment options available for cancer therapy involve the risk of life-threatening side effects so that the search therefore continues to develop the therapy that could selectively act on tumor cells. The principal need in the chemoprevention of cancer remains the discovery of new agents that are effective and safe, and the development of new dose-scheduling paradigms that will allow their beneficial use over relatively long periods of time (but not necessarily constantly) in a manner that is essentially free of undesirable side effects. Cancer chemoprevention can be defined as the prevention, inhibition, or reversal of carcinogenesis by administration of one or more chemical entities, either as individual drugs or as naturally occurring constituents of the diet. Knowledge of chemoprevention science and its application in clinical studies has been growing rapidly over the last few years.

CLASSES OF CHEMOPREVENTIVE AGENTS

Absolute classification of chemopreventive agents is difficult because the precise mechanisms of action are not known for many compounds. In addition, many chemopreventive agents act through more than one mechanism, making it difficult, if not impossible, to establish the most effective mode of action. The classification scheme developed by Wattenberg (1980) is based essentially on the time period during which agents appear to exhibit activity in animal models of carcinogenesis. On this basis, chemopreventive agents are classified as inhibitors of carcinogen formation, blocking agents, and suppressing agents. Blocking agents are inhibitors of tumor initiation, while suppressing agents are inhibitors of tumor promotion/progression.

According to the conventional classification originally proposed by Lee Wattenberg, chemopreventive agents are subdivided into two main categories —Blocking agents and Suppressing agents. Blocking agents prevent carcinogens from reaching the target sites, from undergoing metabolic activation or from subsequently interacting with crucial cellular macromolecules (for example, DNA, RNA and proteins). Suppressing agents, on the other hand, inhibit the malignant transformation of initiated cells, in either the promotion or the progression stage. Chemopreventive agent can block or reverse the pre-malignant stage (initiation and promotion) of multi-step carcinogenesis. They can also halt or at least retard the development and progression of precancerous cells into malignant ones. Recent advances in our understanding of the carcinogenic process at the cellular and molecular level have shown this blocking and suppressing categorization to be an oversimplification, and numerous cellular molecules and events that could be potential targets of chemopreventive agents have been more specifically identified. Therefore, the ability of any single chemopreventive agent to prevent tumour development should be recognized as the outcome of the combination of several distinct sets of intracellular effects, rather than a single biological response. The cellular and molecular events affected or regulated by these chemopreventive agents include carcinogen activation/ detoxification by xenobiotic metabolizing enzymes; DNA repair; cell-cycle progression; cell proliferation, differentiation and apoptosis; expression and functional activation of oncogenes or tumour-suppressor genes; angiogenesis and metastasis; and hormonal and growth-factor activity.

DETOXIFICATION SYSTEM AND CHEMOPREVENTION OF CANCER

Chemopreventive agents, which prevent cancer causing agents (carcinogens) from reaching or reacting with critical targets (cancer initiation) are called blocking agents, whereas those preventing the evolution of the neoplastic process in cells (cancer promotion and progression) are called suppressing agents. Currently, a number of blocking and suppressing agents are being tested using either pharmacological or, less often, nutritional regimens.

Chemopreventive blocking agents can prevent the occurrence of cancer by increasing the detoxification of chemical carcinogens, tumor promoters, and tumor progressors. The pathways of detoxification have been divided into 2 major categories: phase I reactions (oxidations, reductions, and hydrolyses) and phase II reactions (conjugations). Both phase I as well as II reactions are to make the toxin molecule more polar or water-soluble and therefore readily excreted. There has been a growing interest in the inhibitors that induce an increase in activity of phase II detoxifying enzymes, with a special emphasis on glutathione S-transferases. Glutathione S-transferases conjugate ultimate carcinogens, that is, very reactive electrophilic metabolites of carcinogenic compounds, with glutathione. There is also strong evidence for the detoxification of chemical carcinogens or their metabolites and tumor promoters/progressors by UDP-glucuronosyltransferases, which conjugate nucleophilic compounds produced by our bodies (endogenous) as well as foreign chemicals (exogenous), with D-glucuronic acid. Conjugation with D-glucuronic acid, that is, glucuronidation, appears to be the principal conjugation pathway in the tissues of all vertebrate species examined to date. The conjugates are excreted in the bile and urine or transported from one tissue (usually liver) to other tissues. UDP-glucuronosyltransferases are found in the endoplasmic reticulum and nuclear membranes and are markedly inducible by xenobiotic inducers (foreign compounds). High specific activity of nuclear UDP-glucuronosyltransferase in the nuclear membrane in proximity to genetic material suggests a very important role in the deactivation of both foreign and endogenous compounds such as steroid hormones. Thus, this enzyme could also serve as an important barrier for the nuclear genetic apparatus against mutagenic and carcinogenic or otherwise toxic chemical compounds. Carcinogens identified to date, which after metabolic activation are subject to glucuronidation, include polycyclic aromatic hydrocarbons, some nitrosoamines, aromatic amines, and fungal toxins. Certain tumor promoters including steroid hormones also undergo glucuronidation. The importance of conjugating enzymes such as UDP-glucuronosyltransferase, which uses nucleophilic substrates instead of reactive electrophiles, has only recently been recognized. It has become apparent that nucleophilic metabolites are often intermediates in formation of ultimate carcinogens. Furthermore, it is clear that glucuronides may be highly stable forms of these intermediates, which are subject to further activation after hydrolysis by the enzyme β-glucuronidase at sites distant from their site of formation. In fact, the elimination and/or deactivation of potentially damaging chemicals that undergo glucuronidation is limited not only by the rate of conjugation with D-glucuronic acid but also by the rate of de-glucuronidation by this ubiquitous enzyme.

TARGETING BIOTRANSFORMATION: PLEDGING APPROACH TO CANCER CHEMOPREVENTION

Chemopreventive agents influence the carcinogenic process by different mechanisms, however, the most significant impact is a result of effects that occur early in the process (Fig. 17). The initial step of carcinogen activation to reactive electrophiles is catalyzed by phase I enzymes (cytochrome P450 monooxygenase family). This step is followed by detoxification of the biotransformed intermediates by phase II enzymes which are readily inducible by dietary micro-constituents. If phase I enzyme

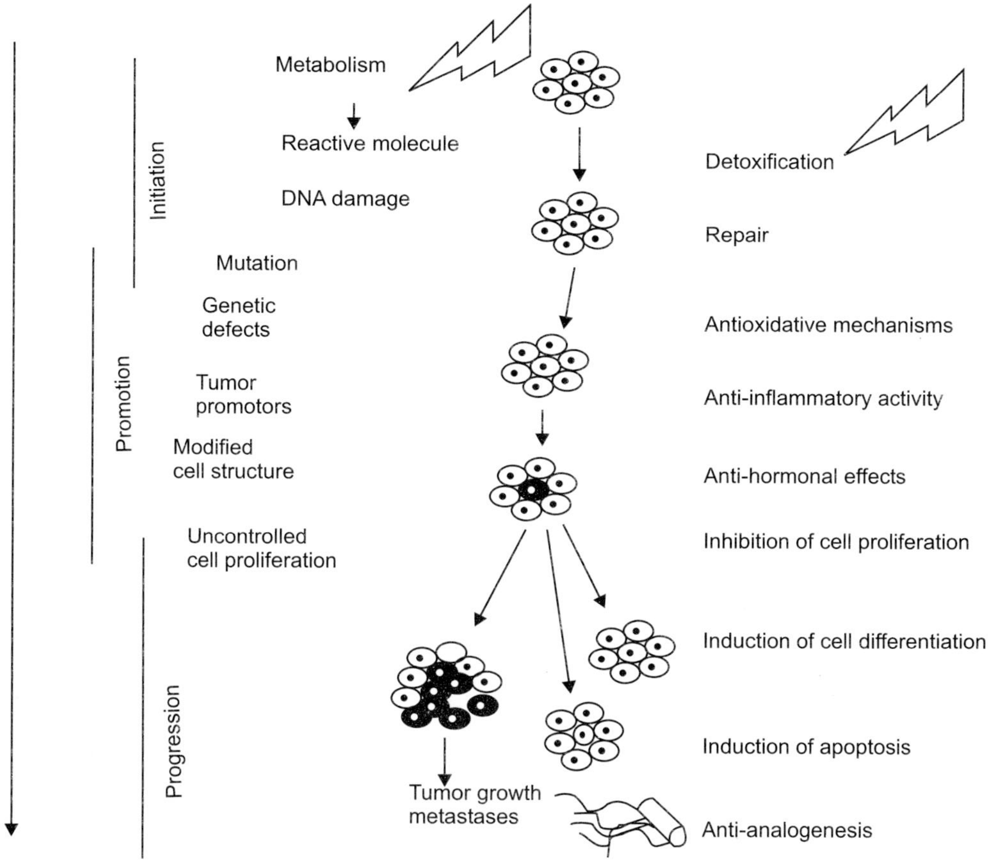

Fig. 17. Targets for chemoprevention of carcinogenesis

activity exceeds that of phase II enzyme activity, excessive covalent binding of the activated intermediates to cellular macromolecules may occur causing cellular toxicity, cellular injury, or carcinogenesis. An imbalance in the coordinated activities of these enzyme systems may result when the burden of exposure to carcinogens is too large to be compensated for by dietary inducers. This may be due either to an excessive exposure, an inadequate intake of dietary inducers, or both. Phase II enzymes are located primarily in the liver where detoxified compounds can be readily eliminated through the portal circulation protecting extra-hepatic tissues from potentially carcinogenic compounds that are absorbed in the gastrointestinal tract. Barrier epithelial tissues also contain phase II enzymes that can be induced to very high levels by dietary factors thus providing the gastrointestinal tract, lung, and skin with protection against incoming harmful exposure from damaging electrophiles. The significance of the effects of these dietary inducers on cancer may be substantial considering that 95% of all cancers are of epithelial origin. Many population-based studies have highlighted the ability of macronutrients and micronutrients in vegetables and fruit to reduce the risk of cancer. Recently, attention has been focused on phytochemicals—non-nutritive components in the plant-based diet that possess cancer-preventive properties. Despite remarkable progress in our understanding of the carcinogenic process, the mechanisms of action of most chemopreventive phytochemicals have not been fully elucidated. Chemopreventive phytochemicals can block initiation or reverse the promotion stage of multi-step carcinogenesis. They can also halt or retard the progression of precancerous cells into malignant ones.

MODULATORS OF BIOTRANSFORMATION ENZYME

One of the first cytochrome P450 inhibitors shown to exhibit chemopreventive activity was disulfiram, which inhibits the activation of dimethylhydrazine and colon cancer induced by this compound. The isothiocyanates are strong P450 inhibitors and among the most potent chemopreventive agents known. For example, dietary phenethyl isothiocyanate, at a concentration of 3 mmol/kg diet, can inhibit 4-(methylnitrosamino)-1-(3-pyridyl)-1-butanone (NNK)-induced lung tumors in F344 rats by approximately 50%. This concentration completely inhibits N-nitrosomethylbenzylamine (NMBA)-induced esophageal tumors in F344 rats. 6-Phenylhexyl isothiocyanate inhibits NNK-induced lung tumorigenicity by > 80% in strain A mice when administered at a dose of 50-fold lower than NNK. Diallyl sulfide, a naturally occurring constituent of Allium vegetables, inhibits carcinogen activation and tumorigenesis in several animal models. Ellagic acid inhibits benzo[a]pyrene metabolism in vitro, NMBA metabolism in vivo and in vitro, and inhibits NMBA-induced esophageal tumors. Enzymes. Another mechanism of action of blocking agents is through induction of cytochrome P450. Inducers of cytochrome P450 act either by increasing the metabolic activation of carcinogens in non-target tissues or by enhancing oxidative detoxification at any tissue site. Indole-3-carbinol (I3C) is a potent inducer of P450 enzymes and has chemopreventive activity in several animal models.

Inducers of phase II detoxifying enzymes are preferred to cytochrome P450 inducers as chemopreventive agents because they are less likely to produce cancers themselves. Sulforaphane, an isothiocyanate found in broccoli, is a potent inducer of GST and inhibits chemically-induced mammary cancer in rats. Another potent inducer of GST is the dithiolethione, oltipraz, which inhibits carcinogen-induced tumorigenesis in a number of animal models. Butylated hydroxyanisole (BHA) stimulates UDP-glucuronyltransferase activity, and this appears to be the mechanism by which BHA inhibits benzo[a]pyrene tumorigenesis in the mouse forestomach. Laboratory studies and epidemiological evidence lead credence to chemoprevention strategy in attenuating the risk of developing cancer in human beings. Many nutrient and non-nutrient dietary constituents of plant origin have evidence of chemoprevention by inhibiting and/or suppressing or reversing cancer incidence trend evoked by multitude of factors including environmental chemical carcinogens. Most of the chemopreventive agents intervene either at initiation, promotion and/or progression stages conceptually associated with ontogeny of carcinogenesis, although the mechanistic detail differs depending on stage at which the chemopreventor acts effectively. The hepatic drug metabolizing system comprising of phase-I and phase-II enzymes are capable of modifying the end result of carcinogenesis at “Initiation stage”.

HERBAL EXTRACTS AS CHEMOPREVENTIVES

In the last few years, considerable interest has been seen in evaluating the chemopreventive potential of different plant extracts. As clarified ahead, the balance between the phase I carcinogen-activating enzymes and the phase II detoxifying enzymes is critical to determining an individual's risk for cancer. There is substantial evidence to indicate that chemopreventive agents including medicinal plants exert their anti-carcinogenic effects by modulation of phase I and phase II xenobiotic biotransformation enzymes. In routine experimental studies plant extracts are screened for their cancer chemopreventive efficacy primarily based upon their potential to modulate biotransformation enzymes. In our research laboratory we have evaluated the effects of aqueous *Azadirachta indica* leaf extract AAILE on certain biotransformation enzymes in certain tissues of mice including liver and forestomach. In our observations AAILE treatment caused a decrease in the contents of cytochrome P450 (in the hepatic tissues) and cytochrome b_5 (in the hepatic as well as forestomach tissues). A significant decrease in cytochrome P450 and cytochrome b_5 content in hepatic tissues of male Balb/c mice upon AAILE treatment has been reported from our laboratory. Significant increase in the activity of GST was observed in the forestomach as well as hepatic tissues and increase in GSH contents in hepatic tissue upon two weeks of AAILE treatment suggesting that *A. indica* has a potential of increasing the metabolic detoxification of carcinogens by increasing the activity of GST and GSH contents. Our results are further supported by earlier investigations, wherein authors reported the increasing effects of *A. indica* on the activity of GST and GSH contents in different experimental models. It was concluded from above studies that *A. indica* mediates its chemopreventive effects through modulation of carcinogen biotransformation.

BENEFICIAL AND ADVERSE EFFECTS OF CHEMOPREVENTIVE AGENTS

Chemopreventive agents are ubiquitously present in commonly consumed foods as well as medicinal plants. It is believed that this type of approach could be promising way of reducing cancer incidence in clinical trials. It is important to note that suitable chemopreventive agents should have little or no toxicity, have a high efficacy, be orally administrable, have a known mode of action and low cost. However, it has been observed that some promising chemopreventive agents, which have been shown to be effective in controlling cancer, have exerted undesirable side effects. Since chemopreventive effects in carcinogenesis are based on the results of *in vitro* studies and animal models and usually involve high doses compared to the low levels in the diet, it may be difficult to adapt these results directly to humans, especially when clinical evaluations on candidate agents are not fully warranted. In addition, a number of chemopreventive agents possess negative effects, such as, mutagenicity, carcinogenicity, and other toxicities. Many antioxidants can, for example, both accept and donate electrons, suggesting their dual potential as antioxidants. Ascorbic acid inhibits the formation of some nitrosamines, but accelerates the formation of others, which might generate carcinogens by transnitrosation. Chemopreventive agents may have an anticancer activity in certain tissues, but at the same time may induce carcinogenic activity in others. Ideally chemopreventive agents should be used to control nutrients and food components, especially micronutrients, at safe doses that effectively interfere with or interrupt the carcinogenic process. The human diet contains a variety of natural carcinogens and many natural chemopreventive agents. The vast majority of chemopreventive blocking agents have been assigned to one or more of the following categories: Phase I enzyme inhibitors as well as inducers, and phase II metabolizing enzyme inducers. The phase II enzymes system is considered to be the main activating machinery, which is capable of converting pre-carcinogens into ultimate carcinogenic forms that can damage DNA, protein and lipids, while phase II enzymes represent the main detoxifying machinery. Many carcinogens may also act through the generation of oxygen radicals. A significant inverse correlation between oxidative damage and plasma antioxidant levels was observed in vitamin E, red ginseng and vitamin C supplement individuals. Some studies have suggested that many antioxidants have carcinogenic and anti-carcinogenic effects. For example, N-acetylcysteine, a known antioxidant, was also shown to be carcinogenic and to act as pro-oxidant. Moreover, several chemopreventive agents that are antioxidants at some concentrations become pro-oxidants at other concentrations. Therefore, dose selection may be important when balancing pro-oxidant and antioxidant effects in antioxidants. Among the plant polysaccharides, aloe polysaccharide has been shown to be the most effective at inhibiting DNA adduct formation, free radical generation, 8-OHdG formation, and cell proliferation. At the same time, aloe can induce angiogenesis and therefore aggravate the progression of carcinogenesis. These results suggest that the chemopreventive role of certain agents may be effective in one stage of carcinogenesis but detrimental in another. Chemopreventive agents are largely identified as enzyme modulators, antioxidants, inhibitors of carcinogen-adduct formation, inhibitors of oncogene activation, or modulators of signaling cascades associated with carcinogenesis, etc. However, the practical application of chemopreventive agents is expected to interfere with not only the carcinogenic process but also with the normal homeostatic process in the body, therefore, many of them may exhibit side effects, including the enhancement of carcinogenicity under certain conditions. It is evident that the overall toxicological effects of potential chemopreventive agents should be thoroughly evaluated and not ignored because of their beneficial effects.

Synergism is another factor that could be considered while assessing chemopreventive efficacy of putative agents. Isolated active components and/or their effective synthetic analogs may show pronounced effects because of their increased potency, but in most of the cases when used for a long duration and in certain cases when used for acute clinical medications may show drastic side effects. On the other hand when taken in crude extract form or as whole dietary supplements they show better compatibility with the human body with minimal side effects. Crude therapeutic products are less toxic than their synthetic counterparts because they contain the total family of medicinal compounds (known and unknown) just as they are found in their natural source and hence offer less risk of side effects. In crude preparations, perhaps, the other components that are present in addition to the active components may be affecting the effects of the active components. These known and unknown components might be acting as synergists for the therapeutic effects and antagonists for the side effects of the active components as well as the other toxic components in the crude preparation. Crude preparations may not be acting like magic bullets of the modern chemically synthesized medicines but these do produce beneficial effects with no or lesser undesirable side effects. There is a need to reassess the therapeutic properties of the crude traditional therapeutic preparations by scientific methodology, which can lead to their rational use. Collection of scientific data supporting the miraculous medicinal properties of these preparations is also required to take maximum benefits without any fear of undesirable side effects.

FUTURE DIRECTIONS

Nowadays, the use of edible phytochemicals with chemopreventive potential is considered to be an inexpensive, readily applicable, acceptable and accessible approach to cancer control as well as management. With healthcare costs being a key issue today, it would be cost-effective to promote the awareness and consumption of phytochemicals as a cancer-preventive strategy for the general population. Many nutrients as well as non-nutritive phytochemicals are being evaluated in intervention trials for their potential as cancer chemopreventive agents. Despite significant advances in our understanding of multistage carcinogenesis, little is known about the mechanism of action of most chemopreventive agents. The chemopreventive effects that most agents exert are likely to be the sum of several distinct mechanisms. Disruption or deregulation of intracellular-signalling cascades often leads to malignant transformation of cells, therefore, it is important to identify the molecules in the signaling network that can be affected by chemopreventive agents to allow for better assessment of their underlying mechanisms. In several cases, the chemopreventive effects of dietary chemo-preventives in cultured cells or tissues are only achievable at supra-physiological concentrations and such concentrations might not be attained when the agents are employed as part of diet. Furthermore, most of the phenolic agents present as glycosides are converted to other conjugated forms after absorption, which might further lower the bioavailablity in various tissues. Both pharmacokinetic properties and bioavailability are key problems in investigating the chemoprevention of cancer and should be assessed carefully before undertaking intervention trials with supplements. The development and use of chemopreventive agents for intervention trials involve many scientific disciplines. With the advances in techniques to assess single nucleotide polymorphisms, researchers/clinicians are now more aware of the specific genes that can directly and indirectly contribute to individual differences in the susceptibility to carcinogenesis. When high-risk groups are identified, practitioners might be able to recommend specific supplements that can modulate or restore the cellular-signalling events that are likely to be disrupted in these individuals. The term 'nutragenomics' has been coined, and much attention is being focused on this relatively new area of research. Tailored supplementation with designer foods that consist of chemopreventive phytochemicals, each having their own distinct anticancer mechanisms, will be available in the near future. These should be developed in line with advances in the genetic and molecular epidemiology of carcinogenesis.

CONCLUSION

It could be summarized that we are unavoidably exposed to numerous chemicals in our environment and most of them have been known to cause cancer. Considering the present scenario, it seems almost impossible to escape exposure to environmental chemical carcinogens. Since the process of chemical carcinogenesis is multi-step in nature, it could be targeted at number of steps by chemopreventive agents. Targeting biotransformation of carcinogens with chemopreventive agents for blocking the initiation of carcinogenesis seems to be an effective approach for cancer chemoprevention. Therefore, the use of chemopreventive agents as a part of diet and/or as supplements should be implicated throughout our life, so that the hazardous effects of environmental chemical carcinogens may be taken care of. Moreover, chemoprevention can be used, especially in some apparently healthy people at risk of cancer, to prevent or reduce their threat of developing cancer. The biomedical community needs to recognize and advocate approaches to prevent cancer with the same enthusiasm that it currently directs towards treating it.

> *"He is a better physician that keeps diseases off us than he that cures them being on us; prevention is so much better than healing because it saves the labour of being sick".*
>
> —Thomas Adams (1618)

SUGGESTED READINGS

1. Pitot HC-III, Dragan YP, Chemical Carcinogenesis. In: Casarett and Doull's Toxicology, The Basic Science of Poisons, eds Curtis D. Klaassen. [6th Ed 1996] The McGraw-Hill Company. INC USA pp 241–319, 2001.
2. *American Cancer Society "Cancer Facts & Figures" American Cancer Society Atlanta* www.cancer.org/downloads/STT/CAFF2006PWSecured.pdf2006
3. Goldstein JA, Faletto MB, Advances in mechanisms of activation and deactivation of environmental chemicals. Environment Health Perspectives 100:169–176, 1993
4. Hanausek M, Walaszek Z, Slaga TJ, Detoxifying cancer causing agents to prevent cancer. Integrative Cancer Therapies 2: 139–144, 2003
5. Lee BM, Park KK, Beneficial and adverse effects of chemopreventive agents. Mutation Research 523–524: 265–278, 2003
6. Surh YJ, Cancer chemoprevention with dietary phytochemicals. Nature Reviews Cancer 3: 768–780, 2003
7. Koul A, Gangar SC, Sandhir V, Pitfalls in journey from traditional to modern medicine. Natural Product Radiance 4: 6–13, 2005
8. Koul A, Mukherji N, Gangar SC, Inhibitory effects of *Azadirachta indica* on DMBA-induced skin carcinogenesis in Balb/c mice. Molecular and Cellular Biochemistry 283: 47–55, 2006
9. Gangar SC, Sandhir R, Rai DV, Koul A, Modulatory effects of *Azadirachta indica* on benzo(a)pyrene induced forestomach tumorigenesis in Balb/c mice. World Journal of Gastroenterology 12: 2749–2755, 2006.
10. Gangar SC, Modulatory effects of *Azadirachta indica* on benzo(α)pyrene induced forestomach tumorigenesis in Balb/c mice, Ph.D. thesis submitted to the Faculty of Science, Punjab University, Chandigarh (2007).

5

Epilepsy: Herbal Remedies

D.V. Rai and Gaurav Kumar

INTRODUCTION

There are various herbal plants in Indian ecosystems, known since ancient days and well popular today also for their medicinal properties. They play a significant role in the therapy of various diseases of human kind. These plants are basis of more than 30% of modern drugs. The earliest evidence of human use of plants of healing dates back to the Neanderthal period. During the dark ages, botanical gardens were created to grow medicinal plants. The use of herbal medications and dietary supplements has increased dramatically in recent times. A significant number of remedies are used to treat various neurological diseases. Herbal medicine practice flourished until the 17th century, when more 'scientific' pharmacological remedies began to flourish. During 19th century, as scientific methods became widely accepted, the practice of botanical healing disappeared from western hemispheres. However, over the last 50 years, the use of herbal medications has increased dramatically again in Europe and North America. According to an estimate, 80% of human population relies primarily on traditional medicines for health care needs. Conversely, side effects from these herbs and dietary supplements may largely go unreported. The medicinal herb has a prolonged effect on various diseases such as cancer, diabetes and various neurological diseases.

Epilepsy is the most common serious neurological disease, affecting approximately one in two hundred people in UK. Onset can occur at any age, although it is the most common in young and old, with an increase in incidence in those aged more than 60 years. Most herbal and complementary medicines are used without a specific evidence of benefit from controlled clinical trials. Some of these products may be anticonvulsant and thus of possible benefit in patients with epilepsy, whereas other may dispose to seizure in those with and without epilepsy. There are more than three thousand of herbal species, which have been recognized to treat various diseases like asthma, Alzheimer's disease, epilepsy and migraine. However, for epilepsy some of the most effective herbal drugs are known to be Bacopa, Canscora, Convolvulus, Elaeocarpus, and Ginkgo.

EPILEPSY

It is the most common chronic neurological condition and is characterized by recurrent unprovoked epileptic seizures. Epilepsy is a general term used for a group of disorders that cause disturbances in electrical signaling in the brain. When this happens a person may suddenly have a seizure. Many people may have undergone a single seizure at some time in their lives. Seizures may occur in any person under certain circumstances, including acute illness, drug overdoses etc. An epileptic seizure, occurs when these energy pulses come much more rapidly as many as 500 per second for a short time due to an electrical abnormality in the brain. This brief electrical surge can happen in just a small area of brain. It falls mainly into two categories:

1. **Partial seizure:** It takes place within discrete regions of brain. If consciousness is fully preserved during the seizures, the clinical manifestations are considered relatively simple and the seizure is called as simple partial seizure. If consciousness is impaired the symptomatology is more complex and is called complex partial seizure. It is of four types:
   (a) Simple partial seizure
   (b) Complex partial seizure
   (c) Aura
   (d) Secondarily generalized seizure

2. **Generalized seizure:** It is the seizure, which arises from both hemispheres simultaneously. It is of six types:
   (a) Absence seizure (Petit mal)
   (b) Tonic seizure
   (c) Clonic seizure
   (d) Myoclonic seizure
   (e) Atonic seizure
   (f) Tonic-Clonic seizure (Grand mal)

The initiation phase of seizure is characterized by two concurrent events in an aggregate of neurons: (a) High frequency burst of action potential and (b) Hypersynchronization. The bursting activity is caused by a relatively long-lasting depolarization of the neuronal membrane due to influx of extra-cellular calcium, which leads to the opening of voltage-dependent sodium channels, influx of Na^+ and generation of repetitive action potentials. This is followed by a hyperpolarizing after potential mediated by Gamma Amino Butyric Acid (GABA) receptors depending on cell type. GABA is a neurotransmitter that slows electrical transmission between the nerve cells. Low levels of GABA in the body have been linked to epilepsy and increased rate for epilepsy. Fig. 1 illustrates how epileptic seizures trigger an activity-dependent sequence of events resulting in neuronal firing.

ENVIRONMENTAL FACTORS CAUSING EPILEPSY

Anything that disturbs the normal pattern of activity in the brain can trigger epilepsy. The various environmental factors, which lead to epilepsy are described below:

1 Environmental and occupational exposure to lead, carbon monoxide and certain chemicals.

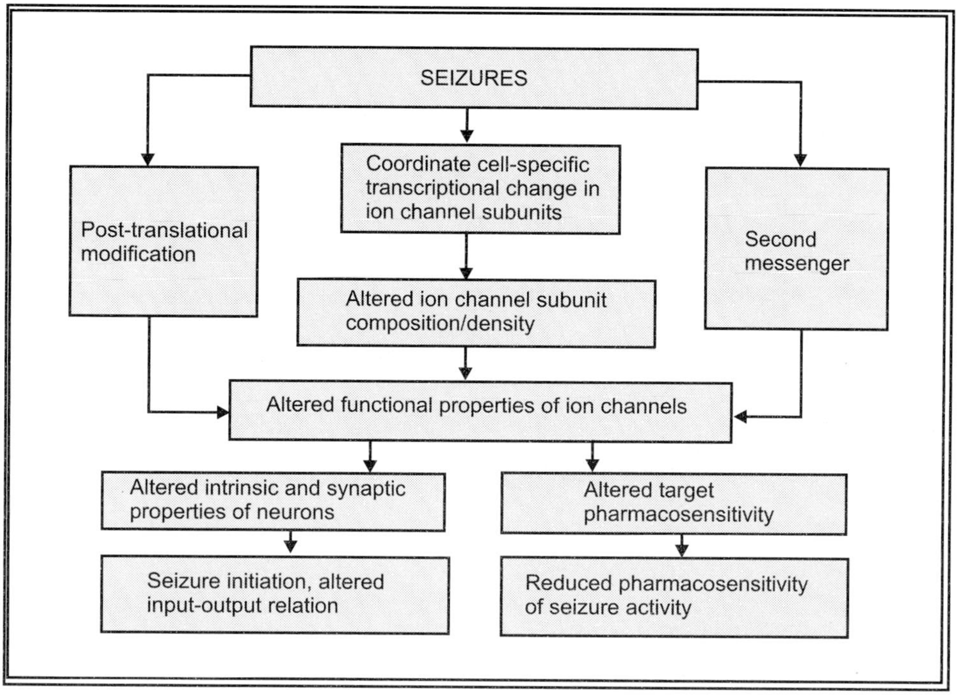

Fig. 1. Mechanism of sequence of events involved in neuronal firing.

2. Use of street drugs and alcohol may cause seizures during intoxication or when offending substance is being withdrawn. Withdrawal of certain medications such as barbiturates and other sedatives can cause epileptic seizure in those who have taken them from long periods.
3. Lack of sleep, stress and hormonal changes.
4. Withdrawal from certain antidepressant and anti-anxiety drugs.

HERBS IN REMEDIAL MANAGEMENT

Herbal medications are being used as anticonvulsants for the past many years. Herbs are used to cure epilepsy frequently as they have, no doubt, slow remedial effects on diseases but they do not have any side effects. More than 3000 herbs have been recognized as perfectly fit for various human diseases like Alzheimer's disease , migraine, nausea and including epilepsy. The herbal plants contain various types of active compounds, which are useful in treatment and management of several human diseases. Many herbs like ashwagandha, shankhpuspi, ginkgo etc. have been used for centuries to control the convulsions, in addition to their myriad effect on central nervous system. However, we have taken only five plants with simple reasons owing to application in curing measures of epilepsy. Currently number of proposed in Ayurveda, which can be utilized in treatment of epilepsy. Some of the herbs, which are greatly used to treat epilepsy are discussed below:

Fig. 2. *Bacopa monnieri.*

Bacopa (Brahmi)

Scientific Name: *Bacopa monnieri*

Family: Scrophulariaceae

English Name: Thyme leaved graticula.

Hindi Name: Brahmi.

General Description: It grows in damp, marshy areas. It is a reputed medicinal herb in Indian systems of medicine. It is also under cultivation as medicinal crop.

Botanical Description: It is a glabrous somewhat succulent, creeping herb, rooting at the nodes, with numerous prostrate branches, each 10–30 cm long; leaves oblong to spathulate, sessile, decussate, rather fleshy, entire, punctuate, obtuse; flowers axillary, solitary, peduncles often much longer and deflexed in fruiting stage; corolla about one centimeter long, pale, lobes 5, oblong, obtuse, subequal, tips purple; capsule ovoid-acuminate or slightly beaked at the apex; seeds oblong, truncate, longitudinally ribbed with transverse striations in between the ribs.

Useful Parts: The whole herb.

Medicinal Uses: According to Ayurveda, it is bitter, pungent, heating, emetic, laxative and useful in ulcers, tumours, ascites, enlargement of spleen, indigestion, inflammations, leprosy, anaemia, biliousness etc. According to Unani system of medicine, it is bitter, aphrodisiac, good in scabies, leucoderma, syphilis etc. It is promising blood purifier and useful in diarrhea and fevers.

Ayurvedic Preparations: Bramhi-Ghrita.

Chemical Constituents: Active principle involved is hersaponin, resembles resperpine and chloromazine in its central action. Famed in Ayurvedic medicine, brahmi has antioxidant properties. It has been reported to reduce oxidation of fats in the blood stream, which is a risk factor for cardiovascular diseases. It has been used for centuries to help benefit epilepsy, memory capacity, increase concentration, and reduce stress-induced anxiety. It is listed as a nootropic, a drug that enhances cognitive ability. In India, this plant has also been used traditionally to consecrate newborn babies in the belief that it will open the gateway of intelligence. Interestingly recent studies suggest bacopa may improve intellectual activity in children.

Canscora (Shankhpushpi)

Family Name: Gentianaceae

Botanical Name(s): *Canscora decussata*

Popular Name(s): Sankh Pushpi, Shankhini, Kambumalini, Samkhapushpi, Sankaphuli

Parts Used: Entire plant and juice

Habitat: This herb is found in India and Burma

Description: Shankhapushpi is a morning glory like perennial herb. The branches of the Shankhapushpi are spread on the ground and can be more than 30 cm long. The flowers are blue in color (5 mm) and the leaves, which are elliptic in shape (2 mm), are located at alternate positions with branches or flowers.

Uses: Primarily, Shankhapushpi is used as a brain tonic. Shankhapushpi is one of the best and prominent natural medicines, which helps in improving memory. Whole plant of Shankhapushpi is used in medical treatment. Various chemical compositions such as glucose, sucrose, glycosides, alkaloids and various acids etc. are found in the plant. Shankhapushpi is also used as one of the most important ingredients in treatment of disorders/syndromes such as hypertension, hypotension, anxiety neurosis, stresses etc.

Convolvulus

Convolvulus is a genus of about 250 species of flowering plants in the bindweed family Convolvulaceae, with a cosmopolitan distribution. Common names include bindweed and morning glory, both names shared with other closely related genera. They are annual or perennial herbaceous vines and (a few species) woody shrubs, growing to 0.3–3 m tall. The leaves are spirally arranged,

Fig. 3. Convolvulus species

and the flowers trumpet-shaped, mostly white or pink, but blue, violet, purple or yellow in some species. Many of the species are problematic weeds, which can swamp other more valuable plants by climbing over them, but some are also deliberately grown for their attractive flowers. This species occurs in many temperate regions. They are mostly slender, creeping winding vines. A few are small perennials. They have simple, alternate leaves and wide funnel-shaped flowers. *Convolvulus* species are used as food plants by the larvae of some *Lepidoptera* species including the leaf-miner *Bucculatrix cantabricella* which feeds exclusively on *Convolvulus cantabricus*.

Scammony (*Convolvulus scammonia*) is a bindweed native to the countries of the eastern part of the Mediterranean basin; it grows in bushy waste places, from Syria in the south to the Crimea in the north, its range extending westward to the Greek islands, but not to northern Africa or Italy. It is a twining perennial, bearing flowers like those of *Convolvulus arvensis*, and having irregularly arrow-shaped leaves and a thick fleshy root. The dried juice, virgin scammony, obtained by incision of the living root, has been used in medicine as *scammonium*, but the variable quality of the drug has led to the employment of *scammoniae resina*, which is obtained from the dried root by digestion with alcohol. The active principle is the glucoside scammonin or jalapin, $C_{34}H_{114}O_6$. The dose of scammonium is 5 to 10 grains, of scammony resin 3 to 8 grains. Like certain other resins, scammony is inert until it has passed from the stomach into the duodenum, where it meets the bile, a chemical reaction occurring between it and the taurocholate and glycocholate of sodium, whereby it is converted into a powerful purgative. Its action is essentially that of a hydragogue, and is exercised upon practically the entire length of the alimentary canal. The drug is not a cholagogue, nor does it markedly affect the muscular coat of the bowel, but it causes a great increase of secretion from the intestinal glands. It acts in about four hours. In large doses it is, of course, a violent gastrointestinal irritant. In consonance with the statement that scammony acts only after admixture with the bile, is the fact that hypodermic or intravenous injection of the drug produces no purgation. The drug frequently kills both roundworm and tapeworm, especially the former, and is therefore an anthelmintic. It is not largely used, but is very effective in the treatment of severe constipation, especially in children.

Elaeocarpus (Rudraksh)

Elaeocarpus is a genus of tropical and subtropical evergreen trees and shrubs. The approximately 350 species are distributed from Madagascar in the west through India, Southeast Asia, Malaysia, southern China, and Japan, through Australia to New Zealand, Fiji, and Hawaii in the east. The islands of Borneo and New Guinea have the greatest concentration of species.

Elaeocarpus ganitrus, the Rudraksha Tree, produces seeds known as Rudraksha. In Sanskrit *rudraksha* literally means "the eye of Rudra". The seeds have been traditionally considered sacred in Hinduism.

Rudraksha is commonly used to mean the seeds of the **rudraksha tree**, *Elaeocarpus ganitrus*. The Rudraksha is a large evergreen broad-leaved tree which grows in the area from the Gangetic Plain to the foothills of Himalayas. Rudraksha trees are also found in middle areas of Nepal. Rudraksha seeds are covered by an outer shell of blue color when fully ripe, therefore, Rudraksha seeds are also

Fig 4. Elaeocarpus species.

called as blueberry beads. Rudraksha is also used for treatment of various diseases in traditional Indian medicine.

Medicinal Value and Biomedical Properties: According to the Ayurvedic medical system, wearing Rudraksha can have a positive effect on the heart and nerves, and relieve you from stress, anxiety, depression, palpitations and lack of concentration. It is also known for its anti-ageing effect, and electromagnetic and inductive properties. People with high blood pressure have been found to have benefited from the use of Rudraksha seeds.

Ginkgo biloba

It is a large shade tree. The species form usually matures at about 80' tall by 60' wide under urban conditions, but specimen trees in open areas can get much, much larger. It is an upright columnar, sparsely branched, with open growth habit when young, usually becoming an upright oval to upright

Fig. 5. *Ginkgo biloba.*

spreading growth habit with maturity, but sometimes becoming highly irregular and picturesque in its branching with age. Its foliage is a medium green and unusually obovate (fan-shaped), up to 3" long, with a petiole that is also up to 3" long; this shape and the elongated petiole cause the foliage to flutter in the slightest breeze. The leaf blade apex is sometimes entire, but often has a single vertical slit in the top center (hence the specific epithet *biloba*); rarely, two slits occur, yielding three lobes. The veins are slightly raised (giving a ribbed appearance to the foliage) and usually dichotomous (forked).

The Ginkgo (*Ginkgo biloba*), frequently misspelled as "Gingko", and sometimes known as the Maidenhair tree, is a unique tree with no close living relatives. It is classified in its own division, the Ginkgophyta, comprising the single class Ginkgoopsida, order Ginkgoales, family Ginkgoaceae, genus *Ginkgo* and just the one species. It is one of the best known examples of a living fossil. In the past it has also been placed in the divisions Spermatophyta or Pinophyta. Ginkgo is a gymnosperm (as opposed to an angiosperm), meaning "naked seed"; its seeds are not protected by an ovary wall and hence, the berry-like structures produced by female ginkgo trees are technically not fruit.

For centuries it was thought to be extinct in the wild, but is now known to grow wild in at least two small areas in Zhejiang province in eastern China, in the Tian Mu Shan Reserve. However, as this area has known human activity for over a thousand years, the wild status of ginkgos there is uncertain.

Medical uses

The extract of the Ginkgo leaves contains flavonoid glycosides and ginkgolides and has been used pharmaceutically. It has many alleged nootropic properties, and is mainly used as memory enhancer and anti-vertigo agent. However, studies differ about its efficacy.

Out of the many conflicting research results, there seem to be basically three effects of Ginkgo extract on the human body: it improves blood flow (including microcirculation in small capillaries) to most tissues and organs; it protects against oxidative cell damage from free radicals (antioxidant); and it blocks many of the effects of PAF (platelet aggregation, blood clotting) that have been related to the development of a number of cardiovascular, renal, respiratory and CNS (Central Nervous System) disorders. Ginkgo can be used for intermittent claudication.

Side effects

Ginkgo may have some undesirable effects, especially for individuals with blood circulation disorders and those taking anticoagulants such as aspirin and warfarin, although recent studies have found that ginkgo has little or no effect on the anticoagulant properties or pharmacodynamics of warfarin. Ginkgo should also not be used by people who are taking monoamine oxidase inhibitors (MAOI) or by pregnant women without first consulting a doctor.

Ginkgo side effects and cautions include possible increased risk of bleeding, gastrointestinal discomfort, nausea, vomiting, diarrhea, headaches, dizziness, and restlessness.

If any side effects are experienced, consumption should be halted immediately. Ginkgo supplements are usually taken in the range of 40–200 mg per day. If the side effects continue usage should be stopped completely.

CONCLUSION

Some herbal medicines may have an anticonvulsant effect; however these herbal remedies are widely promoted and used to treat a diverse array of ailments, including memory loss and dementia, poor concentration, as well as cerebral insufficiency. These herbs described above are preferably prescribed and is gaining worldwide popularity. Though these herbs don't work like bullet but it has complete curable effects on disease with minimal or no side effects.

SUGGESTED READINGS

1. Akah PA, Nwambie AI, Gamaniel KS, *et al*. Experimental study of anticonvulsant plants used for treatment of infantile convulsions in Nigeria. Brain Res Bull, 44: 611–3, 1997.
2. Miller PG, Herbal medicals. Selected clinical considerations focussing on known or potential drug herbs interaction. *Arch Int Med*, 158: 2200–1, 1998.
3. Winslow LC, Kroll DJ. Herbs as medicines. *Arch Intern Med,* 158: 2192–9, 1998.
4. Rastogi RP, Azadirachta In: Rastogi RP and Mehrotra BN (eds.), Compendium of *Indian medicinal plants,* 2: 87–90. 1993.
5. Michael T. Epilepsy—new treatments. *Pharmaceutical Journal* 262(7074), 470–473. 1999
6. Balch J.F, Balch P.A., Prescription for Nutritional Healing. Garden City Park, NY: Avery Publishing Group. 1997.
7. Rai D.V and Gaurav Kumar, Anticonvulsants affecting the mechanical strength of long bones. National symposium on 21st century research in Biochemistry and Biophysics at Kalyani University, Kalyani, Kolkata, pp. 119 – 120, held on Feb. 1–3, 2007.
8. Yanpallewara SK, Raib S, Kumar M, Chauhan S & Acharya SB. Neuroprotective effect of Azadirachta indica on cerebral post-ischemic reperfusion and hypoperfusion in rats. Life Sciences: 76; 1325–1338, 2005.

6

Immunotoxicology and Health

S.K. Arora and Sachin Gupta

INTRODUCTION

Although scientists have known since the early 1900s that occupational exposure to certain chemicals can induce severe immune effects, recent chemical disasters of national and international prominence have renewed interest in the subject. It is generally agreed that human health is influenced by the environment and that many diseases are caused or enhanced by environmental factors, but effects on the immune system have been difficult to delineate. Researchers in the field of immunotoxicology, who study the adverse effects to immune function that result from occupational, inadvertent, or therapeutic exposure to environmental chemicals, biological materials, and drugs, debate, if toxicants damage the immune system and to what extent. Some researchers contend that certain chemicals can affect immunity significantly, increasing an individual's susceptibility to disease, causing hypersensitivity reactions, autoimmunity or immunosuppression in some cases. Others dispute this view, arguing that the evidence for immunotoxicity in humans is greatly overstated. The initial studies of the susceptibility of the immune system and its possible use for detection of sub-clinical toxic states were published in the seventies and early eighties (Vos and van Genderen 1973; Vos 1977; Loose et al. 1978; Faith et al. 1980), when the attention began to focus on the immune system as an important object of toxic action, particularly in connection with the experiments on rodents.

The immune system is known for highly organized cooperation and regulation of various cells, which is ensured on one hand by soluble mediators (immunoglobulins, immunohormones, cytokines) and on the other by intercellular interactions on the level of membrane receptors and anti-receptors. All agents that affect the fine balance mechanisms mentioned can cause agent-specific or species-specific immunity damage, which in majority of cases, results in immunosuppression (e.g. decreased resistance to infectious agents and development of tumors). Manifestations of immunosuppression were observed either on systemic or on the local level (e.g. in lungs or skin, Luster et al. 1996). Pathological increase in the immune response can also occur, such as hypersensitivity, manifested as respiratory tract allergies or contact allergic dermatitis, and some agents can participate in the development of autoimmune diseases.

IMMUNOTOXICOLOGY AS A BRANCH OF SCIENCE

Immunotoxicology is relatively new interdisciplinary scientific field focused on identification and analysis of chemical and, in a broader sense, also physical and biological factors of the environment which can cause unwanted and usually incidental immunomodulations (Dietert et al. 1996). The objective of immunotoxicology is to protect humans and animals against the harmful effects of chemical factors present in the environment, to develop and check immunotherapeutic products and introduce and evaluate the methods intended for determination of interactions between the factors of the external environment and the immune system (Dietert et al. 1996). Harmful effects of environmental factors can arise from (1) direct or indirect action of xenobiotics on the immune system or of products obtained by their biotransformation, (2) induction of the immune response to xenobiotics or their metabolites, (3) modification of self-antigens by chemicals or their metabolites (Berlin 1987).

NATURE OF SUBSTANCES EXHIBITING IMMUNOTOXIC EFFECTS (IMMUNOTOXICANTS)

Immunotoxicants are factors of the external environment, which cause significant changes (modulation) in the immune mechanisms in humans and animals (Dietert et al. 1996). These include **environmental chemical contaminants** (pesticides, industrial emissions and similar), **drugs, biotechnological waste products** (D'Agnolo 1983) and **physical factors** (UV-B light, electromagnetic field—Luster et al. 1996). The action of immunotoxicants results in the reduced reactivity of the immune system—immunosuppression (e.g. PCB, lead, cadmium, Safe 1994; Prochazkova et al. 1990; Skokanova et al. 1993) or, on the contrary, in its excessive reactivity—hypersensitivity (polycyclic aromatic compounds, organophosphate insecticides and others—Luster and Rosenthal 1993). Some substances (e.g. nickel and mercury) can participate in the development of autoimmune diseases (Luster et al. 1996; Henry et al. 1988). The present interest focuses on possible immunologic effects of indoor pollutants. This involves not only the chemicals but also bioaerosols, such as viruses, bacteria, moulds, algae and protozoa, as potential sensitizing agents or mediators of infectious diseases (Luster et al. 1996). Age, genetics, pre-existing disease, lifestyle, diet, drugs and stress may compound the effects of chemical exposure to further compromise immune function and increase the chance of disease. Substances with immunomodulative effects are summarized in Table 1.

MECHANISM OF UV-INDUCED IMMUNOSUPPRESSION

Ultraviolet radiation (UV), in particular the UVB range, suppresses the immune system in several ways. UVB inhibits antigen presentation, induces the release of immunosuppressive cytokines and causes apoptosis of leukocytes. UVB, however, does not cause general immunosuppression but rather inhibits immune reactions in an antigen-specific fashion. UV induced immunosuppression is a highly complex process in which several different pathways are involved (Fig. 3).

Table 1. Xenobiotics inducing immunosuppression or exerting positive immunomodulative effects (Luster et al. 1996; Dietert et al. 1996)

- — polycyclic aromatic hydrocarbons (dimethylbenzanthracene - DMBA)
- — polychlorinated aromatic hydrocarbons (PCB, p-diazine – TCDD)
- — heavy metals (e.g. mercury, lead, cadmium, arsene metalloids)
- — pesticides (e.g. chlordane, organophosphates, carbofuran)
- — organotins (tributyltin oxide)
- — aromatic hydrocarbons (e.g. benzene, toluene)
- — aromatic amines (e.g. benzidine, dimethylnitrosamine)
- — oxidative gases (e.g. O_3, NO_2, SO_2)
- — metabolites of toxigenic moulds (e.g. aflatoxins, ochratoxin, T2 – toxin)
- — drugs of abuse (e.g. cocaine, alcohol, marijuana)
- — environmental particles (e.g. asbestos, silica, beryllium)
- — drugs (e.g. cyclophosphamide, cyclosporine)
- — inducers of allergic reactions (e.g. nickel, palladium) and autoimmune diseases (mercury)
- — immunotoxins (diphtheria toxin, ricin)
- — bioactive substances prepared by recombinant DNA techniques (interleukins, interferons)
- — ultraviolet light (UV-B), electromagnetic field

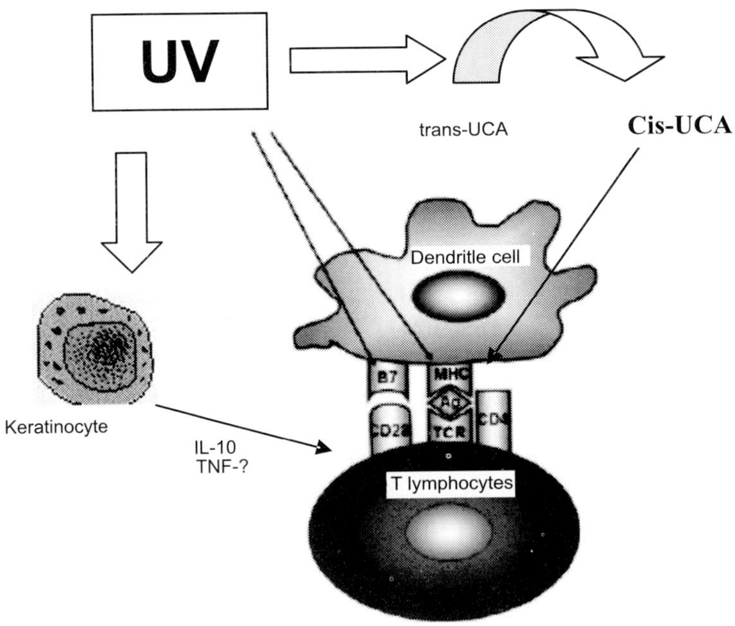

Fig. 3. Various mechanisms involved in UV-induced immunosuppression. Sensitization takes place when an antigen (Ag) is presented by dendritic cells to CD4$^+$ T lymphocytes carrying the appropriate T cell receptor (TCR) in association with major histocompatibility complex class II molecules (MHC II). In addition, the interaction of co-stimulatory molecules e.g. B7 with CD28 is required. UV radiation suppresses the expression of MHC II and of co-stimulatory molecules (e.g. B7). In addition, UV radiation stimulates keratinocytes to release immunosuppressive mediators like IL-10 and TNF-α, which are responsible for systemic immunosuppression. Furthermore, UV radiation converts trans-UCA into cis-UCA, which also exerts immunosuppressive effects.

SPECTRUM OF THE EFFECT OF IMMUNOTOXICANTS

The susceptibility of the immune system to xenogenous substances, which can result in immunosuppression, depends on both the properties of the respective chemical and complex nature of the immune system (antigen recognition and processing, cellular interactions—cooperation, regulation and amplification, activation and differentiation of cells and production of mediators by various cell types). The immunotoxicants can induce changes in highly diverse cellular interactions and reactions in various components of the immune system, depending on its characteristics. The interactions of xenobiotics with the immune system can be affected also by additional factors, such as malnutrition, stress and genetic predisposition (Luster et al. 1993). Moreover, they can participate in the development of allergic reactions and autoimmune diseases (Dietert et al. 1996).

1. Systemic immunosuppressive action

Systemic immunosuppression affects the activity of the immune system as a whole and develops frequently as a result of action of various chemical agents. The most detailed studies of this form of immunotoxicity have been carried out in rodents. The systemic immunosuppression is indicated by: (a) altered weight and histology of lymphoid organs, (b) quantitative changes in cellularity of the lymphoid tissue, peripheral blood leukocytes and bone marrow, (c) impairment of cell function at the effector or regulatory level, (d) increased susceptibility to infectious agents or transplantable tumours (Luster et al. 1996). Some examples of the immunosuppressive mechanisms are presented in Table 2.

Table 2. Possible mechanisms for xenobiotics-induced immunodepression (Spreafico et al. 1987)

- — Affect antigen distribution and/or persistence
- — Influence production-differentiation of precursors immunocytes
- — Affect traffic and/or number of inducer, effector and/or regulatory immune cells
- — Influence activation threshold of immunocytes to antigen and/or growth and differentiation factors (e.g. responsiveness to lymphokines and monokines, expression of receptors)
- — Affect production, release and/or metabolism of lymphomonokines and/or other substances (e.g. α-fetoprotein) modulating immune cell function
- — Influence quantity, quality, persistence of effector products of immune cells (e.g. antibody)
- — Affect number, traffic, functional capacity of other cells (e.g. platelets, neutrophils) and dependent mediators
- — Affect indirect regulatory circuits (e.g. CNS-endocrinological homeostasis)

2. Local immunosuppressive action

Local immunosuppression induced by xenobiotics has most frequently been demonstrated in lungs and on skin in both humans and animals.

(a) Lungs

Pathogenesis of a number of pulmonary diseases, such as fibrosis, granulomatosis and bronchial asthma, is associated with inhalation of pollutants. The pulmonary diseases, which are conditional upon local damage to the immune system, can be induced by various agents, such as oxidizing gases (ozone, NO_2, SO_2) or fine solid particles (silica, beryllium, asbestos, coal dust, sheep wool and others). Numerous studies point to the fact that the progress of pulmonary diseases is related to the post-activation release of cytokines, predominantly interleukin 1 (IL-1), tumor necrotizing factor (TNF-α, TNF-β), platelet-derived growth factor (PDGF), transforming growth factor beta (TGF-β), by pulmonary macrophages (Luster et al. 1996). The suppression of above-mentioned systems can also be caused by pollutant gases, such as ozone and nitrogen oxides, resulting in predisposition of an organism to infectious agents (particularly viruses) and development of tumors (Luster et al. 1996).

(b) Skin

Similar to the lungs, the skin is another potential place of entry of many xenobiotics. Skin reactions can acquire the form of specific immune response (contact hypersensitivity) or non-specific inflammation (contact irritation), both mediated by several similar pro-inflammatory cytokines. Immunologically active cells and their soluble mediators penetrate from the circulatory system into the skin as a response to the stimulus induced by xenobiotics. Besides, other dermal cells (such as Langerhans cells), which participate in the immune reactions, can be activated in response to dermal stimuli. Under the action of exogenous factors, such as ultraviolet radiation (UV) or some chemicals (dimethyl benzanthracene), they can disappear or lose their function (Luster et al. 1996). Keratinocytes induced by UV-B radiation, chemical irritants, sensitizing compounds and some pharmacological substances produce a wide scale of cytokines: IL-1, granulocyte and macrophage colony stimulating factor (GM-CSF), IL-6, IL-8, TGF-α, TGF-β, TNF-α and IL-3. The basic response of keratinocytes consists of the production and secretion of IL-1α and TNF-α which results in the expression of leukocyte adhesion molecules on the surface of dermal endothelial cells (e.g. VCAM-1). Important is also their paracrine effect in which TNF-α and IL-1 also play role as autocrine mediators inducing release of IL-8 by keratinocytes. IL-8 is an important chemotactic and activating substance affecting polymorphonuclear neutrophils (PMNL). In case of antigenic character of the stimulant (e.g. nickel sulphate), an increased number of mononuclear cells and subsequent participation of T-helper cells is characteristic of the response. This leads to significant release of IFN-γ and TNF-α with a resultant enhancement of the response (Luster et al. 1996).

3. Exclusive effect on target cells

Some xenobiotics exhibit more intensive selective effect on certain type of cells or some cells exhibit primary susceptibility to the action of such substances. Table 3 shows the examples of the effect of

Table 3. Selective immunosuppressive effect of some xenobiotics on monocyte–macrophage functions (Spreafico et al. 1987)

Direct cytotoxicity: Azathioprine, Steroids, Dacarbazine, Actinomycin D, Vinblastine

Responsiveness—production of lymphokines: Steroids, Azathioprine, Cyclosporin,Vinblastine

Antigen processing—phagocytosis: Steroids, Actinomycin D, Azathioprine

Chemotaxis—migration: Steroids, Azathioprine

immunosuppressive substances on the function of macrophages and monocytes (Spreafico et al. 1987).

Hypersensitivity mediated by xenobiotics

Chemically induced hypersensitivity, as a manifestative stage of immunotoxicity, aroused great interest of clinicians (Berlin 1987). Contact hypersensitivity reaction is the principal demonstration of hypersensitivity induced by xenobiotics (Asherson 1987). Such a reaction develops after the contact of various substances with body surface during which xenobiotics bind to skin proteins and change from incomplete antigens (haptens) to complete ones capable of inducing the immune response. Eczema are produced clinically in the place of contact with an allergen. The acute phase is manifested as erythema with swelling and formation of vesicles while production of papulae and scales and focal pruritus predominates in the acute phase (Buc 1997). There are records of contact dermatitis caused by various components of paint coats (polycyclic aromatic compounds), pesticides (organophosphates, Luster and Rosenthal 1993), different chemicals (e.g. formaldehyde, epoxide, Peruvian balm, lanolin, stains and others), drugs (neomycin) and plants, such as poison ivy (Buc 1997). Particular attention was paid to contact dermatitis caused by nickel, which is a frequent constituent of various articles (costume jewellery, jewellery, metal eyeglass frames, dog collars, etc.). Nickel dissolves in weakly acidic environment of sweat and its salts have high affinity to various proteins, including albumin and proteins of the complement system which creates conditions for the development of hypersensitivity reaction (Dolovich et al. 1984; Fishelson et al. 1983; Asherson 1987).

Autoimmune diseases

One of the characteristic features of the immune system is the recognition and the development of immune response to various xenogenous antigens present in the external environment and the absence of immune response to the own tissues. An impairment of the immunotolerance caused by various factors results in autoimmune diseases (Klein 1991). One of the possible ways of the development of autoimmune disease is the reaction of autoantibodies with xenogenous substances, which are bound to own cells or structures of the organism. In this way, for example, autoimmune hemolytic anemia can develop as a result of interaction of antibodies with erythrocyte-bound penicillin (Jeffe 1979) or pesticide dieldrin. Autoimmune glomerulonephritis, which develops as a result of deposition of circulating immune-complexes (mercury and gold) in the basal glomerular membrane, can serve as another example.

WAYS OF DETERMINATION OF IMMUNOTOXIC PROPERTIES OF XENOBIOTICS

The immunotoxic mechanism of xenobiotics is implemented either through direct action of the chemical with the immune cells and their membrane receptors, affecting the production of cell mediators and antibodies, or indirectly, through secondary effects on important central organs such as liver (hepatotoxicity—production of hepatoproteins with immunomodulative properties) and the neuroendocrine system (neuroendocrine toxicity—chemically induced changes in hormones or neuroactive substances with immunomodulative properties). The direct and indirect action can produce a combination of both effects (Dietert et al. 1996). The methods commonly used to evaluate immunotoxic effect of xenobiotics recognize non-functional and functional tests (Vandebriel et al. 1995). Non-functional tests (Table 4) provide information mainly about the changes in the lymphoid tissue, number of peripheral lymphocytes and monocytes, level of total globulins, cytokines, etc. (Vos 1977, 1987; Dean et al. 1987; Vandebriel et al. 1995; Luster et al. 1996). Functional tests (Table 5) reflect in greater detail the situation *in vivo* because they focus on the direct assessment of phagocytic and antigen-specific components of immunity. The evaluation of immunotoxicological

Table 4. Review of non-functional tests used in immunotoxicology (Luster et al. 1993; Vandebriel et al. 1995; Dietert et al. 1996)

Weights of lymphatic organs
- thymus, spleen
- lymph nodes (after oral exposure the mesenteric and after inhalatory exposure the bronchial nodes)

Histopathology (hematoxylin-eosin staining)
- thymus, spleen, lymph nodes
- mucosal immune system (Peyer's patches in the gut and lymphoid tissue in the respiratory tract)

Immunohistochemical methods (immunoperoxidase)
- differentiation cell antigens (CD)
- adhesion cell molecules (e.g. ICAM-1)
- activation cell markers (e.g. CD25 - IL-2R)

Basal immunoglobulin level
- total antibody levels
- total IgM, IgG, IgA and IgE levels

Bone marrow
- qualitative and quantitative analysis of immunocompetent cell populations

Enumeration of leukocytes
- peripheral blood
- bronchoalveolar and peritoneal lavage, skin

Fluorescence-activated cell scanner (FACS) analysis
- the CD4/CD8 ratio
- T cells (e.g. CD3, CD4+ CD8-, CD4-CD8+, CD4+CD8+)
- B cells (surface Ig)

Soluble factors
- complement (total and fractions)
- cytokine determination
- biological assays (cytokine-dependent cell lines, anti-cytokine monoclonal antibodies)
- ELISA (commercially available for mouse and rat)
- quantification of mRNA (blotting and subsequent hybridization or reverse transcription and subsequent PCR)

Table 5. Review of functional tests used in immunotoxicology (Luster et al. 1993; Vandebriel et al. 1995; Dietert et al. 1996)

Functional tests of phagocytes
- phagocytic activity (PHA) and index of PHA
- metabolic and bacterial activity of phagocytes

Antigen-specific antibody responses
- sensitization to LPS, ovalbumin (ELISA)
- sheep red blood cells immunization (ELISA, plaque forming cell assay PFC)

Mitogen responsiveness to B-cell mitogens
- the proliferative response of B cells to LPS

Mitogen responsiveness to T-cell mitogens
- the proliferative response of T cells to concanavalin A (Con A) and phytohaemagglutinin (PHA)

Mixed lymphocyte reaction (MLR)
- difference in cell proliferation between lymphocytes cocultured with allogeneic cells and those cocultured with syngeneic cells

Cytotoxic T-lymphocyte assay
- differentiation of T-cells into cytotoxic effector cells

Natural killer (NK) activity
- cell populations are cultured together with NK-sensitive target cells (e.g. YAC lymphoma cell line)

Delayed-type hypersensitivity responses (DTH)
- evaluating cell-mediated immunity
- sensitization to PPD, ovalbumin, keyhole limpet hemocyanin (KLH) and *L. monocytogenes*

Host resistance models
- *Listeria monocytogenes, Streptococcus pneumoniae, Pasteurella multocida*
- Cytomegalovirus, Influenza virus
- *Trichinella spiralis, Plasmodium yoelii*

Tumour models
- B16F10 melanoma, P4B6 fibrosarcoma, MADB106 adenocarcinoma

Autoimmune and allergy models

action of xenobiotics requires a spectrum of methods that can provide as detailed as possible information about the immune system status.

SUGGESTED READINGS

1. Asherson, G. L. Basic mechanisms of hypersensitivity: An update. Immunotoxicology, eds. A.Berlin, J. H. Dean, M. H. Draper, M. B. Smith, F. Spreafico. Martinus Nijhoff Publishers, Dordrecht, pp. 411–425, 1987.
2. Berlin, A. Synopsis, conclusions and recommendations. Immunotoxicology, eds. A. Berlin, J. H. Dean, M. H. Draper, E. M. B. Smith, F. Spreafico. Martinus Nijhoff Publishers, Dordrecht, pp.XI–XXVII, 1987.
3. Buc, M. Clinical immunology (in Slovak). VEDA, publishing house of Slovak Academy of Sciences, Bratislava, pp. 54–77, 1997.
4. D'agnolo, G. The control of drugs obtained by recombinant DNA and other biotechnologies. In: Current problems in drug toxicology. Eds. G. Zbinden, F. Cohadon, J. Y. Detaille and G. Mazue. John Libbey Eurotext: Paris and London, pp. 241–247, 1983.
5. Dean, J. H., Lauer, L. D., House, R. V., Ward, E. C., Murray, M. J. Experience with validation of methodology for immunotoxicity assessment in rodents. Immunotoxicology, eds. A. Berlin, J. H. Dean, M. H. Draper, E. M. B. Smith, F. Spreafico. Martinus Nijhoff Publishers, Dordrecht, pp.135–157, 1987.

6. Dietert, R. R., Golemboski, K. A., Kwak, H., Miller, T. L., Davison, T. F. Environment – immunity interactions. T.F. Davison, T.R. Morris, L.N. Payne (eds.): Poultry immunology, Carfax Publish. Co. Abingdon, UK, pp. 343–356, 1996.
7. Dolovich, J., Evans, S. L., Niebor, E. Occupational asthma from nickel sensitivity: I. Human serum albumin in the antigenic determinant. Brit. J. Ind. Med. 41:51–55, 1984.
8. Faith, R. E., Luster, M. L., Vos, J. G. Effects on immunocompetence by chemicals of environmental concern. In: Hodgson, E., Bend, J. R., Philpot, R. M.(eds.): Reviews in Biochemical Toxicology 2, Elsevier, Holland, pp. 173–211, 1980.
9. Fishelson, Z., Pangburn, M. K., Muller-Eberhard, H. J. C3 convertase of the alternative complement pathway. J. Biol. Chem. 258: 7411–7415, 1983.
10. Jaffe, I. A. Penicillamine in rheumatoid arthritis: Clinical pharmacology and biochemical properties. Scand. J. Reumatol. (suppl.) 28: 58–64, 1979.
11. Klein, J. Immunology. Blackwell Sc. Publ. Inc. New York, pp. 477–481, 1991.
12. Loose, L. D., Pittman, K. A., Benitz, K. F., Silkworth, J. B., Mueller, W., Coulston, F. Environmental chemical-induced immune dysfunction. Ecotoxicol. Environ. Safety **2**: 173–198, 1978.
13. Luster, M. I., Germolec, D. R., Kayama, F., Rosenthal, G. J., Comment, C. E., Wilmer, J. L. Approaches and concepts in immunotoxicology. Experimental immunotoxicology, eds. R. J. Smialowicz and M. P. Holsapple, CRC Press. Inc., pp.13–27, 1996.
14. Luster, M. I., Portier, C., Pait, D. G., Rosenthal. G. J., Germolec, D. R., Corsini, E., Blaylock, B. L., Pollock, P., Kouchi, Y., Craig, W., White, K. L., Munson, A. E., Comment, C. E. Risk assessment in immunotoxicology. II. Relationships between immune and host resistance tests. Fundam. Appl. Toxicol. 21: 71–82, 1993.
15. Luster, M. I., Rosenthal, G. J. Chemical agents and the immune response. Environ. Hlth. Perspect. 100: 219–236, 1993.
16. Prochazkova, J., Jilek, P., Vancurov A, M. Immunotoxicology of xenogenous substances (in Czech). Ceskoslov. Hyg. 35: 102–107, 1990.
17. Safe, S. H. Polychlorinated biphenyls (PCBs): Environmental impact, biochemical and toxic responses, and implications for risk assessment. Crit. Rev. Toxicol. 24: 87–149, 1994.
18. Skokanova, V., Kriz, J., Kodl, M. The influence of lead on selected immunity parameters in children I. Relationship between the values of immunity parameters investigated and the level of lead in blood (in Czech). Ceskoslov. Hyg. 38:142–150, 1993.
19. Spreafico, F., Meredino, A., Braceschi, L., Sozzani, S. Immunodepressive drugs as prototype immunotoxicants. Immunotoxicology, eds. A. Berlin, M. H. Draper, J. H. Dean,.M. B. Smith, F. Spreafico. Martinus Nijhoff Publishers, Dordrecht, pp.192–208, 1987.
20. Vandebriel, R. J., Garsen, J., Van Loveren, H. Methods in Immunotoxicology. Methods in Neurosciences. 24: 151–169, 1995.
21. Vos, J. G. Immune suppression as related to toxicology. CRC Crit. Rev. Toxicol. 5: 67–101, 1977.
22. Vos, J. G. The role of histopathology in assessment of immunotoxicology. Immunotoxicology, eds. A. Berlin, J. H. Dean, M. H .Draper, M. B. Smith, F. Spreafico. Martinus Nijhoff Publishers, Dordrecht, pp.125–134, 1987.
23. Vos, J. G., Van Genderen, H. Toxicological aspects of immunosuppression. In: Deichmann W.B. (ed.). Pesticides and the Environment: A Continuing Controversy. Intercontinental Medical Book Corp., New York, pp. 527–545, 1973.

7

Stem Cells: New Perspective in Biomedicine

R.C. Sobti, D.V. Rai, Gaurav Kumar, R.S. Verma and Raj Bahadur

INTRODUCTION

Recently, the work of medical scientists doing research in the field of stem cell biology have drawn considerable attention due to a variety of reasons; one of the reasons is that it may have applications in modern medicine. Secondly expanding fields of scientific inquiry in the research on stem cells raises fundamental ethical questions as rapidly as it generates new discoveries. Research on stem cells is advancing knowledge about how an organism develops from a single cell and healthy cells replace damaged cells in adult organisms. The promising area of science is investigating the possibility of cell-based therapies to treat diseases, and is often referred to as regenerative or reparative medicine.

Stem cells are the potentially immortal cells capable of self-renewal and also give rise to differentiated cell types of specialized tissues. A putative stem cell can both self renew and give rise to differentiated progeny Medical researchers believe that research in this area has the potential to change the way human disease will be treated in future. Medical researchers anticipate being able to use technologies derived from stem cell research to treat cancer, Parkinson's disease, spinal cord injuries, muscle damage, corneal degeneration, and diabetes etc. amongst a number of other diseases, impaired conditions of organs. However, there still exists a great deal of social and scientific uncertainty surrounding stem cell research, which could possibly be overcome through public debate and future research. There exits a widespread controversy over stem cell research that emanates from the techniques used in the creation and usage of stem cells. This present document on stem cells is intended for anyone who wishes to learn more about the biological properties of stem cells, the important questions about stem cells that are the focus of scientific research, and the potential use of stem cells in research and in treating disease. The primer includes information about stem cells derived from the embryo and adult. Much of the information included here is about stem cells derived from human tissue, but some studies of animal-derived stem cells are also described.

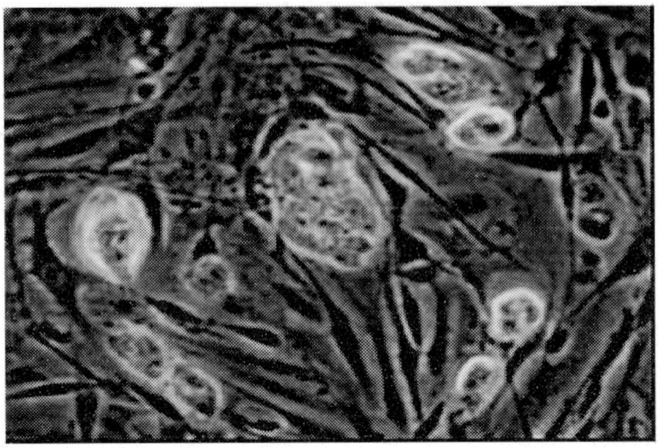

Fig. 1. Human stem cells.

STEM CELLS

Stem cells are primal cells that retain the ability to renew themselves through cell division and can differentiate into a wide range of specialized cell types. Stem cells have two important characteristics that distinguish them from other types of cells. First, they are unspecialized cells that renew themselves for long periods through cell division. The second is that under certain physiologic or experimental conditions, they can be induced to become cells with special functions such as the beating cells of the heart muscle or the insulin-producing cells of the pancreas. Photograph of stem cells is shown in Fig. 1.

The rigorous definition of a stem cell requires that it possesses two properties:

- Self-renewal—the ability to go through numerous cycles of cell division while maintaining the undifferentiated state.
- Unlimited potency—the capacity to differentiate into any mature cell type. In a strict sense, this makes stem cells either totipotent or pluripotent, although some multipotent and/or unipotent progenitor cells are sometimes referred to as stem cells.

Potency definitions

Pluripotent, embryonic stem cells originate as inner mass cells within a blastocyst. The stem cells can become any tissue in the body, excluding a placenta. Only the morula's cells are totipotent, able to become all tissues and a placenta.

Potency specifies the differentiation potential (the potential to differentiate into different cell types) of the stem cell.

- Totipotent stem cells are produced from the fusion of an egg and sperm cell. Cells produced by the first few divisions of the fertilized egg cell are also totipotent. These cells can differentiate into embryonic and extraembryonic cell types.
- Pluripotent stem cells are the descendants of totipotent cells and can differentiate into cells derived from the three germ layers.
- Multipotent stem cells can produce only cells of a closely related family of cells (e.g. hematopoietic stem cells differentiate into red blood cells, white blood cells, platelets, etc.).

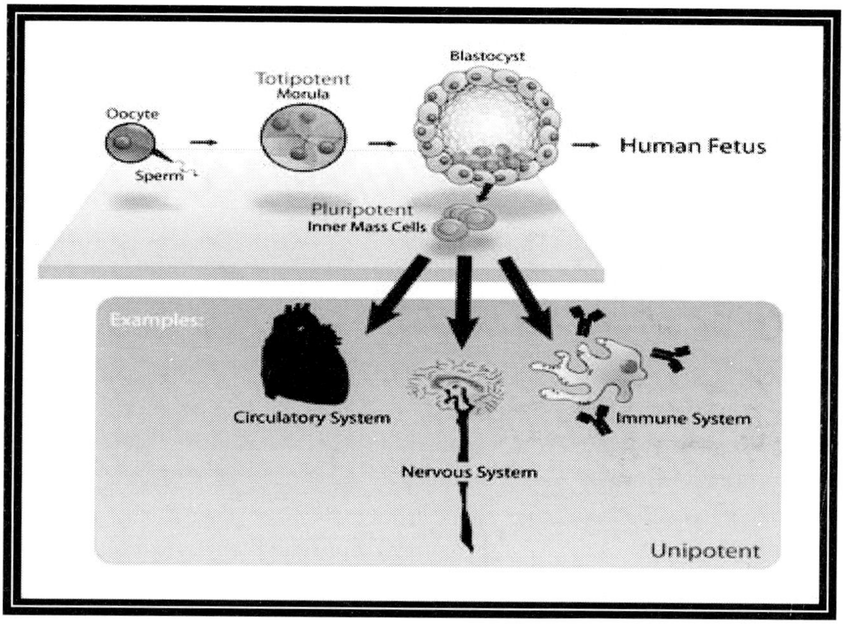

Fig. 2. Pluripotency of Embryonic Stem Cells.

- Unipotent cells can produce only one cell type, but have the property of self-renewal which distinguishes them from non-stem cells.
  There are two types of stem cells
  1. Embryonic stem cells (ES cells).
  2. Somatic stem cells (adult stem cells).

(1) Embryonic stem cells (ES cells)

Embryonic stem cells (ES cells) are stem cells derived from the inner cell mass of a blastocyst. A blastocyst is an early stage embryo—approximately 4 to 5 days old in humans and consisting of 50–150 cells. ES cells are pluripotent, and give rise during development to all derivatives of the three primary germ layers: ectoderm, endoderm and mesoderm. In other words, they can develop into each of the more than 200 cell types of the adult body when given sufficient and necessary stimulation for a specific cell type. They do not contribute to the extra-embryonic membranes or the placenta. When given no stimuli for differentiation, they will continue to divide in vitro and each daughter cell will remain pluripotent. The pluripotency of these cells has been rigorously demonstrated *in vitro* and *in vivo,* thus they can be indeed classified as stem cells. Because of their unique combined abilities of unlimited expansion and pluripotency, embryonic stem cells are a potential source for regenerative medicine and tissue replacement after injury or disease.

A. Generation of embryonic stem cells: Embryonic stem cells, as their name suggests, are derived from embryos. Specifically, embryonic stem cells are derived from embryos that develop from eggs that have been fertilized *in vitro*—in an *in vitro* fertilization clinic—and then donated for research purposes with informed consent of the donors. They are not derived from eggs fertilized in a woman's body. The embryos from which human embryonic stem cells are derived are typically four or five days old and are a hollow microscopic ball of cells called the blastocyst. The blastocyst includes three structures: the trophoblast, which is the layer of cells that surrounds the blastocyst; the blastocoel, which is the hollow cavity inside the blastocyst; and the inner cell mass, which is a group of approximately 30 cells at one end of the blastocoel.

B. Culture in the laboratory: Growing cells in the laboratory is known as cell culture. Human embryonic stem cells are isolated by transferring the inner cell mass into a plastic laboratory culture dish that contains a nutrient broth known as culture medium. The cells divide and spread over the surface of the dish. The inner surface of the culture dish is typically coated with mouse embryonic skin cells that have been treated so they will not divide. This coating layer of cells is called a feeder layer. The reason for having the mouse cells in the bottom of the culture dish is to give the inner cell mass cells a sticky surface to which they can attach. Also, the feeder cells release nutrients into the culture medium. Recently, scientists have begun to devise ways of growing embryonic stem cells without the mouse feeder cells. This is a significant scientific advancement because of the risk that viruses or other macromolecules in the mouse cells may be transmitted to the human cells.

Over the course of several days, the cells of the inner cell mass proliferate and begin to crowd the culture dish. When this occurs, they are removed gently and plated into several fresh culture dishes. The process of replating the cells is repeated many times and for many months, and is called subculturing. Each cycle of subculturing the cells is referred to as a passage. After six months or more, the original 30 cells of the inner cell mass yield millions of embryonic stem cells. Embryonic stem cells that have proliferated in cell culture for six or more months without differentiating, are pluripotent, and appear genetically normal are referred to as an embryonic stem cell line. Once cell lines are established, or even before that stage, batches of them can be frozen and shipped to other laboratories for further culture and experimentation.

C. Laboratory tests used to identify embryonic stem cells: At various points during the process of generating embryonic stem cell lines, scientists test the cells to see whether they exhibit the fundamental properties that make them embryonic stem cells. This process is called characterization. As yet, scientists who study human embryonic stem cells have not agreed on a standard battery of tests that measure the cells' fundamental properties. Also, scientists acknowledge that many of the tests they do use may not be good indicators of the cells' most important biological properties and functions. Nevertheless, laboratories that grow human embryonic stem cell lines use several kinds of tests. These tests include: growing and subculturing the stem cells for many months. This ensures that the cells are capable of long-term self-renewal. Scientists inspect the cultures through a microscope to see that the cells look healthy and remain undifferentiated using specific techniques to determine the presence of surface markers that are found only on undifferentiated cells. Another important test is for the presence of a protein called Oct-4, which undifferentiated cells typically make. Oct-4 is a transcription factor, meaning that it helps turn genes on and off at the right time, which is an important part of the processes of cell differentiation and embryonic development. This is a method to assess whether the chromosomes are damaged or if the number of chromosomes has changed. It does not detect genetic mutations in the cells.

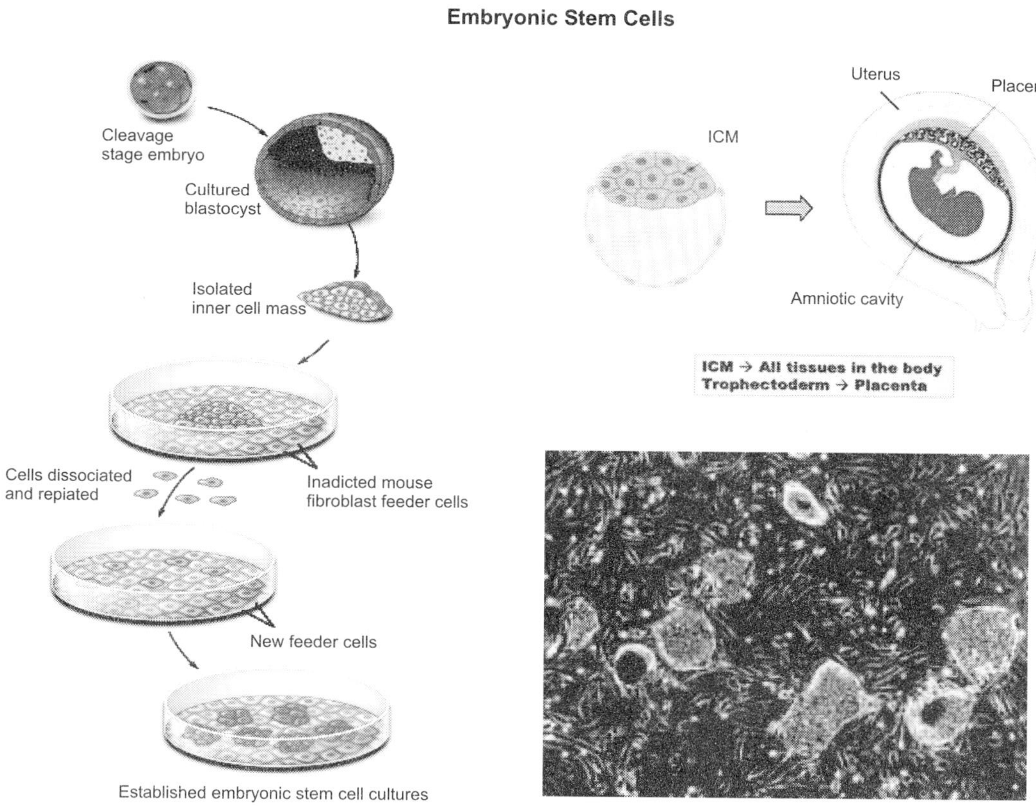

Fig. 3. Culture of ESCs in Laboratories.

D. How are embryonic stem cells stimulated to differentiate: As long as the embryonic stem cells in culture are grown under certain conditions, they can remain undifferentiated (unspecialized). But if cells are allowed to clump together to form embryoid bodies, they begin to differentiate spontaneously. They can form muscle cells, nerve cells, and many other cell types. Although spontaneous differentiation is a good indication that a culture of embryonic stem cells is healthy, it is not an efficient way to produce cultures of specific cell types. So, to generate cultures of specific types of differentiated cells—heart muscle cells, blood cells, or nerve cells, for example—scientists try to control the differentiation of embryonic stem cells. They change the chemical composition of the culture medium, alter the surface of the culture dish, or modify the cells by inserting specific genes.

(2) Adult stem cells

Adult stem cells are undifferentiated cells found throughout the body that divide to replenish dying cells and regenerate damaged tissues. Also known as somatic stem cells, they can be found in children, as well as adults. A great deal of adult stem cell research has focused on clarifying their capacity to divide or self-renew indefinitely and their differentiation potential. Many ASCs may be better classified as progenitor cells, due to their limited capacity for cellular differentiation. The primary role of these cells in a living organism are to maintain and repair the tissue in which they are found. Some scientists now use the term somatic stem cell instead of adult stem cell. Unlike embryonic stem cells, which are defined by their origin (the inner cell mass of the blastocyst), the origin of adult stem cells in mature tissues is unknown.

A. Location of ASCs: The main locations of adult stem cells are:

- HEMATOPOIETIC – Yolk sac blood islands, fetal liver spleen, bone marrow, circulating umbilical cord blood.
- LIVER – Portal zone near bile duct.
- INTESTINE – Crypt.
- EPIDERMAL – Basal layer.
- RETINAL – Pigmented ciliary margins.
- BREAST EPITHELIAL – End buds and basal layers of mammary glands.
- PANCREAS – Pancreatic islets and ducts.
- MESENCHYMAL – Bone marrow stroma and adipose tissue.

B. Methods for isolating stem cells from adult tissue:

- FLOW CYTOMETRY:
  a) Hematopoeitic stem cells.
  b) Neural crest stem cells.
  c) Epidermal stem cells.
- MICRO DISSECTION or COLONY FORMING ASSAY:
  a) Retinal stem cells.
  b) Hair follicle stem cells.
- CELL ADHESION or LACK OF ADHESION:
  a) Human epidermal stem cells adhered most rapidly to Type IV collagen, fibronectin.
  b) Late attachment of human breast epithelial stem cells on plastic.
- CONTACT INSENSITIVE GROWTH ON CELL MAT:
  a) Putative human kidney epithelial stem cells.

C. Tests used for identifying adult stem cells:

Scientists do not agree on the criteria that should be used to identify and test adult stem cells. However, they often use one or more of the following three methods: (1) labeling the cells in a living tissue with molecular markers and then determining the specialized cell types they generate; (2) removing the cells from a living animal, labeling them in cell culture, and transplanting them back into another animal to determine whether the cells repopulate their tissue of origin; and (3) isolating the cells, growing them in cell culture, and manipulating them, often by adding growth factors or introducing new genes to determine what differentiated cells types they can become. Also, a single adult stem cell should be able to generate a line of genetically identical cells—known as a clone—which then gives rise to all the appropriate differentiated cell types of the tissue. Scientists tend to show either that a stem cell can give rise to a clone of cells in cell culture, or that a purified population of candidate stem cells can repopulate the tissue after transplant into an animal. Recently, by infecting adult stem cells with a virus that gives a unique identifier to each individual cell, scientists have been able to demonstrate that individual adult stem cell clones have the ability to repopulate injured tissues in a living animal.

D. Adult stem cell differentiation

Studies have reported that adult stem cells occur in many tissues and that they enter normal differentiation pathways to form the specialized cell types of the tissue in which they reside. Adult stem cells may also exhibit the ability to form specialized cell types of other tissues, which is known as transdifferentiation or plasticity.

Normal differentiate pathway in a living animal, adult stem cells can divide for a long period and can give rise to mature cell types that have characteristic shapes and specialized structures and functions of a particular tissue. The following are examples of differentiation pathways of adult stem cells (Fig. 4). Hematopoietic stem cells give rise to all the types of blood cells: red blood cells, B lymphocytes, T lymphocytes, natural killer cells, neutrophils, basophils, eosinophils, monocytes, macrophages, and platelets. Bone marrow stromal cells (mesenchymal stem cells) give rise to a variety of cell types: bone cells (osteocytes), cartilage cells (chondrocytes), fat cells (adipocytes), and other kinds of connective tissue cells such as those in tendons. Neural stem cells in the brain give rise to its three major cell types: nerve cells (neurons) and two categories of non-neuronal cells—

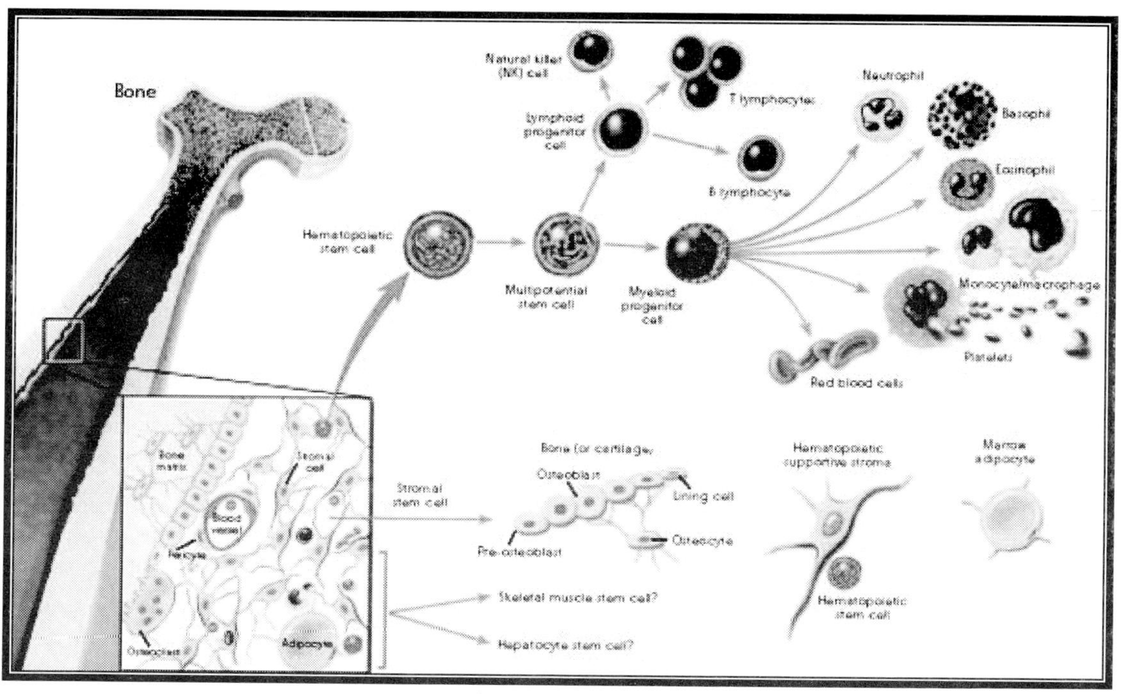

Fig. 4. Hematopoietic and stromal stem cell differentiation.

astrocytes and oligodendrocytes. Epithelial stem cells in the lining of the digestive tract occur in deep crypts and give rise to several cell types: absorptive cells, goblet cells, Paneth cells, and enteroendocrine cells. Skin stem cells occur in the basal layer of the epidermis and at the base of hair follicles. The epidermal stem cells give rise to keratinocytes, which migrate to the surface of the skin and form a protective layer. The follicular stem cells can give rise to both the hair follicle and to the epidermis.

Adult stem cell plasticity and transdifferentiation: A number of experiments have suggested that certain adult stem cell types are pluripotent. This ability to differentiate into multiple cell types is called plasticity or transdifferentiation. The following list offers examples of adult stem cell plasticity that have been reported during the past few years. Hematopoietic stem cells may differentiate into: three major types of brain cells (neurons, oligodendrocytes, and astrocytes); skeletal muscle cells; cardiac muscle cells; and liver cells. Bone marrow stromal cells may differentiate into: cardiac muscle cells and skeletal muscle cells. Brain stem cells may differentiate into: blood cells and skeletal muscle cells. Current research is aimed at determining the mechanisms that underlie adult stem cell plasticity. If such mechanisms can be identified and controlled, existing stem cells from a healthy tissue might be induced to repopulate and repair a diseased tissue.

SIMILARITIES AND DIFFERENCES BETWEEN EMBRYONIC AND ADULT STEM CELLS

Human embryonic and adult stem cells each have advantages and disadvantages regarding potential use for cell-based regenerative therapies. Of course, adult and embryonic stem cells differ in the number and type of differentiated cell types. Embryonic stem cells can become all cell types of the body because they are pluripotent. Adult stem cells are generally limited to differentiating into different cell types of their tissue of origin. However, some evidence suggests that adult stem cell plasticity may exist, increasing the number of cell types a given adult stem cell can become. Large numbers of embryonic stem cells can be relatively easily grown in culture, while adult stem cells are rare in mature tissues and methods for expanding their numbers in cell culture have not yet been worked out. This is an important distinction, as large numbers of cells are needed for stem cell replacement therapies. A potential advantage of using stem cells from an adult is that the patient's own cells could be expanded in culture and then reintroduced into the patient. The use of the patient's own adult stem cells would mean that the cells would not be rejected by the immune system. This represents a significant advantage as immune rejection is a difficult problem that can only be circumvented with immunosuppressive drugs. Embryonic stem cells from a donor introduced into a patient may cause transplant rejection. However, whether the recipient would reject donor embryonic stem cells has not been determined in human experiments.

APPLICATIONS IN MEDICINE

Diabetes

Prevalence of diabetes (WHO Health reports) says that worldwide nearly 8% out of 194 million number of people are affected from diabetes, in which ~90% are affected from Type 2 and ~ 4–8%

are from Type 1, the data estimated in 2006. Diabetes is characterized by hyperglycemic conditions associated with decreased insulin secretion or resistance with progressive loss in islets mass. Islet regeneration is an important issue, which can compensate loss of functional beta cells and thus is attractive strategy to cure diabetes.

- Both type 1 and type 2 diabetes are characterized by a marked deficit in β cell mass causing insufficient insulin secretion.
- Type 1 diabetes is caused by autoimmune-mediated destruction of beta-cells and drugs—STZ / alloxan/caerulein, Dietary Cu deprivation, Hemipancreatectomy, viruses. Once hyperglycemia develops as much as 90% of beta-cells have been lost.
- Type 2 diabetes is mainly because of the impaired peripheral uptake of glucose from adipocytes and skeletal muscle. Studies in animal models have shown that fetal deficiency of macro/micro nutrients results in Type 2 diabetes. It is also characterized by a 65% decrease in beta-cell mass associated with a 10-fold increase in beta-cell apoptosis.
- Taken together, these data highlight the importance of beta-cell mass for the maintenance of normoglycemia.
- Therefore, beta-cell replacement will clinically be a feasible approach in the management of diabetes.

Transplantation of insulin producing islet cells,

a) Pancreas: syngenic/allogenic and xenogenic
b) Islet cell transplantation (Cadaver)
c) Stem cell: embryo, fetus and adult tissue.
d) Genetic engineering.

HEMATOLOGICAL DISORDERS

Pluripotent hematopoietic stem cells (PHSCs) are responsible for multilineage hematopoietic cell differentiation and maturation. These cells possess specific homing properties and can be identified by certain surface markers. PHSCs give rise to common myeloid and lymphoid progenitors, the parent cells of different types of blood cells. Qualitative and quantitative abnormalities of embryonic stem cells results in various blood disorders like leukemias, immunodeficiency syndromes etc. Presence or absence of stem cell markers help to diagnose leukemias and lymphomas accurately.

Hematological disorders are of various types as mentioned below:

a) Malignant: These are of two types:
   - Leukemia
   - Lymphomas
b) Pre-malignant: These are of three types;
   - Myelodysplastic syndromes (MDS)
   - Aplastic anemia
   - Paroxysmal nocturnal hemoglobinuria (PNH)
c) Non-malignant: These are of three types:
   - Inherited hemolytic anemia

  - Immuno-deficiency syndromes
  - Metabolic disorders
- d) Diagnosis
- e) Prognosis
- f) Treatment: They are of following types:
  - Conventional
    - Supportive
    - Specific
  - Aggressive
    - Chemotherapy
    - Molecular targeted therapy
    - Stem cell transplantation
    - Gene therapy

Classification of Clonal Hemopoietic Disorders:

- a) Acute myeloblastic leukemia (AML - M0-8)
- b) Chronic myeloproliferative disorders (CMPD)
- c) Myelodysplastic syndromes (MDS)
- d) Acute lymphoblastic leukemia (ALL - L1-3)
- e) Chronic lymphocytic leukemia
- f) Hematopoietic Stem Cell Transplantation (HSCT)

Sources

- Bone marrow
- Peripheral blood (mobilized)
- Umbilical cord blood
- Fetal liver cells

Types

- Allogeneic
  - VUD
  - Sibling
- Syngeneic
- Autologous

HSC Transplantation: Phases

- Preparative Phase: eliminate tumour cells, create space, prevent graft rejection, less non-hematopoietic toxicity
- Transplant
- Early post-transplant:
  - Blood product and nutritional support
  - Anti-microbial prophylaxis

  - Prevention of GvHD
  - Treatment of Acute GvHD
  - Management of other complications
- Late phase management
- Manipulations:
  - RBC and/or plasma depletion
  - T cell depletion

Complications of HSCT:

- Infectious
- Non-infectious
  - GvHD (acute and chronic)
  - Direct toxicities
  - Supportive Rx.

PARKINSON'S DISEASE

People who have Parkinson's disease experience difficulty with movement, balance and speech. These problems result from the death of specialized brain cells called dopamine neurons. These cells produce dopamine, a chemical that helps control muscle movements. The effects of the disease can be treated with drugs that help increase dopamine in the brain, but there is no known cure. Researchers studying Parkinson's disease knew that the problem was to replace the dead dopamine neurons with healthy dopamine-producing cells. Fig. 5 shows the pathway of causing Parkinson's disease, and the treatment strategy planning to use for treatment.

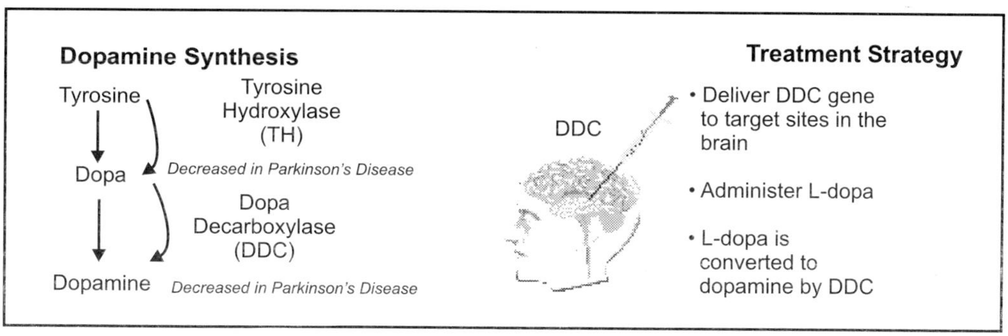

Fig. 5. Pathway and new coming treatment strategy.

Human stem cells have potential uses and obstacles that must be overcome before their uses.

There are many ways in which human stem cells can be used in basic research and in clinical research. However, there are many technical hurdles between the promise of stem cells and the realization of these uses, which will only be overcome by continued intensive stem cell research. Studies of human embryonic stem cells may yield information about the complex events that occur during human development. A primary goal of this work is to identify how undifferentiated stem cells become differentiated. Scientists know that turning genes on and off is central to this process. Some of the most serious medical conditions, such as cancer and birth defects, are due to abnormal cell division and differentiation. A better understanding of the genetic and molecular controls of these processes may yield information about how such diseases arise and suggest new strategies for therapy. A significant hurdle to this use and most uses of stem cells is that scientists do not yet fully understand the signals that turn specific genes on and off to influence the differentiation of the stem cell. Human stem cells could also be used to test new drugs. For example, new medications could be tested for safety on differentiated cells generated from human pluripotent cell lines. Other kinds of cell lines are already used in this way. Cancer cell lines, for example, are used to screen potential anti-tumor drugs. But, the availability of pluripotent stem cells would allow drug testing in a wider range of cell types. However, to screen drugs effectively, the conditions must be identical when comparing different drugs. Therefore, scientists will have to be able to precisely control the differentiation of stem cells into the specific cell type on which drugs will be tested. Current knowledge of the signals controlling differentiation fall well short of being able to mimic these conditions precisely to consistently have identical differentiated cells for each drug being tested. Perhaps the most important potential application of human stem cells is the generation of cells and tissues that could be used for cell-based therapies. Today, donated organs and tissues are often used to replace ailing or destroyed tissue, but the need for transplantable tissues and organs far outweighs the available supply. Stem cells, directed to differentiate into specific cell types, offer the possibility of a renewable source of replacement cells and tissues to treat diseases including Parkinson's and Alzheimer's diseases, spinal cord injury, stroke, burns, heart disease, diabetes, osteoarthritis, and rheumatoid arthritis.

HEART INJURIES

It may become possible to generate healthy heart muscle cells in the laboratory and then transplant those cells into patients with chronic heart disease. Preliminary research in mice and other animals indicates that bone marrow stem cells, transplanted into a damaged heart, can generate heart muscle cells and successfully repopulate the heart tissue. Other recent studies in cell culture systems indicate that it may be possible to direct the differentiation of embryonic stem cells or adult bone marrow cells into heart muscle cells (Fig. 6). In people who suffer from type I diabetes, the cells of the pancreas that normally produce insulin are destroyed by the patient's own immune system. New studies indicate that it may be possible to direct the differentiation of human embryonic stem cells in cell culture to form insulin-producing cells that eventually could be used in transplantation therapy for diabetics. To realize the promise of novel cell-based therapies for such pervasive and debilitating diseases, scientists must be able to easily and reproducibly manipulate stem cells so that they possess the necessary characteristics for successful differentiation, transplantation and engraftment. The following is a list of steps in successful cell-based treatments that scientists will have to learn to precisely control to bring such treatments to the clinic. To be useful for transplant purposes, stem cells must be reproducibly made to:

- Proliferate extensively and generate sufficient quantities of tissue.

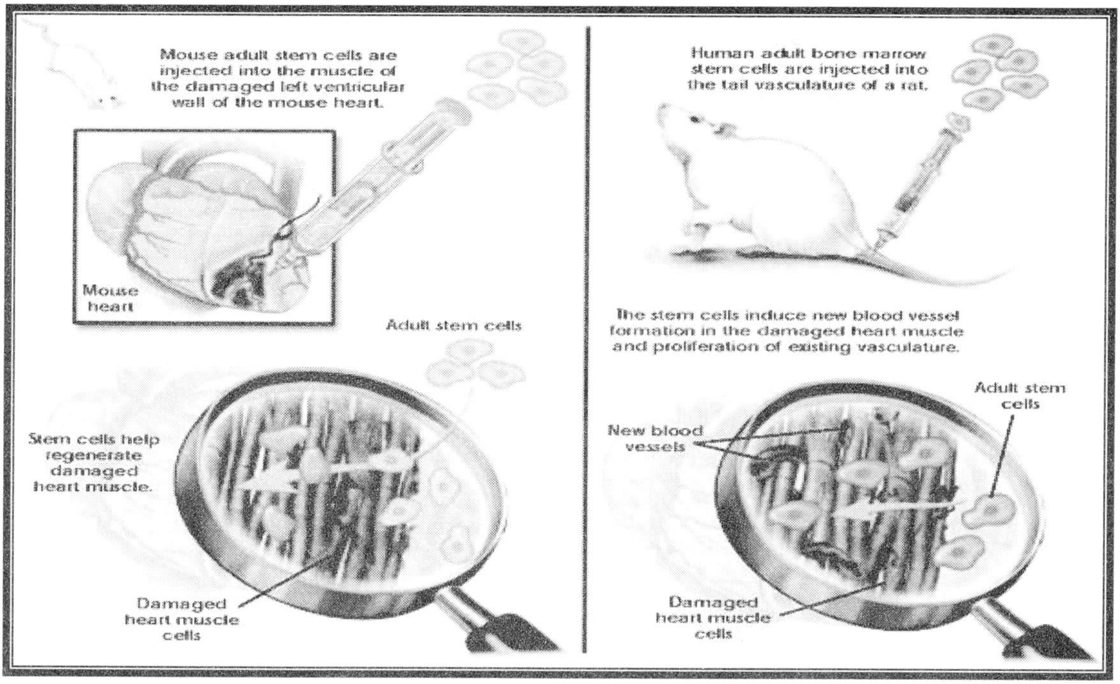

Fig. 6. Heart muscle repair with adult stem cells.

- Differentiate into the desired cell type(s).
- Survive in the recipient after transplant.
- Integrate into the surrounding tissue after transplant.
- Function appropriately for the duration of the recipient's life.
- Avoid harming the recipient in any way.

Also, to avoid the problem of immune rejection, scientists are experimenting with different research strategies to generate tissues that will not be rejected.

Spinal Cord Injuries

Spinal Cord Injury (SCI) is damage to spinal cord that results in a loss of function such as mobility or feeling. Frequent causes of damage are trauma (car accident, gunshot, falls, etc.) or diseases (polio, spina bifida, Friedreich's ataxia, etc.). Approximately 32 injuries per million population or 7800 SCIs occur in the USA each year. SCI is very different from back injuries such as ruptured disks, spinal stenosis or pinched nerves. The spinal cord does not have to be severed for a loss of function to occur. In fact, in most people with SCI, the spinal cord is intact, but the damage to it results in loss of functioning. Acute spinal cord damage occurs during the first few weeks of the injury. In turn the chronic period begins after a few months. Due to the scarring in chronic phase, transplanting adult stem cells into a damage spinal cord does not spur nerve growth. Currently, drug treatment is given during the acute phase to help stabilize the injury site, but it provides only a very mild benefit, and do not foster regeneration of injured tissue. It is anticipated that the stem cell treatment in humans will occur during spinal stabilization at the acute phase, when screws and rods are placed in the spinal column to restabilize it after injury. A treatment derived from human embryonic stem cells has shown to improve mobility in patients with spinal cord injuries, providing the first physical evidence that the therapeutic use of these cells can help restore motor skills lost from acute spinal cord tissue damage.

ETHICAL ISSUES

a) Sources of ES cells

- ES cell lines
- Excess embryos from IVF clinics
- Embryos created for research by IVF
- Therapeutic cloning

b) Ethical debate

- Harvesting ES cells destroys the blastocyst.
- Is this a murder
- ES cell research requires human cells
- Could create a commercial market for human cells
- This devalues life
- If excess IVF embryos are being discarded anyway, they should be put to good use"
- Therapeutic cloning is a slippery slope—it will lead to reproductive cloning.

CONCLUSION

It is to be believed now that stem cell research has the potential to completely change the treatment of human diseases. Such treatment for a number of current treatments already exists, although the majority of them are not commonly used because they tend to be experimental and not very cost-effective. Medical researchers anticipate being able to use technologies derived from stem cell research to treat cancer, Parkinson's disease, spinal cord injuries, and muscle damage, amongst a number of other diseases and impairments. However, there still remains a great deal of social and scientific uncertainty surrounding stem cell research, which could possibly be overcome through public debate and future research. Stem cells, however, are already used extensively in research, and some scientists do not see cell therapy as the first goal of the research, but see the investigation of stem cells as a goal worthy in itself. If scientists can reliably direct the differentiation of embryonic stem cells into specific cell types, they may be able to use the resulting, differentiated cells to treat certain diseases at some point in the future.

SUGGESTED READINGS

1. Axllrad, A.A., D.L. Mcleod, M.M. Shreeve and D.S. Heath. Properties of cells that produce erythrocytic colonies *in vitro*. In: Hemopoiesis in culture.W.A. Robinson, ed., pp. 226–237, 1974.

2. Byron, J.W. Manipulation of cell cycle of hemopoietic stem cells. Exp. Hemat. 3: 44–53, 1975.
3. Metcalf, D. and M.A.S. Moore, Hemopoietic cells, 1971.
4. Lima C et al. Olfactory mucosa autografts in human spinal cord injury: A pilot clinical study, Journal of Spinal Cord Medicine 29, 191–203, 2006.
5. Spees JL, Gregory CA, Singh H, Tucker HA, Peister A, Lynch PJ, et al: Internalized antigens must be removed to prepare hypoimmunogenic mesenchymal stem cells for cell and gene therapy. *Mol Ther*, 9(5):747–756, 2004.

8

Microbes and Human Health

S.K. Soni, Raman Soni and D.K. Rahi

INTRODUCTION

Microorganisms have lived on Earth for billions of years and are essential to our very existence. They are ubiquitous and represent a huge biological diversity in soil, marine and aquatic ecosystems. A variety of microbes are found in all common environments such as soil, water and air as well as exotic locales as diverse as deep sea hydrothermal vents and soda lime lakes. These are essentially bags of enzymes which along with minerals are some of the most reactive, catalytic surfaces in soils and sediments of the Earth's crust. Microbes crave energy, and while acquiring it, they chemically transform many important elements through oxidation and reduction reactions. Consequently, bacteria are responsible for most of the global decomposition of organic matter and the production of greenhouse gases. The biogeochemical cycles of the majority of biological elements are controlled over geologic time by microbially-mediated processes. A variety of microorganisms are found in the diverse environment. Their number and types vary significantly and determined by various factors including the amount and type of nutrients, moisture content, degree of aeration, temperature, pH, existence of roots of plants, salinity, hydrostatic pressure, intensity of light and turbidity etc. Microorganisms play a critical role in animal and human health. Bacteria, for example, play an important role in digestion. They help in synthesizing vitamin K and absorb certain nutrients and help to convert bile and acids in the intestines. Some microorganisms also help to prevent other harmful microorganisms from invading the intestines. Microorganisms that are normally found in animal and human bodies are referred to as normal flora. Most normal flora are found in the gastrointestinal tract, but are found throughout the body, including the eye. Bacteria commonly found in the body include *Streptococcus* and *Staphylococcus,* bacteria which can also cause serious infections. The discovery of the role of microorganisms or germs, in causing disease was the beginning of a great revolution in health care. Although Antony van Leeuwenhoek first observed bacteria in the late 1600s, it was not until late in the 19th century that the germ theory of disease became generally accepted. The research of French scientist Louis Pasteur provided persuasive evidence that certain microorganisms were responsible for human illness. Among his other findings, he discovered three bacteria, *Staphylococcus, Streptococcus* and *Pneumococcus* that can become pathogenic. However, pathogenicity among microorganisms is the exception, not the rule. Considering the huge population of bacteria and viruses in our environment and in our own bodies, it is a relatively rare occurrence that these typically symbiotic relationships become harmful. After all, microorganisms are neutral or have little to gain, from killing their host. As physician and microbiologist Lewis Thomas reminded us, "The man who catches a meningococcus is in considerably less danger for his life, even without chemotherapy than meningococci with the bad luck to catch a man." Human disease is in fact the body's immunologic reaction to the presence of a bacterium or virus. It is not yet completely understood why some immunological reactions occur. Symptoms of infection are the result of the immune system's response to the presence of potential threats. Fever, for example, is part of the body's natural defence mechanism. Higher temperatures reduce the ability of viruses and bacteria to replicate. In most cases, the body's immune system is successful in preventing serious harm from infections. Indeed, infections bestow a benefit on human health in building up the body's immunity to pathogens so that they do not cause any harm. In the last several decades, as people have increased their use of antibiotics to treat routine infections, a serious unintended consequence has been seen i.e. an increase in the populations of microorganisms that are resistant to antibiotics and difficult to treat.

Although modern science and medicine has made vast improvements in human and animal health, it is remarkable how much remains to be understood. No one, for example, knows within an order of magnitude how many microorganisms there are. New infectious viruses, such as West Nile virus and the Sudden Acute Respiratory Syndrome (SARS), appear from time to time, posing a threat to human health. The origin of some of these viruses is unknown. There also remains much to be understood about the common cold and the flu. The present chapter describes both the positive and negative roles of microorganisms in human health.

MICROORGANISMS AND FOOD PRODUCTION

Microorganisms are associated with all the foods that we eat and are responsible for the formation of certain foods through the process of fermentation. Microbes may also be directly used as a source of food in the form of single cell proteins and food supplements for human health in the form of pigments, amino acids, vitamins, organic acids, enzymes and lipidic substances and inhibit other microbes which may produce the toxins. The exploitation of microorganisms in food industry is as old as the history of civilization. The traditional application has been in the production of various indigenous fermented foods by involving the natural microflora coming from the environments. The current applications involve the production of various foods and food additives by the process of fermentation with genetically modified microbes (GMO's) Indigenous fermented legume and cereal products form an important part of human diet in South-East Asia including India, the Middle East and Africa. They are also becoming popular in the developed world due to their high nutritional value and organoleptic characteristics.

Beer, wine and bread making by yeasts, the processing of milk to dairy products by lactic acid bacteria and the fermentation of vinegar by acetic acid bacteria show that these microbes are among the oldest domesticated microorganisms. The reports about the origin of fermented foods are lacking but it is believed that the discovery of fermentations was purely by chance. The transition from hunting and gathering food to organized food cultivation and production is believed to have taken place some 10000 to 15000 years ago in the Middle East. It is believed that it was purely by chance that the fermentation became a method of preservation. Some of the common indigenous fermented foods consumed in different parts of the world are listed in Table 1 which indicates the geography, substrates employed, the nature and uses of various products. Indigenous fermented products form an important part of human diet in South-East Asia including India, the Middle East and Africa. They are also becoming popular in the developed world due to their high nutritional value and organoleptic characteristics. People of Asia are pioneers in the development of means of fermenting vegetable proteins to produce meat-like flavours. Indonesians developed the fermentation methods for introducing meat-like texture into vegetable products. Koreans are credited for the production of acid fermented vegetables. People of Egypt developed wheat breads leavened with yeasts while Indians discovered methods for souring and leavening cereal-legume batters. Generally, the traditional methods of preparing indigenous fermented foods are simple and inexpensive. However, the ancient methods of making such foods are changing rapidly through modern microbial technology. Some of the fermented products including dairy products, coffee, cocoa and bread have been studied extensively for many decades and their production methods have been modernized keeping pace with the advancement in modern science. During the last few years much interest has been generated in the fermented foods of Asian and African countries including India where such foods are still being manufactured according to the traditional, technologically less advanced methods using simple equipment at cottage industry scale by means of natural micro-flora from the staples and the surroundings.

MICROORGANISMS AS THE SOURCE OF FOOD

Though every bite of food contains microbes, we are eating them all the time and there is a constant stream of microbes into animal intestines. Some of the microorganisms or their fruiting bodies are used directly as a source of single cell proteins, mushrooms and truffles are a source of proteins, generally termed as mycoproteins. The term Single cell proteins (SCP) refers to dead and dried cells of microorganisms such as bacteria, algae, molds, and yeasts. SCP from bacteria, yeasts, and fungi are produced by growing them on cheaper feedstocks such as molasses, methane, methanol, ethanol, cheese whey, cassava starch, a range of agricultural, forestry wastes and other industrial wastes. Generally, four groups of microorganisms including bacteria, filamentous fungi, yeasts and algae are used as source of single cell proteins (Table 2). Among these yeasts are best due to their high nutritive values, ability to grow on low pH, vast substrate acceptability, less chances of contamination and are easy to harvest. Next to yeasts are the filamentous fungi. Bacteria may not be safe as food or feed as many of them produce toxins. While algae need large surface area for cultivation, economically and technically are difficult to grow.

A very important group of microorganisms used directly as a source of food are the mushrooms, most of which belong to basidiomycetes and a few belong to ascomycetes. These fungi produce a fleshy fruiting structure containing the sexually produced spores, basidiospore or ascospores respectively. The major, but relatively rare ascomycete fruiting structure eaten are morels (*Morchella esculenta*). Most market mushrooms belong to group basidiomycetes. Mushrooms are consumed not

Table 1. Some important indigenous fermented foods of the world

	Fermented food	Country/place	Ingredient(s)	Nature of product	Product use
Legume products					
1.	*Bhallae*	India	Black gram	Deep fried patties	Snack after soaking in curd, water
2.	*Chee-fan*	China	Soybean, whey curd	Solid, cheese like	Eaten fresh
3.	*Khaman*	Western India	Bengal gram	Spongy cake	Breakfast food
4.	*Kenima*	Nepal, Sikkim, Darjeeling district of India	Soybeans	Solid	Snack
5.	*Ketjap*	Indonesia	Black soybeans	Syrup	Seasoning agent
6.	*Meitauza*	China, Taiwan	Soybean cake	Solid	Fried in oil or cooked with vegetables
7.	*Meju*	Korea	Soybeans	Paste	Seasoning agent
8.	*Miso*	Japan, China	Soybeans/Soybeans and rice	Paste	Soup base, seasoning agent
9.	*Natto*	Northern Japan	Soybeans	Solid	Roasted or fried in oil used as a meat substitute
10.	*Oncom*	Indonesia	Peanut press cake	Solid	Roasted or fried in oil, used as a meat substitute
11.	*Papadam*	India	Black gram and spices	Circular tortilla like wafers	Condiment
12.	*Soybean milk*	China, Japan	Soybeans	Liquid	Drink
13.	*Sufu*	China, Taiwan	Soybean, whey curd	Solid	Soybean cheese, condiment
14.	*Soy sauce*	Japan, China, Philippines	Soybeans/Soybeans and wheat	Liquid	Seasoning agent for meat, fish, cereals and vegetables
15.	*Tempe*	Indonesia and vicinity	Soybeans	Solid	Fried in oil, roasted or used as meat substitute in soup
16.	*Vadai*	India	Black gram	Deep fried patties	Snack
17.	*Warries*	Northern India	Black gram and spices	Ball like hollow, brittle	Spicy condiment, eaten with vegetables, legumes, rice
Legume-cereal products					
18.	*Dhokla*	Western India	Rice, Bengal gram	Spongy cake	Condiment
19.	*Dosa*	Southern India	Rice, black gram	Spony, pan cake	Staple food
20.	*Hama natto*	Japan	Wheat flour, whole soybeans	Raisin like,	Flavouring agent for meat and fish, eaten as snack
21.	*Idli*	Southern India	Rice, Black gram	Spongy	Breakfast food
22.	*Kecap*	Indonesia and vicinity	Wheat, soybeans	Liquid	Condiment, seasoning agent
23.	*Tao-si*	Philippines	Wheat flour, soybeans	Semi-solid	Seasoning agent
24.	*Taotjo*	East Indies	Roasted wheat meal or glutinous rice, soybeans	Semi-solid	Condiment

(Contd.)

Table 1. (Contd.)

Fermented food	Country/place	Ingredient(s)	Nature of product	Product use
Cereal products				
25. *Ang-Kak (Chinese red rice)*	China, Taiwan, Philippines, Thailand	Rice	Solid	Colouring other foods
26. *Ambali*	India	Millet flour	Semi solid	All time food
27. *Bhatura*	India (North)	White wheat flour	Deep fried	Breakfast bread
28. *Fermented rice*	India	Rice	Semi solid	Breakfast
29. *Hopper (Appa)*	Sri Lanka	Rice or white wheat flour and cocconut water	Semi solid	Breakfast
30. *Jalebie*	India and Pakistan	White wheat flour	Deep fried pretzel	Confection food
31. *Kenkey*	Ghana	Maize	Semi solid	Breakfast, lunch and supplement to fish stews
32. *Kisra*	Sudan	Sorgham flour	Spongy bread	Staple food
33. *Kulcha*	India (North) and Pakistan	White wheat flour	Flat bread	Staple food
34. *Mahewu*	South Africa	Maize meal	Liquid	Beverage food
35. *Nan*	India (North), Pakistan, Iran and Afghanistan	White wheat flour	Flat bread	Staple food
36. *Ogi*	Nigeria	Maize, Millet or Sorghum	Semi solid	Breakfast and Infant food
37. *Pozol*	Mexico	White maize	Liquid	Beverage or Porridge
38. *Puto*	Philippines	Rice	Semi solid	Breakfast and snack food
39. *Shamsy bread*	Egypt	Wheat flour	Spongy bread	Staple food
40. *Uji*	Kenya, Uganda and Tanzania	Maize, Sorghum or Millet flour	Semi solid	Breakfast and lunch
Fermented Milk Products				
41. *Cultured cream*	United States, Russia Central Europe	Cow's milk cream	Sour gel/cream containing butter-like aromatic flavour	Toppling on vegetables, salads, fish, meats, fruits, filling in cakes and in soups
42. *Cultured butter milk*	United States	Cow's skim milk, cream	Viscous fluid milk	
43. *Yoghurt*	United States, Europe, Asia	Cow/goat/sheep's milk, condensed skim milk, cream, non-fat dry milk, sucrose, fruits	Flavoured curd	Beverage food
44. *Acidophilus milk*	United States, Russia	Cow's milk	Acid fermented fluid milk	All time food
45. *Kefir*	Russia	Cow/goat/sheep's milk	Acid and alcoholic milk	Beverage food with therapeutic value
46. *Koumiss*	Russia	Mare's milk	Acid and alcoholic milk	Beverage food
47. *Cottage cheese*	United States	Cow's milk	Natural unripened soft cheese	Beverage food, generally used in pulmonary tuberculosis, All time food

Table 2. Microorganisms and substrates used for production of single cell proteins

Microorganisms	Substrates for production
Bacteria	
Aeromonas hydrophylla	Dairy wastes
Achromobacter delvacvate	Hydrocarbons
Acinetobacter calcoacetious	Hydrocarbons
Bacillus megaterium	Meat industry wastes
Bacillus subtilis	Forest, agricultural, horticultural wastes
Cellulomonas species	Forest, agricultural, horticultural wastes
Flavobacterium species	Forest, agricultural, horticultural wastes
Lactobacillus species	Starch wastes
Methylomonas methylotrophus	Hydrocarbons
Methylomonas clara	Hydrocarbons
Pseudomonas fluorescens	Manures
Rhodupseudomonas capsulata	Starch wastes
Thermomonospora fusa	Lignocellulosic wastes
Filamentous fungi	
Aspergillus niger	Lignocellulosic wastes
Aspergillus fumigatus	Starch wastes
Aspergillus oryzae	Lignocellulosic wastes
Cephalosporium eichorniae	Starch wastes
Chaetomium cellulolyticum	Lignocellulosic wastes
Penicillium cyclopium	Dairy wastes
Rhizopus chinencis	Starch wastes
Scytaldium acidiphilum	Lignocellulosic wastes
Trichoderma viridae	Lignocellulosic wastes
Trichoderma album	Lignocellulosic wastes
Yeasts	
Amoco torula	Hydrocarbons
Candida tropicalis	Starch wastes
Candida utilis	Lignocellulosic wastes
Candida novellas	Hydrocarbons
Candida intermedia	Dairy wastes
Saccharomyces cerevisae	Lignocellulosic wastes
Algae	
Chlorella pyrenidosa	Utilizes CO_2 through photosynthesis
Chlorella sorkiniana	Utilizes CO_2 through photosynthesis
Chondrus chrispum	Utilizes CO_2 through photosynthesis
Scenedesmus species	Utilizes CO_2 through photosynthesis
Spirulina species	Utilizes CO_2 through photosynthesis
Porphyrium species	Utilizes CO_2 through photosynthesis

only for their flavours and taste but also for their nutritive values. The nutritive value of edible mushrooms nearly equals to that of milk. The protein content of mushrooms are above most vegetables and somewhat less than most meats and milk. Protein content can vary from 10 to 50 % on dry weight basis. They contain high amount of amino acids (lysine, valine, and leucine), carbohydrates, vitamins, unsaturated fatty acids with a mineral content exceeding that in fish and meat and twice that in most vegetables. Fruiting bodies in general, on dry weight basis, contain 32% proteins, about 55% carbohydrate, 2% fat and rest is ash constituting the minerals. Besides, mushrooms are a good source of vitamins like B-complex vitamins including riboflavin (B_2), niacin, pantothenic acid, thiamin (B_1), biotin, folate and vitamin B_{12}. Mushrooms are a particularly rich source of riboflavin. Mushrooms are unique in that they contain vitamin B_{12}, something that vegetables can't produce at all. Since B_{12} is mainly of animal origin, deficiency is commonly associated with vegetarian diets. Mushrooms were found to contain 0.32-0.65 mg per gram of B_{12}, pro-vitamin D is present in some mushrooms, particularly shiitake, and can be converted to vitamin D by ultraviolet irradiation (sunlight).Vitamin A is uncommon although several mushrooms contain detectable amounts of pro-vitamin A measured as the β-carotene equivalent. Most cultivated mushrooms are believed to contain low amounts of the fat-soluble vitamins, K and E, and only a small contribution to the daily requirement of vitamin C. More than 2000 species of mushrooms are reported to be edible throughout the world and about 150 species belonging to 30 genera are prominent among them. Over 60 species are cultivated commercially and 10 have reached an industrial scale in many countries. The prominent edible mushroom species cultivated throughout the world are enlisted in Table 3.

There are several edible mycorrhizal mushrooms including the species of *Cantharellus, Russula, Lactarius* which cannot be cultivated on organic substrate due to their obligatory mycorrhizal nature and are known as mycorrhizal mushrooms.

Truffles are a group of edible *mycorrhizal* (subterranean) *fungi*, the *ascoma* (*fruiting body*) of which is highly prized as food. These are generally served uncooked and shaved over steaming pasta or salads. The production of truffle is done by digging tree seedling in a mycorrhizal slurry before planting. After several years, under favourable growing conditions for both, the tree and the fungus, truffles form underground fruiting bodies that roughly resemble potatoes. These range from the size of a pea to that of a fist and give off a distinctive odor. Since these mushrooms do not completely emerge from the ground, they have traditionally been sniffed out by pigs or trained dogs. The requirements for growing the black truffle, *Tuber melanosporum* Vitt., include choosing an appropriate host plant (usually oak or hazel nut), inoculating its roots with spawn, and planting it.

MICROORGANISMS AS A SOURCE OF FOOD SUPPLEMENTS

A food supplement, mycoprotein, is the processed cellular mass that is obtained from the filamentous fungus *Fusarium venenatum.* The filamentous nature of the cells gives the cellular mass a meat-like texture that makes it suitable for a variety of applications in food, including use as a muscle replacer in meat-alternative products, as a fat replacer in certain dairy products and as a cereal replacer in products such as breakfast cereals or puffed snacks.The protein content of mycoprotein typically ranges from 42–50% on a dry weight basis. Mycoprotein is a high quality protein because mycoprotein contains all of the essential amino acids. The fat content of mycoprotein typically ranges from

Table 3. Prominent cultivated edible mushrooms

Mushroom species	Common name	Substrates for cultivation
Agaricus bisporus, *A. bitorquis*	Button mushroom	Wheat straw
Pleurotus ostreatus, *P. florida, P. sajor-caju*	Oyster mushroom	Wheat straw, bean straw, coir, corn cobs, paper wastes, cotton wastes (textile industry), crushed bagasse, etc.
Volvariella volvacea	Paddy straw mushroom	Paddy straw, oil palm pericarp waste, banana leaves, water hyacinth
Lentinus edodes	Shiitake or Oak mushroom	Sawdust, coffee pulp, sawdust-rice bran
Pholiota nameko	Nameko or viscid mushroom	Logs, sawdust-rice bran
Auricularia spp.	Jew's ear mushroom	Sawdust
Hericium erinaceus	Lion's head or Pom-Pom mushroom	Sawdust distillers grain waste
Grifola frondosa	Maitake or sitting hen mushroom	Sawdust
Strophoria rugosonannulata	—	Paper waste
Morchella sp.	Morels	*see footnote
Flammulina velutipes	Winter mushroom	Sawdust rice-bran

*Commercial production of morels is currently not a practical option. Morels are being grown using a patented process, at only one production facility in North america (in Atlanta). The patent and facility are owned by Terry Farms and represent the only successful commercial process for cultivation of this ascomycetous mushroom.

12–14% on a dry weight basis and this fat content is more like vegetable fat than animal fat because it has a low proportion of saturated fatty acids and a high proportion of mono- and polyunsaturated fatty acids.The fungus is grown aerobically under steady-state conditions maintained by a continuous feed of nutrients and simultaneous removal of the culture. Quorn is the trademark of a fungus-based food product, sold (largely in Europe) as a meat substitute or imitation meat. It is marketed to the health-conscious and to vegetarians. Quorn is the first commercial mycoprotein to be used in general food supplies. A wide range of metabolites, produced by microorganisms, in the form of organic acids, amino acids, vitamins, lipids, proteins, enzymes etc. are also used as food supplements for the purpose of preservation, flavour enhancement, stabilization and an improvement in digestibility and nutritive value (Table 4).

Bacteria and fungi are producers of a variety of organic acids including lactic acid, citric acid, acetic acid, gluconic acid and malic acid which are important in food industry. The traditional organisms for citric acid fermentation are selected strains of *Aspergillus niger,* though citric acid has also been produced with several yeasts, mainly species of *Candida,* especially, *Candida lipolytica.* Citric acid provides a pleasant tartness and complements fruit, berry, and other flavors. Blends of citric acid combined with malic acid, lactic acid or fumaric acid are also used in beverage formulations. Other functions of citric acid in these applications rely on its ability to prevent metal-catalyzed off-flavors and the deterioration of colour. This important food additive also serves as an antimicrobial preservative, retarding the growth of spoilage organisms. Acetic acid fermentation is applied in the aerobic, two-step process for the production of vinegar. However, in the last decade a significant amount of research has been devoted to the anaerobic production from biomass and its potential to

Table 4. Food additives of microbial origin

Additive function	Product	Producing microorganisms
Acidulants	Citric acid	*Aspergillus niger, Candida guilliermondii, Candida lipolytica*
	Acetic acid	*Acetobacter aceti, clostridia*
	Lactic acid	*Lactobacillus delbrueckii, L. bulgaricus*
	Malic acid	*Brevibacterium flavum*
Amino acids	Glutamic acid	*Corynebacterium glutamicum, Brevibacterium flavum*
	Lysine	*Corynebacterium glutamicum, Brevibacterium flavum*
	Tryptophan	*Klebsiella aerogenes*
Antimicrobial agents	Nisin	*Lactobaciluss lactis*
	Natamycin	*Streptomyces natalensis, S. chattanoogensis*
Antioxidants	Ascorbic acid	*Acetobacter suboxidans*
Colours	β-Carotene	*Blakeslea trispora, Dunaliella salina*
	Lycopene	*Blakeslea trispora, Streptomyces chattanoogensis*
	Zeaxanthin	*Flavobacterium* species
	Lutein	*Spongiococcum excentricum, Chorella pyrenoidosa*
	Cantaxanthin	*Cantharellus cinnabarinus, Rhodococcus maris*
	Astaxanthin	*Mycobacterium lacticola, Xanthophyllomyces dendrorbous*
Emulsifiers	Various polysaccharides	Many organisms
	Glycolipids	*Rhodococcus erythropolis*
Flavours and flavor enhancer	Vanillin	*Saccharomyces* species
	Nucleosides	*Bacillus subtilis, Brevibacterium ammoniagenes*
	Monosodium glutamate	*Corynebacterium glutamicum, Brevibacterium flavum*
Humectants	Glycols	*Bacillus* species
Lipids and fatty acids	Glycolipids	*Bacillus subtilis*
	γ-linoleic acid	*Mucor* species
Non-nutritive sweetners	Aspartame	From microbial production of aspartic acid and phenylalanine
Stablizers and thickners	Glycan	*Saccharomyces* species
	Pulullan	*Auerobasidium*
	Xanthan	*Xanthomonas campestris*
Vitamins	Ascorbic acid (C)	*Acetobacter suboxidans*
	Cyanocobalamin (B_{12})	*Propionibacteria* species, *Pseudomonas denitrificans, Streptomyces griseus*
	Riboflavin (B_2)	*Ashbya gossypii, Ermothecium ashbyii*, recombinant *Bacillus subtilis*
Enzymes	α-Amylases	*Aspergillus flavus, A. oryzae*
	Amyloglucosidase	*Aspergillus oryzae, Rhizopus niveus*
	Catalase	*Micrococcus lysodeikticus*
	Cellulase	*Aspergillus niger, Trichoderma viridae*
	α-Galactosidase	*Mortierella vinacea*
	Glucose isomerase	*Actinoplanes missouriensis, Streptomyces olivaceous, Bacillus coagulants*
	Glucose oxidase	*Aspergillus niger, Penicillium chrysogenum*
	Invertase	*Saccharomyces cerevisae*
	Lactase	*Candida pseudotropicalis, Kluyveromyces lactis*
	Lipase	*Aspergillus oryzae*
	Nuclease	*Penicillium citrinum*
	Pectinase	*Aspergillus niger*
	Urease	*Lactobacillus fermentum*
	Protease	*Bacillus licheniformis, Bacillus subtilis, Aspergillus niger*
	Rennet	*Bacillus cereus, Endothia parasitica, Rhizomucor miehei,* Recombinant *E. coli*

Table 5. Applications of microbial enzymes in food industry.

Enzymes	Source		Functions
	Bacteria	Fungi	
Proteases	*Bacillus subtilis, Lactobacillus acidophilus, L. plantanum, L. lactis, L. delbrueckii, L. helveticus*	*Aspergillus oryzae*	Baking, brewing, cheese production, meat tenderization, flavorants
Lipases	*Bacillus* spp.	*Aspergillus oryzae, A. niger, Rhizopus* sp., *Rhizomucor miehei, Rhodotorula* sp.	Flavorants, in dairy industry
α-amylases	*Bacillus* spp.	*Aspergillus* spp. *(Aspergillus oryzae)*	Baking, brewing, starch processing
β-amylases	*Bacillus* spp.	*Rhizopus* sp.	Brewing, maltose syrup production
Amyloglucosidases (glucoamylase)	—	*Aspergillus niger, Rhizopus niveus*	Brewing, baking, giucose syrup production
β-galactosidases (lactases)	*Lactobacillus* spp.*Bacillus* spp., *E. coli, Streptococcus lactis*	*Kluyveromyces* sp., *Candida* spp.	Processing of dairy products, whey-fructose syrup production
Glucose isomerases	*Actinoplane* sp., *Arthrobacter* spp., *Streptomyces* spp.	—	High-fructose syrup production
Glucose oxidases	—	*Aspergillus niger, Penicillium notatum*	As antioxidant, used alongwith catalase to remove oxygen from food products
Pullulanases	*Klebsiella pneumoniae, Bacillus acidopullulutícus*	—	Brewing, debranching of starch in sugar syrup manufacture
Pectinases		*Aspergillus niger, Penicillium* spp.	Extraction of plant and fruit juices, coffee bean fermentation
Hemicellulases & β-glucanases/ cellulases	*Bacillus* spp.	*Trichoderma reesei, Penicillium funiculosum, P. emersonii, Aureobasidium pullulans, Cryptococcus* sp., *Trichosporon* sp.	Juice production, alcoholic beverages, baking, brewing, nutraceutics
Urease	*Lactobacillus fermentum*	—	Wine production
Invertase	—	*Saccharomyces cerevisae*	Variously in food processings

compete with chemical synthesis. Vinegar is a liquid condiment used as a food flavoring and as a preservative. It is produced by the action of aerobic bacteria of the genus *Acetobacter* on dilute solutions of ethanol such as cider, wine, or diluted distilled alcohol solutions. The ethanol solutions utilized are produced microbiologically by the fermentation of sugars by *Saccharomyces cerevisiae*. When starch materials are used (as in malt or rice vinegar), the enzymatic saccharification of starch is required prior to the alcohol fermentation. Lactic acid has been produced from various lactic acid bacteria in indigenous fermented foods including dairy products and is mainly responsible for the precipitation of milk proteins for yoghurt and cheese production. *Aspergillus niger* is used to produce gluconic acid at large scale. A portion of the gluconic acid is converted to δ-gluconolactone, a slow-acting acidulant which is used with sodium bicarbonate for controlled leavening of baked goods. The lactone is also used in the dairy industry for cheese curd formation and for the improvement of the heat stability of milk. In addition, it serves as an additive in cured meat products, luncheon meats, tofu, and packet desert mixes. L-malic acid obtained with fungal fermentations plays a role similar to that of citric acid as a food ingredient. As an acidulant, it has a smooth, long-lasting flavor profile. Malic acid exhibits a synergy with aspartame permitting a 10% reduction in the usage of this sweetner in beverages.

Amino acids include an important class of compounds required by living cells as growth factors. Production of amino acids by fermentation processes was initiated in 1956 when L-glutamic acid was discovered as a product in the spent medium of *Corynebacterium glutamicum*. About 66% of amino acids are used in food industry and 31% used as feed additives. In food industry amino acids in the form of sodium glutamate, sodium aspartate, D,L-alanine, glycine are used to enhance flavours. L-cysteine improves quality of bread during baking process and acts as antioxidant in fruit juices. L-tryptophan + L-histidine act as antioxidants in powdered milk. L-phenylalanine and L-aspartic acid are used as the starting molecules for aspartame (L-aspartyl-L-phenylalanine methyl ester) which is used as a low calorie sweetner. Lysine is added in bread in Japan and in some countries soya products are enhanced by addition of methionine, L-lysine. D,L-methionine used to upgrade animal feeds. The principal microorganisms capable of producing glutamic acid include *Corynebacterium, Brevibacterium, Microbacterium, Arthrobacter,* however, *Corynebacterium glutamicum* is mainly used commercially. L-methionine is synthesized with immobilized *Aspergillus oryzae* aminoacylase. L-lysine is produced with L-amino-caprolactum hydrolase from *Cryptococcus laurentii* and L-phenylalanine is produced with immobilized *Rhodotorula rubra* cells bearing phenylalanine ammonia lyase. Phenylalanine is in demand for the synthesis of aspartame, a low-calorie, artificial sweetner used in soft drinks and desserts. Tryptophan, lysine and methionine all essential amino acids in the human diet are used to enrich foods. Certain fungi are able to produce substantial amounts of these amino acids but are surpassed by improved industrial strains of bacteria. Strains of *Pichia anomala* are prolific producers of tryptophan. Strains of the yeasts *Candida utilis* and *Saccharomyces cerevisae* and the molds *Ustilago maydis* and *Gliocladium* spp. produce significant quantities of lysine. Mutants of the yeasts *Candida tropicalis, Pichia pastoris* and *C. utilis* have an increased methionine content and may have potential as food supplements.

Vitamins are essential mediators of metabolism and play an important role in the human diet. They are often supplements in foods and animal feeds. Many microorganisms are the potential producers of various vitamins. Many fungi and bacteria can produce riboflavin. The two ascomycetes *Ermothecium ashbyii* and *Ashbya gossypii* are the potential producers of riboflavin. Industrial fermentation with *A. gossypii* is now preferred and generates about 30% of world output. Sterols are widely distributed among the fungi. Ergosterol, a commercially important lipid, extracted from dehydrated yeast can be converted into vitamin D by irradiation with ultraviolet light. Such yeast strains may serve as alternative sources of vitamin D in future and can be used in various dairy products, like milk and margarine in breakfast cereals and baby foods. There are certain fungal species including *Blakeslea trispora*, a heterothallic fungus of the Mucorales group and *Phycomyces blakesleeanus* are extremely rich in mycelial β-carotene, as a potential source of carotenoides which act as precursor for vitamin A production.

Microbial polysaccharides in the form of pullulan, scleroglucan and agar have attracted considerable interest as industrial fungal polymers with unique properties with food application. Pullulan is a neutral, linear homopolysaccharide composed of maltotriose units is produced by the dimorphic black yeast *Aureobasidium* (*Pullularia*). Its ability to form strong transparent films with low oxygen permeability enhances its usefulness as a food coating and packaging agent. Scleroglucan, a highly branched, neutral homopolysaccharide with a backbone of about 90 D-glucose units is produced by *Sclerotium glucanicum* or *S. rolfsii*. Scleroglucan exhibits pseudo-plasticity over a broad range of temperature, pH, and salt conditions, making it suitable for use as a food additives. Agar, a linear galactan sulphate produced by Red Sea weed (*Gelidium algas*), having an ability to absorb water-forming gels, is used as a stabilizer in foods.

About 70% of the world's supply of fats and oils is derived from plants and currently no food lipids are produced by microbiological means. It is notable that processes for fat production with yeasts *Rhodotorula gracilis, Lipomyces lipofer, Lipomyces starkeyi, Endomyces vernalis, Cryptococcus terricolus* and *Candida* species, which accumulate significant levels of fat, have been developed. The molds and fungi like *Mortierella vinacea, Mucor circinelloides, Aspergillus ochraceous, A. terreus*, and *Penicillium lilacinum Hansenula, Chaetomium, Cladosporium, Malbranchea, Rhizopus* and *Pythium* are also good produces of fat. Fatty acids for the food, medical, and cosmetics industries could be standard products of fungal fermentation in coming years. A wide range of microbial enzymes are also widely used in food processing and the manufacture of beverages (Table 5) which help to control texture, appearance and nutritive value of food as well as to generate desirable flavours and aromas.

Many foods function as probiotics which are an important part of diet good for health. Probiotics are foods that contain live bacteria (Table 6). It is the bacteria and metabolites which they produce that give these probiotics their health promoting properties. The best known example of a probiotic is yoghurt. The consumption of yoghurt shows increased milk digestibility, quicker recovery from certain types of diarrhoea, enhanced immune function, reduction in certain cancers and possible lowering of blood cholesterol levels. The probiotics exert the beneficial effect by producing certain metabolites known as prebiotics. Products that combine prebiotics with probiotics for enhanced health benefits are called as synbiotics and is the major new trend in the dairy sector. Synbiotic yoghurt has promising enhanced health benefits, such as improved immune function and increased calcium absorption in addition to overall health benefits. These products are at the forefront in providing consumers with a combination of health benefits in the tasty and convenient forms. Other emerging domestic products incorporating prebiotic fibers include kefir, yoghurt and other dairy drinks, sprouted products, functional waters, nutrition bars, weight loss products, soymilk, green foods, probiotic supplements, mineral supplements, medical foods and pet foods. These products incorporate prebiotic fibers for various health benefits, including enhanced mineral and isoflavone absorption, fiber contribution, gut integrity, immune function and cholesterol control.

Table 6. Important bacterial strains used as probiotics

Bacterial strains	Beneficial effects
Lactobacillus acidophilus/L. bulgaricus	Balance intestinal flora, immune enhancement
Mixture of *L. lactophilus, L. plantarum* and *Bacillus bifidum*	Prevention and treatment of antibiotic and rotavirus associated diarrhea
Lactibacillus casei	Prevention of harmful intestinal microbiota
Bacillus subtilis	Used in bacteriotherapy, immunostimulatory agent
Bifidobacterium bifidium	Prevention of viral diarrhea
Propionibacterium fredenreichii	Growth stimulation of other useful bacteria

MICROORGANISMS IN PHARMACEUTICAL INDUSTRY

Microorganisms also find a wide application in the pharmaceutical industry and are the source of life-saving drugs in the form of antibiotics, alkaloids, vaccines, bacteriophages, compounds of therapeutic value including steroids, microbial probiotics, prebiotics, enzymes nutriceuticals recombinant proteins and peptides.

Antimicrobial agents from microorganisms: Antibiotics are probably the most important group of compounds synthesized by microorganisms. Most antibiotics are secondary metabolites produced by actinomycetes, filamentous fungi and bacteria (Table 7). Apart from bacteria, actinomycetes and fungi, algae also produces certain compounds which possess pharmaceutical properties such as anaesthetic, antipyretic, cough remedy, wound healing, thirst quenching and treatments for gout, gallstones, goitre, hypertension, bowel movement and skin diseases (Table 8).

Alkaloids are a diverse group of small nitrogen containing organic compounds produced by certain plants and microorganisms. Many are toxic, but some have various therapeutic properties. Species of filamentous fungus, *Claviceps*, produce a range of alkaloids. These compounds are produced within the sclerotia (fruiting bodies) of *Claviceps purpurea* that develop naturally when this organism infects developing cereal grains. Infected grains become black and are referred to as ergots. These structures contain indole alkaloids, derived from a tertracyclic ergoline ring system, which are classified into two groups. The members of first group are based on clavin and contain no peptide component. Clavin-based alkaloids are also produced by other groups of fungi, including species of *Aspergillus* and *Penicillium*. Some possess antibiotic and anti-tumor activity. The second group, based on lysergic acid (LSA) and containing a tripeptide or an amino alcohol, are found only in *Claviceps* species. Examples include ergometrine, ergocristine, ergosine, and ergotamine. These LSA-based alkaloids have medical roles as analgesics in migraine therapy, as hallucinogens and for treating circulation problems. Others have particular uses in obstetrics, for inducing the smooth muscle of the uterus to contract during labour and after childbirth. Earlier, the alkaloids were extracted from ergots that developed within the infected cereal crops, usually rye. Most are now produced by fermentation of *Claviceps fusiformis, C. paspali* or *C. purpurea* in surface, submerged or immobilized cell culture.

Table 7. Antibiotics of microbial origin

Microbial genera	Microbial species	Antibiotics
Bacteria		
Pseudomonas	*P. aeruginosa*	Pyocyanin
	P. viscose	Viscosin
Streptococcus	*S. lactis*	Nisin
Diplococcus	*Diplococcus* X-5	Diplomycin
Chromobacterium	*C. prodigiosum*	Prodigiosin
Escherichia	*E. coli*	Coliformin
Proteus	*Pr. vulgaris*	Protaptin
Bacillus	*B. brevis*	Gramicidin
	B. subtilis	Subtilin
	B. polymyxa	Polymixin
	B. licheniformis	Bacitracin
Actinomycetes		
Streptomyces	*S. griseus*	Streptomycin
	S. aureofaciens	Tetracycline
	S. rimosus	Tetracycline
	S. spheroids	Novobiocin
	S. antibioticus	Actinomycin
	S. flavus	Actinomycin
	S. flaveolus	Actinomycin
	S. parvus	Actinomycin
	S. parvulus	Actinomycin
	S. chrysomallus	Actinomycin
	S. faradiae	Actinomycin
	S. melanochromogenes	Actinomycin
	S. erythreus	Erythromycin
	S. hygroscopicus	Hygromycin
	S. rimosus	Oxytetracycline
	S. griseofulvus	Oxytetracycline
	S. armilatus	Oxytetracycline
	S. aureofaciens	Oxytetracycline
	S. venezuelae	Chloramphenicol
	S. faradiae	Neomycin
	S. albogriseolus	Neomycin
	S. kanamyceticin	Neomycin
Nocardia	*N. sulphurea*	Tetracycline
Fungi		
Imperfect fungi		
Penicillium	*P. chrysogenum*	Penicillin
	P. griseofulvum	Griseofulvin
Trichothecium	*T. roseum*	Trichothecin
Basidiomycetes		
Lenzites (thermophilic fungi)	*L. thermophia*	Lenzitin
Ascomycetes		
Chaetomium	*C. cochloides*	Chetomin
Lichen	*Lichen*	Usnic (Usninic) acid
Algae		
Chlorella	*C. vulgaris*	Chlorellin

Table 8. List of the algal species with respective medicinal properties

Organisms	Compound
Lyngbya majuscula	Terpenes, Carbohydrate
Trichodesmium erythraeum	Terpenes, Carbohydrate
Fragilaria sp.	Peptides
Chaetoceros lauderi	Polysaccharides
Asterionella japonica	Nucleotides fatty acids
Gonyaulax tamarensis	Terpenes, Carbohydrate
Isochrysis sp.	Terpenes, Carbohydrate
Phaeocystis pouchetii	Acrylic acid

Many therapeutic steroids that are used for the treatment of allergies, inflammation, skin diseases and as oral contraceptives are prepared by microbial transformations (Table 9). Initially, they were prepared by extraction from animal tissues or via complex chemical synthesis, both of which were extremely costly. Many steroids are now manufactured using combination of chemical and microbial transformation steps. These processes employ relatively cheap sterols as the starting materials, often diosgenin extracted from the Mexican yam (*Dioscorea composite*), and stigmasterol, a byproduct of soybean oil manufacture. The microorganisms involved are mostly filamentous fungi (*Rhizopus, Curvularia, Fusarium* and *Aspergillus* species), and *Mycobacteria*, in the form of suspensions or immobilized growing cells, resting cells, spores and cell-free extracts. They perform key reactions to modify the basic steroid structure, a cyclopentanoperhydrophenanthrene, including hydroxylations at positions 2 and 17. Various side chain cleavages, hydrogenations and dehydrogenation and ring expansions, from a five-membered to a six-membered ring.

Table 9. Some steroids whose production involves microbial transformation

Biotransformation	Substrate	Product	Organisms
1-Dehydrogenation	Hydrocortisone	Prednisolone	*Arthrobacter simplex*
11α-Hydroxylation	Progesterone	11α-Hydroxyprogesterone	*Rhizopus nigricans*
11β-Hydroxylation	Reichstein compounds	Hydrocortisone	*Curvularia lunata*
Side chain cleavage	β-sitosterol	Androstadienedione	*Mycobacterium* species

Vaccines, another pharmaceutical preparation derived from suspensions of killed or attenuated microorganisms (bacteria, viruses, fungi, protozoa, or rickettsiae) or antigenic proteins derived from them are administered in the animals for the prevention, amelioration, or treatment of infectious and other diseases (Table 10).

Recombinant DNA technology which employs the potential use of recombinant *E. coli*, has allowed the production of many recombinant therapeutic proteins and peptides with potential medical application (Table 11). Microbial vaccines are produced using either attenuated microbial cells, in which the organisms are adapted or engineered to delete pathogenicity while retaining immunogenicity, or live microbial vectors in which the microbial cells carrying added genes for an immunogenic protein are used.

Table 10. Microbial vaccines for medical and veterinary use

Live attenuated bacterial vaccines	
Bacillus anthracis (spore vaccine)	Anthrax
Brucella abortus S19	Brucellosis
Salmonella typhi	Typhoid
Shigella sonnei	Shigelosis
Inactivated bacterial vaccines	
Capsular antigens	
Escherichia coli K88, K99 and 987P	Diarrhoea and food poisoning
Neisseria meningitides A & C	Meningitis
Pasteurella multocida	Pasteurellosis
Cell wall antigens	
Bordetella pertussis	Whooping cough (pertussis)
Salmonella paratyphi	Paratyphoid
Salmonella typhi	Typhoid fever
Vibrio cholera	Cholera
Toxoid antigens	
Clostridium perfringens	Food poisoning
Clostridium tetani	Tetanus
Corynebacterium diphtheria	Diphtheria
Recombinant vaccines	
Bacterial	
Chlamydia	Chlamydia (pelvic inflammatory disease)
Clostridium tetani	Tetanus
Corynebacterium diphtheria	Diphtheria
Protozoal	
Plasmodium vivax	Malaria
Viral	Hepatitis B Herpes simplex Influenza Measels (rubeola) Poliomyelitis Rabies
DNA vaccine	Malaria

Due to increasing problems caused by antibiotic-resistant bacterial pathogens, alternatives to antibacterial chemotherapy using antibiotics are now being sought. The benefits of bacteriophage therapy are that bacteria do not develop resistance to viruses, as they too evolve to continue invading their host bacteria. Also, they are specific and do not kill beneficial bacteria of the human or animal natural microflora. This therapy appears to be useful particularly for the treatment of cholera, dysentery, meningitis, septicaemia, tuberculosis and infections caused by *S. aureus* and species of *Campylobacter* and *Pseudomonas*.

Table 11. Important recombinant proteins and peptides for therapeutic use

Antibodies
Tissue necrosis factor-α antibody (rheumatoid arthritis treatment)
Cancer and viral diseases
Interferons
Interleukins
Tissue necrosis factors
Transforming growth factors
Cardiovascular diseases
Erythropoietin (boosts red blood cell proliferation)
Hirudin (thrombin inhibitor from leech)
Urokinase (thrombosis treatment)
Tissue plasminogen activator (clot dissolution)
Hormones
Human growth hormone
Insulin
Neurological diseases
Endorphins
Nerve growth factors
Neuropeptides
Vaccines
Foot and mouth disease
Hepatitis B
Wound healing and blood clotting factors
Epidermal growth factors
Fibroblast growth factor
Clotting factor VIII (haemophilia treatment)

MICROORGANISMS AS PROBIOTICS

A probiotic is an organism which contributes to the health and balance of the intestinal tract. A probiotic is also referred to as the 'friendly', 'beneficial', or 'good' bacteria which when ingested acts to maintain a healthy intestinal tract and helps fight illness and disease. Probiotic is a live microbial feed supplement, which beneficially affects the host by improving its intestinal microbial balance. Probiotics enhance the *immune system* by favorably altering the gut micro-ecology and preventing unfriendly organisms from gaining a foothold in the body. They prevent the overgrowth of yeast and fungus and produce substances that can lower cholesterol. Probiotics are widely recommended for the treatment of candidal infection. Probiotics are also essential in the treatment and prevention of thrush, vaginal yeast infections, and athlete's foot. The important bacteria and yeasts used as probiotics include:

Bifidobacterium: They are saccharolytic organisms that produce acetic and lactic acids without generation of CO_2, except during degradation of gluconate. They are also classified as lactic acid bacteria (LAB). Bifidobacteria used as probiotics include *Bifidobacterium adolescentis, Bifidobacterium bifidum, Bifidobacterium animalis, Bifidobacterium thermophilum, Bifidobacterium breve, Bifidobacterium longum, Bifidobacterium infantis* and *Bifidobacterium lactis.*

Lactobacillus: Lactobacilli are normal inhabitants of the human intestine and vagina. Lactobacilli are gram-positive, facultative anaerobes, non-spore forming and non-flagellated rod or coccobacilli. Lactobacilli are also classified as lactic acid bacteria (LAB). Lactobacilli used as probiotics include *Lactobacillus acidophilus, Lactobacillus brevis, Lactobacillus bulgaricus, Lactobacillus casei, Lactobacillus cellobiosus, Lactobacillus crispatus, Lactobacillus curvatus, Lactobacillus fermentum, Lactobacillus gasseri, Lactobacillus johnsonii, Lactobacillus plantarum, Lactobacillus salivarus, Lactobacillus reuteri, Lactobacillus johnsonii, Lactobacillus delbrueckii* subspecies *delbrueckii, Lactobacillus delbrueckii* subspecies *bulgaricus* and *Lactobacillus paracasei* subsp. *paracasei.*

Lactococcus: Lactococci are gram-positive facultative anaerobes. They are also classified as lactic acid bacteria (LAB). Lactococci that are used or are being developed as probiotics include *Lactococcus lactis, Lactococcus lactis* subspecies *cremoris (Streptococcus cremoris), Lactococcus lactis* subspecies *lactis.*

Saccharomyces: The principal probiotic yeast is *Saccharomyces boulardii. Saccharomyces boulardii* is also known as *Saccharomyces cerevisiae* Hansen CBS 5296 and is used to treat diarrhea associated with antibiotic use.

Streptococcus: One of the probiotic member of this genus is *Streptococcus thermophilus* which is a gram-positive facultative anaerobe and is nonmotile, non-spore forming and homofermentative. It is also classified as a lactic acid bacteria (LAB). It is found in milk and milk products. *Streptococcus salivarus* subspecies *thermophilus* is another probiotic strain.

Enterococcus: Enterococci are gram-positive, facultative anaerobic cocci of the Streptococcaceae family and are part of the intestinal microflora of humans and animals. *Enterococcus faecium* SF68 is a probiotic strain that has been used in the management of diarrheal illnesses.

The probiotic bacteria exert various beneficial effects on health, both through their direct interactions with the gut, and in their metabolic products, known as prebiotics which enhance health in a number of ways through the support of probiotic bacteria in the gut. One important example of prebiotic metabolite is short chain fatty acids (SCFAs) which are the products of prebiotic fermentation by these gut microflora and are crucial for human health and well-being. The probiotic bacteria and their SCFAs restrict the growth and activity of less beneficial species, including putrefactive and disease-causing bacteria and yeast. The effect of probiotics crowding out bad or less beneficial species is known as competitive exclusion which is essentially the good bacteria restricting the bad ones. The SCFAs produced by probiotic strains make the gut environment less friendly to harmful strains. In addition, there is limited food and space in the colon, so the provision of a food source that is preferred by the good bacteria helps ensure they will have the competitive advantage. The SCFAs, in particular, are crucial to gut integrity and function, immune system modulation, calcium absorption and cholesterol maintenance. In fact, most of the energy required by the colon is provided directly by SCFAs. If the colon does not have enough SCFAs to provide energy, decreased integrity and function may follow. Various 'starved bowel' disorders, known collectively as irritable bowel syndrome (IBS), are generally thought to be linked to inadequate SCFA production and a poor balance of gut microflora. The important point about SCFAs is that they are totally provided by probiotic fermentation of prebiotic fibers in the gut.

MICROBIAL ENZYMES IN PHARMACEUTICALS

Enzymes as drugs have two important features that distinguish them from all types of drugs. First, enzymes often bind and act on their targets with great affinity and specificity. Second, enzymes are catalytic and convert multiple target molecules to desired products. These two features make enzymes specific and potent drugs that small molecules cannot. These characteristics have resulted in the development of many enzymes as drugs for a wide range of disorders (Table 12).

Mushroom nutriceuticals

Mushrooms were treated as a special kind of organisms which were known to possess all the essential curative ingredients in adequate quantities. The prominent health giving beneficial ingredients of mushrooms are mostly polysaccharides which are different in their composition, mode of action, degree of ramification and polymerization etc. yet they all are immuno stimulants with no side effects and all of them strengthen health and increase immunity (Table 13). Thus, mushroom nutriceutical is a refined/partially defined food extractives which is consumed in the form of capsules or tablets as a dietary supplement (not a food), and which has potential therapeutic applications. A regular intake of mushroom nutriceutical enhances the immune response of the human body thereby increasing resistance against disease and in some cases causes regression of a disease state.

CONCLUSION

Greater awareness regarding health of the mankind has brought about a more intensive search for alternative cleaner bioprocess technologies using microorganisms and their enzymes for the production of various value-added products from renewable resources. Many of these products including industrial enzymes, fermented foods, vaccines, antibiotics, bioplastics and biopolymers etc. have become indispensable, hence their supply has to be increased to meet the demands of ever-increasing population. Consumer demand for natural, safe and environmentally friendly food products is increasing. The 21st century has dawned with renewed hope for a better livelihood for the population of this earth and the science of microorganisms has also grown up and well established in the form of microbial biotechnology by its robust performance in the commercial and industrial sectors. This is primarily due to the blending in classical, basic microbiology with several new techniques that are likely to influence the development of the bioindustries of future with lot of applications. In spite of having a lot many discoveries, inventions and developments in microbial science, a serious attention is required to use microbial labour and capabilities more efficiently in various sectors. The first,

Table 12. Microbial enzymes used for various therapeutic purposes

Enzymes	**Source**	**Applications**
α-amylase	*Aspergillus* and *Bacillus* species	Digestive aids
Alcohol dehydrogenase or alcohol oxidase	*Saccharomyces cervisae, Candida boidnii* and *Pichia pastoris*	Ethanol testing
Asparaginase	*Escherichia coli, Serratia marcesens* and *Erwinia carotovora*	Maintains low asparagines levels in treating cancers whose cells cannot synthesize the amino acid
Catalase	*Aspergillus niger, Corynebacterium glutamicum* and *Micrococcus lysodeikticus*	Contact lens cleaning systems
Cholesterol esterase	*Pseudomonas fluorescens*	Monitoring serum cholesterol levels
Creatininase	*Pseudomonas putida* and recombinant *E. coli*	Determination of serum creatinine levels
Glucose oxidase	*Aspergillus niger*	Analysis of blood glucose levels
β-lactamase	*Bacillus cereus* and *E. coli*	Treatment of penicillin allergy
Proteases (various)	Several bacterial and fungal sources	Used as digestive aids, for wound debridement and contact lens cleaning
Rhodanase (thiosulphate sulphur transferase)	*Trichoderma* species	Treatment of cyanide poisoning
Streptokinase (a cysteine protease)	Various haemolytic streptococci	To break down blood clots
Urease	*Lactobacillus fermentum*	Removal of bacteria from blood in renal failure
Uricase (ureate oxidase)	*Arthrobacter globiformis* and *Candida utilis*	Used in the treatment and diagnosis of gout

Table 13. Nutriceutically important mushrooms with curative properties.

Mushroom species	**Pharmacodynamics**	**Compounds**
Ganoderma lucidum, Lentinus edodes	Immunomodulatory, Anticancer, Hepatoprotective, Hypoglycemic, Antihistamine release, Anti-angiogeinic, Antiviral	Polysaccharides/Proteins
Ganoderma lucidum	Cytotoxic tumor, Anti-HIV, Hypolipidemic, Hypotensive, Antiplatelet aggregation, Cardioactive, Immunomodulatory	Triterpenes
Ganoderma lucidum	Anti-hypersensitivity, Anti-autoimmune diabetes, Anti-hepatitis B, Immunomodulatory	Amino acids
Ganoderma lucidum	Antiplatelet aggregation	Adenosine and derivatives
Ganoderma lucidum	Antitumor	Organic germanium
Many species	Antibacterial	Hirustic acid
Odemansiella radiate	Antibiotic effect	E-b methoxy acrylate
Volvariella	Cardiac tonic	Volvatoxin/Flammutoxin
Collybia velutipes	Decrease cholesterol	Eritadenine
Ganoderma lucidum, Grifola frondosa	Decrease blood sugar	Peptide/Glycogen/ Ganodran/Glucan
Marasmius androsaceus	Analgesic/Sedative effects	Marasmic acid

important among them is health sector, which still needs to develop newer, effective, products of healthcare for combating diseases like AIDS, tuberculosis, and polio; second, in food sector, there is need to put more efforts on encircling and developing microbial source of nutriceuticals. Above all, the indiscriminate use of microbial biotechnology should be avoided.

SUGGESTED READINGS

1. Chang, S.T. and Miles, P.G. Edible Mushrooms and their Cultivation. CRC Press, Boca, Raton Fl, 1991.
2. Godfrey, T. and West, S. Industrial Enzymology, Second edition, Stockholm Press, New York, 1996.
3. Huis In't Veld, J.H.J., Bosschaert, M.A.R. and Shortt, C. Health aspects of probiotics. Food Science & Technology Today, 12 : 46-49, 1998.
4. Leuchtenberger, W., Huthmacher, K. and Drauz, K. Biotechnological production of amino acids and derivatives: current status and prospects. Appl. Microbiol. Biotechnol. 69:1-8, 2005.
5. Pelczar (Jr), M.J., Chan, E.C.S., and Kreig, N.R. Microbiology. 5th ed. McGraw Hill Book Company, Singapore, 1986.
6. Prescott, L.M., Harley, J.P. and Klein, D.A. Microbiology. 6th ed. McGraw Hill Companies, Inc. New York, 2005.
7. Rahi, D.K., Rajak,R.C., and Pandey, A.K. (2005). Mushroom Nutriceuticals : An Emerging Health Care Aid. In : Frontiers in Plant Sciences. (Eds. K.G. Mukerji, K.V.B.R. Tilak, S.M. Reddy, L.V. Gangwane, P.Prakash, I.K. Kunwar). I.K. International Pvt. Ltd., New Delhi 110016, India.
8. Soni, S.K. (2007). Microbes: a source of energy for 21st century. New India Publishing Agency, New Delhi, India
9. Soni, S.K. and Gupta, J.K. Enzymes in agro-processing: Trends and Future Prospects. In: Biotechnological Strategies in Agro-Processing (S.S. Marwaha and J.K. Arora, Eds), Asiatech Publishers Inc., New Delhi, India. pp. 113-139, 2003.
10. Soni, S.K. and Sandhu D.K. Microbiology of Fermentations. In: Biotechnology: Food Fermentation Microbiology, Biochemistry and Technology. (Eds. V.K. Joshi and A. Pandey) Volume I: Basic, Educational Publishers and Distributors, New Delhi, India. 25-85, 1999.
11. Soni, S.K. and Sandhu, D.K. Indian fermented foods: Microbiological and Biochemical aspects. Indian J. Microbiol. 30: 135-157, 1990.
12. Soni, S.K. and Sandhu, D.K. Fermented cereal products. In: Biotechnology: Food fermentation Microbiology, Biochemistry and Technology. Vol 2. (Edited by V.K. Joshi and A. Pandey). Educational Publishers and Distributors, New Delhi. pp. 895-950, 1999.
13. Stanier, R.Y., J. L. Ingraham, M. L. Wheelis, and P. R. Painter. General Microbiology. 5th ed. Upper Saddle River, New Jersey: Prentice Hall, 1986.

9

Shilajit: Health Rejuvenant

D.V. Rai and Mehak Goel

INTRODUCTION

Shilajit is a reputed remedy of Ayurveda, the ancient system of medicine. Charaka Samhita, the classical Ayurvedic treatise on medicine, has described physical properties and therapeutic uses of Shilajit. The ancient medical literature attributes many health benefits to it such as general physical strengthening, anti-ageing, blood sugar stabilization, injury healing, urinary tract rejuvenation, enhancing brain function, bone healing, kidney rejuvenation, immune system strengthening, management of arthritis and hypertension and treating obesity. It also relieves digestive problems, increases sex drive, improves memory, counters diabetes, improves pancreas function, purifies the blood, improves spleen function, strengthens digestion, improves metabolism, dissolves tumors, reduces tiredness, counteracts thirst, and is a powerful rejuvenator. Modern clinical studies have shown anti-diabetic and antioxidant activity of shilajit. Shilajit has been used in treatment of Alzheimer's disease. In many parts of the world, shilajit is known for its ability to increase vitality, sexual energy, and overall physical and mental well-being. It is a supplement that, with the aid of fulvic acids, can dig deep into the body's hiding places for toxins and bore them out. It is like an empty vessel that travels through your body to collect bad free radicals and toxins that bog you down over time. Our body can't make minerals, so we must get them from food or supplements. Supplements like Shilajit are extremely important if you want to rid your body of the toxins that cause arthritis, high blood sugar levels, mental slowness, and low organ function. In addition, Shilajit works to purify the blood and clear the mind. It thus helps in obtaining overall good health. Its overall effects are alternative, tonic, slightly laxative, respiratory stimulant, disinfectant, expectorant, intestinal antiseptic, diurnal and lithotriptic.

Native Shilajit (Sanskrit/Indian name is Shilajit, scientific name (Latin) is Asphaltum puniabiunum and other common names are mineral wax, ozokerite silajatu, silaras, shilajeet, asphalt, mineral pitch) is a blackish-brown exudation, of varying consistency, from steep rocks of different formations. It is commonly found in the Himalayas primarily in India, at altitudes between 1000–5000 m, from Arunachal Pradesh in the East to Kashmir in the West. It is also found in other mountain ranges of world, e.g. Afghanistan (Hindukush, Badakh-shan), Australia (Northern Pollock ranges) Bhutan, China, Nepal, Pakistan, Tibet (Himalayan belt) and the USSR (Tien Shan, Caucasus, Pamir, Ural). The word 'shila' signifies rock. In Sanskrit the literal meaning of Shilajit is "Rock Like"—the power to make our body like a rock enabling it to withstand the ravages of time. Use of Shilajit makes the human body robust and rock-like. Shilajit has the unparalleled powers of arresting and reversing the ageing process. Ancient Indian *yogis* and practitioners of Ayurvedic system considered Shilajit to have divine powers capable of healing the body of virtually any ailments and above all preserve youthfulness.

The genesis of Shilajit is linked to the formation of mountains. The Himalayan mountains have been formed from the collision of the Sub-Asian and Indian continents about 50 million years ago. During the collision dense vegetation comprising entire rain forests are understood to have been trapped under millions of tons of pressure, which gradually transformed into humus, a rich organic mass that is food for new plant life. The flora of the Himalayas is rich and varied, and for thousands of years the plants have come to life, absorbed nutrients from the soil, and then died out. This is a process which has been repeated again and again countless times, and continued for millennia. It is believed that Shilajit found in the Himalayas are the fossilized form of those plants, and the particular biosphere of the Himalayas created them and bestowed medicinal qualities to them. Due to microbial action and the tremendous pressure from the weight of the mountains the humus was transformed into a dense, viscous, mineral rich mass, this ultimately leads to a thick fluid Shilajit. Thus Shilajit represents originality of ancient vegetation and purity of non-exposure to any harmful fertilizer, pesticide, herbicide and other pollutants. The trapped layers of Shilajit become exposed due to the freezing winters, hot summer sun and erosion from monsoon rains. Shilajit "flows" out from between the cracks in the layers of rock during the summer when the temperature of the mountains gets warm enough to make Shilajit less viscous. In the raw form (before refinement), Shilajit is a semi-hard, brownish black to dark, greasy, black resin that has a distinctive smell and taste.

It has been variously described a bitumen (asphalt), a mineral resin, a plant fossil, a substance of mixed plant and animal origin, or an inorganic substance. Pure Shilajit is in the form of a thick paste that oozes between rocks due to geothermal processes. Shilajit found in the higher altitudes of the Himalayas are collected in raw form during summer months when the ice melts. Shilajit so collected is purified and processed by several drug manufactures and presented in capsule form for human consumption.

ORIGIN

About 200 million years ago, India was a large island off the Australian coast separated from the Eurasian continent by the Tethys Sea. The Indian continent drifted north at a rate of about 9 metres a century. This movement led to the eventual disappearance of the Tethys Sea. Fifty million years ago, the Indian continent collided with the Asian continent. This caused the sea bed of the Tethys Sea to be pushed up and keep moving up to eventually form the Himalayan mountains. The Himalayan mountains continue to rise more than one centimetre a year. During this transition the mineral rich and fertile soil of the sea bed gave rise to a lush and dense tropical jungle. As the ground continued to be pushed up to become mountains a lot of plants became trapped by layers of rock and soil and remained preserved for thousands of years. They are gradually transformed into humus, a rich organic mass that is food for new plant life. Today, millions of years later the monsoon rains and extreme freeze and thaw conditions work together to crack large rock formations, exposing the precious shilajit. Because of its ancient nature, the vegetation was never exposed to any type of fertilizer, pesticide, herbicide, or pollution.

Ancient Sanskrit holy texts, over 3,000 years old, make reference to a mysterious substance called shilajit, which they describe as the "destroyer of weakness". The sacred substance was prescribed for thousands of years for many different health problems and became a powerful tool in Ayurvedic medicine. There is some indication that shilajit may have been the priceless *soma* of the Eastern alchemists. The ancient Vedic text Rig Veda states that *soma* "has mountains and stones for its body" and "dwells within the mountainous rock where it grows". Mountainous rocks are the "abode of *soma*," and it is "plucked from between the rocks by mountain dwellers and brought to the priests-alchemists who prepared the *soma* by washing and grinding and cooking". *Soma* was considered the elixir of immortality, the secret substance used by alchemists to perfect both body and mind. This ancient wisdom was passed from generation to generation among the Indian and Nepali alchemists and holy men, but it escaped the notice of the Western medical establishment until the last days of the twentieth century, when explorer John Anderson heard the amazing benefits of this substance and refused to give up the search until he found its source. He journeyed throughout India and Nepal until he learned the perilous harvesting of the raw shilajit from the cliffs. He also documented the realms of Sanskrit studies showing the rare plant's benefits. He spoke first-hand with more than fifty Indian and Nepalese researchers that have been studying the wonderful effects of shilajit and perfecting the processes for delivering the purest, most concentrated shilajit ever known to man. In the modern time, Sir Martin Edward Stanley, a British explorer during his expeditions to the Himalayas, in 1870, observed that monkeys on the gangetic plains became old by the time they were 10 years old, whereas the monkeys on the higher altitudes continued to be highly active and agile even during old age. He noticed that the monkeys at the higher altitudes were eating a rock-like melted material oozing out from rock crevices during the summer months when the ice cover melted.

NATIONAL AND INTERNATIONAL STATUS

A lot of research has been done in this area at both national and international level. At national level there are many institutes which are exploring the benefits of shilajit. Studies have been done to show the medicinal properties of this herbomineral. Many workers in India are working on possible antiulcerogenic and anti-inflammatory activities of shilajit and its effect on changes in liver and serum lipids produced by carbon tetrachloride, antioxidant defence by both native and processed shilajit. Shilajit has been used by Indian and Nepalese alchemists for centuries but it first came into notice of westerners around the end of 19th century. They were amazed by the medicinal properties of this drug. Thereafter, a large amount of work was done on this blackish-brown exudation. Laboratories in many countries contributed in the research of Shilajit. For example, in South Korea, the spermatogenic and ovogenic effects of chronically administered shilajit to rats were shown. Likewise other counties like Jordan, Kuwait, Germany, and Britain also explored shilajit. The various research done in other countries include seeing effects of shilajit on the development of tolerance to morphine in mice (Jordan), preliminary evaluation of shilajit as a suspending agent in antacid suspensions (Safat, Kuwait), systemic administration of refined extracts from *Withania somnifera* (Indian Ginseng) and Shilajit differentially affects cholinergic but not glutamatergic and GABAergic markers in rat brain (University of Leipzig, Germany), etc.

COMPOSITION AND STRUCTURE

Shilajit consists of a very broad spectrum of organic compounds and minerals. It contains more than 85 minerals in ionic form and a large number of organic compounds like fulvic acid, humic acid, hippuric acid and benzopyrones. Traces of benzoic acid are also present. The standardisation of shilajit is based on fulvic acid.

Shilajit contains two classes of organic compounds: humic substances and non-humic organic metabolites. Humic substances are produced by interaction of plants, algae, mosses (bryophytes) and microorganisms by a process known as humification. Humification of latex- and resin-bearing plants is primarily responsible for the production of the water-soluble humic substances. The phytochemistry of vegetation around shilajit-bearing rocks is therefore very important. Humus is formed when soil microorganisms decompose animal and plant material into elements usable by plants. Plants are the source of all our food and humus is the source of plant food. Humus consists of organic residues that have lost their original structure following rapid decomposition in the environment. They are the major organic constituents of native shilajit, present in an amount of about 80–85%; these substances have molecular weights ranging from several thousand for humic acids (HAs), to up to several million for polymeric humins (HMs) and only a few hundred for its fulvic acid component. Humic substances also are found in soils and sediments distributed over the earth's surface, occurring in almost all terrestrial and aquatic environments. The humus extracts

Fig. 1. Structure of natural Shilajit.

possess a lattice-like structure, perforated by voids of varying dimensions (0.01–0.05 mm). These are filled with organic molecules or metal complexes, which may be responsible for the therapeutic effects. However, shilajit humus is not special; degrading humus samples and analyzing the products spectroscopically shows that it is structurally similar to ordinary soil. Humus contains two components of interest: fulvic acid and humic acid. The former has the lowest molecular weight components, containing uronic acids, phenolic glucosides and amino acids, while the latter is composed of high molecular weight compounds and contains a high proportion of phenolics.

Fulvic acid is rapidly being recognized as one of the key elements in many outstanding health and scientific breakthroughs of the 21st century. Scientists and doctors throughout the world are beginning to discover fulvic acid and are beginning to recognize its extraordinary potential. Fulvic acid has always occurred naturally in organic plants and soils, yet its recent discovery and tremendous value is now just beginning to be recognized. It can balance and energize cell life and enhance biological properties. Fulvic acid is created in soil in the presence of microorganisms and humus. It helps in transporting minerals from soil to the plants in faster and easier manner. The percentage of fulvic acid is very important in shilajit. The fulvic acid in shilajit is absorbed almost completely and more effectively than fulvic acid in pure form it is because shilajit contains around 85 minerals which makes the transportation and absorption of fulvic acid by body cells maximum. The chemical formula, structure and characteristics of fulvic acid have been determined by NMR.

Humic acid is present in fractions in shilajit and is insoluble in water. It is light brown to yellow brown in color. Humic acid, along with fulvic acid, supports free-radical scavenging and antioxidant activities. It can acquire positive ions (free radicals) under one condition and release them when conditions change. It picks up ions depending on the availability of different ions to replace the ones released. Positive ions are called cations and the pickup and release property is called cation exchange capacity. Humic acid holds cations so they can be absorbed, improving micronutrient cellular exchange and transference to the body's circulation system.

Shilajit contains significant amounts of free form hippuric acid ($C_6H_5CONHCH_2COOH$) and benzoic acid. Consumption of cranberries is associated with augmented hippuric acid production and has shown significant benefits in maintaining good urinary tract health and weight loss. Although the mechanisms are not yet clearly understood, when the body metabolizes benzoic acid (that naturally occurs in cranberries and in shilajit), the result is elevated amounts of hippuric acid, which in turn is believed to inhibit microbial growth. It is interesting that shilajit is used traditionally for genitourinary and kidney conditions and part of this effectiveness is related to the presence of free form hippuric acid and benzoic acid (which is transformed in the body to hippuric acid).

The therapeutic actions of the raw material vary by the region it is harvested from. There are other substances that contain humic and fulvic acids, but true shilajit has a very important therapeutic, bioactive ingredient that is not present in other "shilajit-like" substances. The authenticity and therapeutic quality of shilajit is identified by the inclusion of oxygenated dibenzo-alpha-pyrones. While there are several areas from which the raw material is collected, the highest levels of therapeutic ingredients come from specific areas in the Himalayan mountains in Nepal at 10,000–12,000 feet above sea level. The processing of the raw shilajit is very important as it contains free radicals and may also contain mycotoxins and fungal toxins. The processing removes these free radicals, polymeric quinone radicals, toxins, mycotoxins, and other inactive ingredients.

The non-humic substances of shilajit are low molecular weight compounds of marine fossil, plant and microbial origin, occurring in and around shilajit-bearing rocks. They comprise a mixture of low molecular weight aromatic, aliphatic, alicyclic, and heterocyclic (N- and S-containing) compounds. Of particular biological interest are oxygenated dibenzo-alpha-pyrones (DBP) and hydroxyacetophenones (HAPs). The most common (Fig. 2) low molecular weight compounds present are oxygenated dibenzo-α-pyrones (6), (7) and (8). They are probably produced from the naturally occurring macro cycle (9), via an intermediate such as (10), (Scheme 1). Solvent extraction of shilajit results in a compound similar to (10), which can be autoxidised to (6) and (8) when exposed to light.

Benzopyrones are groups of compounds related to flavonoid compounds. Flavonoids, such as the nutritional bioflavonoids as found in vitamin C-containing foods, have antioxidant activity. Benzopyrones are also contained in the proanthocyanidins in pycnogenol and grape seed, and related compounds such as silymarin and rutin, and primarily act to reduce edema caused by protein leaking out of capillaries. In addition, benzopyrones are known to enhance diuretic action important for reducing edema and water weight related to obesity. Benzopyrones have been recommended for the treatment of viral infections. Dibenzo-alpha-pyrones amount to around 1% (Table 1).

Around 85 trace minerals are found in shilajit. These minerals occur in natural ionic form which is very vital for maintaining the equilibrium of energy metabolism in body. The minerals in shilajit are not similar to the mineral supplements available in market. These minerals are in ionic form and have previously been absorbed by rich plant life and returned back to earth. So they are easily absorbed by the body cells. These large numbers of minerals present in shilajit are Aluminium, Antimony, Barium, Bismuth, Bromine, Cadmium, Calcium, Cobalt, Cerium, Cesium, Cobalt, Copper, Fluorine, Gallium, Germanium, Gold, Indium, Iron, Lanthanium, Lead, Lithium, Magnesium,

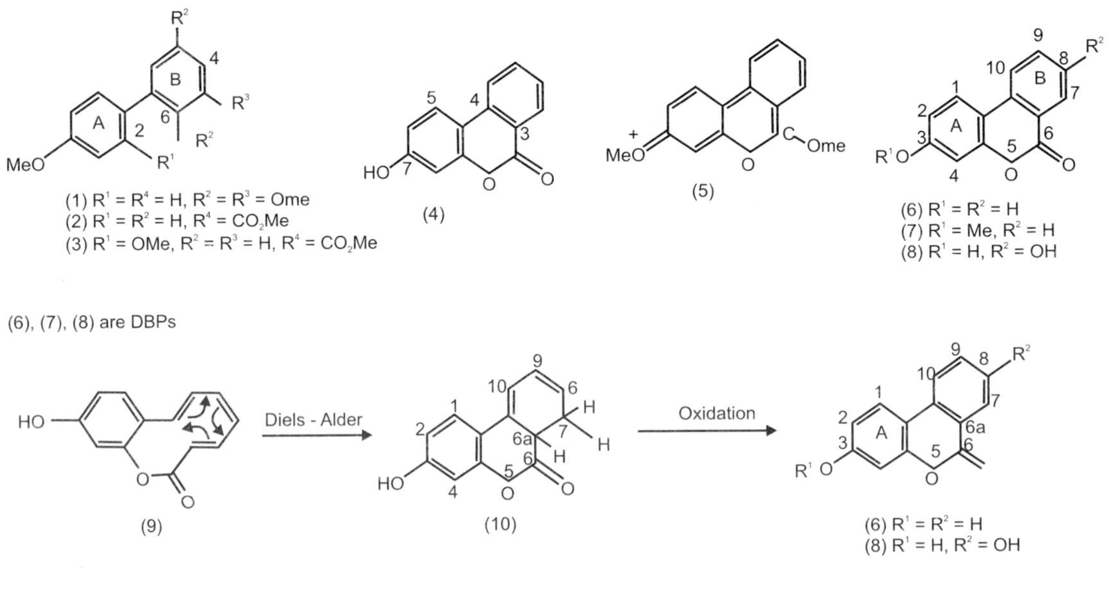

Scheme 1- A possible route to making dibenzo-α-pyrones (6) and (8)

Fig. 2. Some of the compounds that may be present in samples of shilajit.

Manganese, Molybdenum, Mercury, Niobium, Phosphorus, Protactinium, Rubidium, Silica, Sodium, Strontium, Scandium, Silver, Tellurium, Thorium, Tin, Titanium, Tungsten, Yttrium and Zinc.

Table 1. Relative abundance of bioactive constituents in purified shilajit composition.

Bioactive constituents	Amount (%)
Dibenzo-alpha-pyrone & di- or tetramers	0.3
3-OH & 3,8-$(OH)_2$Hydroxy (mono-di)	0.1–0.4
Benzoic acid or its N,S,-heterocyclic	3–12
Dibenzo-[alpha]-pyrone acetophenones, long-chain esters other aromatic or its esters	0.5–1
Fulvic acids	60

The presence of additional compounds (Table 2) and their amounts, in the purified shilajit composition has been established by comprehensive chromatographic (HPTLC, HPLC, GLC) and spectroscopic (GC-MS, NMR) analyses using specific markers.

Table 2. Characteristics of Purified Shilajit

Appearance	Fine powder
Color	Dark brown
Taste	Bitter
Odor	Characteristic
pH of 2% aqueous solution	7.01
Water soluble ext.	92.0 (% w/w)
Alcohol soluble ext.	26.8 (% w/w)
Fulvic acid	65.3%
Total Dibenzo-alpha-pyrones	0.427%
Total mineral matter	18.48%
Acid insoluble mineral matter	4.18%

ACTIVE COMPONENTS

The biological effects of shilajit are believed to be due to following bioactive compounds :

1. Fulvic acids (FAs)

Fulvic acid ($C_{14}H_{12}O_8$, is a yellow acidic metabolite) is a small, water-soluble, natural organic substance of low molecular weight, which is derived from humus. Fulvic acid is absorbed by plants that need the trace minerals and other nutrients for their own growth. Whenever one eats plants that have been produced in the presence of fulvic acid, the body can then assimilate more of nature's building blocks to build and maintain the body. Fulvic acid is the active ingredient of shilajit. Fulvic acid is credited with revitalizing the electrical energy of body cells. Porous by nature, fulvic acid neutralizes toxins. It is the most powerful electrolyte found in nature, and it can quickly cure food poisoning in minutes. It does this by supplying the body with nutrients and minerals and maximizing cell absorption. This nutrient works throughout the body to boost immunity and supercharge cell function. Basically, fulvic acid optimizes body's defences by ridding it of toxins. It contains an array of naturally-occurring phytochemicals, biochemicals, strong antioxidants, nutrients, enzymes, amino acids, antivirals, antifungals, and much more. This increases blood circulation, immunity, energy, vitality and sexual function. The many benefits of fulvic acid are well documented and touted by ayurvedic experts throughout India and all over the globe. For centuries, humans have obtained this nutrient in small amounts in food. This naturally-occurring substance is found in soil and mineral deposits, and it works to transport vital nutrients and minerals throughout the soil to plants. Over time, the acid becomes supercharged with natural scavengers of free radicals, detoxifying agents and cell-boosting electrolytes. Porous by nature, this substance seeks out all the bad free radicals in body and carries them out through waste. When this happens, body's immune system gets a load lifted off its shoulders and is able to fight other issues like arthritis, diabetes, and osteoporosis. It is an excellent bone-healing agent, and is easily accepted by the body for maximum absorption.

The cells (particularly the protoplasm in the cell) are the basic building blocks of body and life. Our body is made of millions and millions of cells. It is these cells that process food and oxygen to build up the body and give energy. It can be seen that with aging there occurs a continuous death and decay of internal body cells, which manifest in weakness of internal organ systems. All kinds of ailments creep in body with age. Our skin cells die and wrinkles are formed. The fulvic acid and the other minerals bring about rejuvenating changes in our body cells to restore youthfulness using following ways:

- The porous fulvic acid and humic acids in shilajit carry herbal compounds deeply into the tissues of the body. These porous cavities also hook toxins (chelates heavy metals and body toxins, removing them from the system) and escort them out of the body.
- Fulvic acid acts as a source of protection from environmentally-induced oxidation. Free-radical damage (oxidation) has been implicated in over 60 disorders. Antioxidants are extremely beneficial to fight disease. Fulvic acid is a very powerful antioxidant so it has the ability to dramatically reduce the oxidative effects of free radicals. This means that fulvic acid could potentially help your body ward off disorders such as cancer, premature aging, wrinkling of the skin and arthritis.
- Fulvic acid potentiates the availability of essential nutrients. Potentiation is a term that describes how long an ingredient can remain active. It functions as a carrier for the bioactive DBPs and trace elements. This "carrier" action is responsible for part of fulvic acid's health benefits. The trace minerals are needed as cofactors for enzymes, play important roles in turning food into energy, maintain the electrical balance in bodily fluids, carry oxygen in the body, are part of blood and bone, allow nerves to transmit messages and more.
- On eating food, body sets into motion a complex series of actions, which are designed to break down the foods with the help of enzymes. It then converts these digested foods into energy and building materials. Since there are only a finite number of enzymes that are available to the body, supplementing them from outside sources (such as with raw fruits and vegetables and enzyme supplements) is an absolute necessity. The presence of fulvic acid actually increases the activity of several enzymes, including transaminase, invertase and alkaline phosphates.
- Speaking of the potentiation of certain substances, fulvic acid may also have a positive effect on RNA and DNA. According to several researchers, the actual content of DNA in cells and the

synthesis of RNA are enhanced by the presence of fulvic acid. Furthermore, when supplemented with traditional vitamins and minerals, their bioavailability may be enhanced by the addition of fulvic acid.

- The only way that the body can benefit from nutrients is after they have been absorbed. The same is true of plants. Before a plant can properly process minerals, vitamins and other nutrients, they must be incorporated into their structure. Fulvic acid enhances this absorption process. Prior to elemental minerals being converted and stored by a plant as plant tissue, fulvic acids begin to ionize them. Without fulvic acid, the substances that surround plants would not be absorbed since the root hairs are single ion chambers that pull nutrients into them ion by ion. However, when fulvic acid acts upon a substance its molecular size and weight are altered. This enables it to pass through cell membranes—including the semi-permeable membranes that line intestine.
- It is a powerful natural electrolyte and helps to restore electrochemical balance. As the most powerful natural electrolyte, fulvic acid restores electrical balance to damaged cells, neutralizes toxins and can eliminate food poisoning within minutes.
- Fulvic acid acts as an important protective agent. An important aspect of humic substances is related to their absorptive interaction with environmental chemicals, either before or after they reach concentrations toxic to living organisms. The toxic herbicide known as paraquat is rapidly detoxified by humic substances (fulvic acid). Fulvic acids have a special function with respect to the demise of organic compounds applied to soil as pesticides. When it encounters free radicals with unpaired positive or negative electrons, it supplies an equal and opposite charge to neutralize the free radical. Fulvic acid acts as a refiner and transporter of organic materials and cell nutrients. It has the ability to dramatically detoxify herbicides, pesticides, and other poisons that it interacts with, this includes many radioactive elements. This detoxification process may extend to animals and humans.

2. Dibenzo-alpha-pyrone chromoproteins (DCPs)

Oxygenated DCPs are isolated from shilajit. DCPs are composed of oxygenated dibenzo-alpha-pyrone (DBPs) or its conjugates, phosphor-creatine, proteins, fatty acid esters of glycerol and other small ligands, e.g., carotenoids, sterols and aromatic acids, as core structural fragments. DCPs are the major bioactives of shilajit. DCPs comprise organo-mineral constituents and exhibit orange, purple and yellow colors contributed by oxygenated carotenoids known as xanthophylls and indigoids. The DCPs exhibit absorption maxima in the UV and visible regions at λ_{max} about 225, 275, 320, 392, 470, 492, 500–535, 620–660 nm.

The DCPs participate in a wide variety of functions in animal biological systems. DCPs have been encountered in the lowest form of animal organisms (foraminifera, a marine invertebrate, and in haemolymph of termites), in higher animals (rodents, beaver, chimpanzee, sheep) and in man. DCPs play part in treating chronic stress disorders, in increasing cognition learning, skin care, hair care, pharmaceutical or nutritional or veterinary formulations. Also, the skin care or protection formulation can be in the form of a lotion, cream, gel or spray. Pharmaceutical formulation is in the form of a tablet, syrup, capsule.

3. Dibenzo-Alpha-Pyrones (DBPs)

DBPs (Dibenzo-Alpha-Pyrones, along with triterpenes and phenolic lipids) both mono- and bis-compounds are present in free and metal-ion conjugated forms. DBPs are able to pass the blood-brain barrier and act as a powerful antioxidant protecting the brain and nerve tissue from free radical damage. It also inhibits the enzyme acetylcholinesterase, which breaks down acetylcholine. This will increase the levels of acetylcholine. Low levels of acetylcholine are associated with Alzheimer, poor memory and concentration. The low molecular weight chemical markers viz. aucuparins, oxygenated dibenzo-K-pyrones and triterpenic acids of the tirucallane type (free and conjugated) occurring in the core structure of shilajit humus are the other major active constituents of shilajit.

SOURCES

Shilajit drips from the cracks of the rocks during hot weather. It is formed due to decomposition of the plant matter in the rocks. The bio-transformed plant matter is extruded from the rocks by geothermal pressures. Chemical research indicated that shilajit is quite recent in origin. Humification of latex and resin-bearing plants was responsible for the major organic mass of shilajit (about 80–85%). Shilajit samples collected from different places exhibit variations in chemical characteristics and bioactivities. The factors that cause variations in shilajit humus are: (i) altitude and the nature of shilajit-bearing rocks; (ii) atmospheric conditions (e.g. alternate wetting and drying); (iii) pH and moisture content of the rock source; (iv) activity of the rhizospheric microorganisms and their exo-enzymes.

The common plant sources of humus, in mountain soils, are the perennial grasses and legumes, which possess finely branched root systems capable of regeneration. Other important sources of humus are the litter and latex of plants. In its native state, shilajit is a compact mass of vegetable organic matter composed of a dark gummy matrix—thought to be largely the result of humification of the plant *Styrax officinalis* Linn and/or *Euphorbia royleana* Boiss (Euphorbiaceae). *Euphorbia royleana* Boiss is a rich latex producing plant. Some other latex and resin producing common species are the legumes, e.g. *Trifolium* and *Rhus* (family: Anacardiaceae), *Ficus* (Moraceae), and *Juniperus* (Cuprassaceae). *T. ripens* (Leguminosae) collected from different places in the Himalayan belt yielded several phenylpropanoid-acetate-derived metabolites. Many of the shilajit-bearing mountains have been found to be rich storehouses of marine invertebrate fossils, such as of the phyla of Arthropoda, Brachiopoda and Mollusca, of the Phanerozoic era. This co-occurrence of shilajit and the invertebrate fossils is a consistent phenomenon. Marine invertebrates contribute to the formation of shilajit. DBP-carotenoproteins and other low molecular weight organic and coloring constituents (e.g., indigoids) were found in the marine invertebrate samples. DCP-fraction comprises astaxanthin fatty acyl derivatives and canaxanthin. 3,8-Dihydroxy dibenzo-alpha-pyrone and the amino acids have also been isolated from DCPs. The coloured molecules comprising carotenoids, indigoids, and glycation of protein may form stable complexes by coordination with metal ions. Such intra-crystalline biomolecules act as a nucleation site for biomineralization. When limb muscles of dead marine animals decay, the vacated spaces are filled with minerals, such as pyrite (FeS, CaSiO) before the thin organic cuticles that surround them have time to collapse or decay. Precipitation is ensued as a

Fig. 3. Natural Shilajit dripping out from the cracks of the rocks during hot weather.

result of diffusion of Fe and S into the cell. Pyrite does not replace the tissue directly but precipitates on surfaces and within spaces.

MEDICAL APPLICATIONS

Shilajit has been attributed with many miraculous healing properties. The underlying power of shilajit comes from its property of restoring the electrical potency of the body cells, reducing the stress on the internal organs and effecting balanced energy metabolism. Shilajit is useful in as aphrodisiac, rejuvenator, nervine tonic, respiratory stimulant, internal antiseptic, diuretic, lithotriptic, piles, adiposity, renal and bladder stones, nervous diseases, amenorrhea, dysmenorrhea, menorrhagia, eczema, anorexia, fracture of bones, etc.

- Shilajit is a reputed rejuvenator and immunomodulator claimed to arrest the process of ageing and prolong life. It rejuvenates and heals every cell of the body and corrects all the imbalances. By restoring balanced energy metabolism, it keeps us young for a very long time. A younger body obviously can resist any type of ailments. If you are a woman, shilajit will ensure that you remain in youthfulness for a long time. The ageing process involves the action of highly reactive free radicals, produced systemically, which interact with other cellular compounds and produce oxidative damages and eventually kill cells and tissues and impairs the immune function of the organism. Such free radical damage accumulates and increases with age, creating degenerative diseases, such as Alzheimer's, cardiovascular, arthritis, and cancer and over a hundred other diseases. It stimulates the immune system and reduces chronic fatigue. It counteracts debility and general fatigue. It is an adaptogen that helps to combat immune disorders.

- It helps to accelerate processes of protein and nuclei acid metabolism and stimulates energy providing reactions. DNA, the cellular building block of the body, is very sensitive to oxidative stress. Although repairs to damaged DNA are constantly being made, the cell's mechanism cannot keep up with the number of mutations that occur in the organism, particularly in the aged. Mitochondria, which is responsible for producing cellular energy has its own DNA, but it does not have a repair mechanism to give it protection against free radical induced damage. The mutation of mitochondrial DNA therefore produces a greater adverse effect than DNA mutation elsewhere in the system. Researchers in recent years have shown that certain individual natural supplements, such as omega-3-polyunsaturated fatty acids and metabolites thereof, oxygenated dibenzo-alpha-pyrones, and their O-acyl-esters, as well as hydroxyacetophenones can protect against oxidative damage to mitochondrial DNA. Accordingly, it is desired in this invention to provide a purified composition of bioactive agents to protect the body against free radical damage.
- Shilajit helps in supplementing the minerals (like phosphorus, calcium and magnesium) and vitamins in body. It also promotes the movement of minerals, especially calcium, phosphorous, and magnesium, into muscles, tissue and bone. It reduces recovery time in muscles, bones and nerve injuries. It promotes strong bones and heals damaged muscle tissues. It also helps to cure arthritis. The antioxidant and anti-inflammatory properties of shilajit help to decrease and relieve joint inflammation and pain. The effects on the neurotransmitters in the brain also seem to help relieve joint pain. It stimulates the immune system and improves restoration (recovery) after exercise.
- Natural electrolyte, fulvic acid, assists every stage of cellular metabolism and is the most powerful, natural electrolyte known. It restores electrical balance to damaged cells, neutralizes toxins and can eliminate food poisoning within minutes. When it encounters free radicals with unpaired positive or negative electrons, it supplies an equal and opposite charge to neutralize the free radicals. Fulvic acid acts as a refiner and transporter of organic materials and cell nutrients. Whenever minerals come into contact with fulvic acid, in a water medium, they are naturally dissolved into an ionic form. These minerals literally become part of the fulvic acid itself. Once the minerals melt into the fulvic acid complex, they become bio-active, bio-available, and organic. Thus, when elemental minerals are transformed into an organic state, through a natural chemical process involving fulvic acid and photosynthesis, they are safe to be used by both humans and animals
- Organic tissue growth: A modern day medical wonder is our ability to actually remove damaged tissues and bones from the human body and replace them with healthy bones and tissues. Unfortunately, such transplantation is also very risky. Transplants of this type are sometimes rejected. To lower the incidence of rejection the most desirable donor tissues come from alternate sites of the same body. Unfortunately, this requires a separate surgery. Extra incisions mean more discomfort and a longer period of healing. The only other alternative is finding a donor from other sources. Due to the lack of human donor tissues, animal bone in the form of bovine calcium hydroxyapatite, an inorganic calcium compound, are used. Two problems with such a procedure are that neither animal bones nor inorganic calcium are readily absorbed by the human body. However, these problems were circumvented with the use of fulvic acid. When fulvic acid was inserted into the animal bone before replacement, the patients experienced dramatically

improved regeneration of the transplanted bones. The fulvic acid was so readily accepted and used by the patients that their bodies became highly osteoconductive. This means that new bone tissue began to form at an accelerated pace, thereby enhancing growth and healing. The inorganic calcium was also absorbed by the body due to the fact that the fulvic acid had transformed it into an organic compound.

- Reproductive system: Shilajit is commonly referred to as the Indian Viagra. It helps in increasing strength and endurance power in men. It works as an aphrodisiac agent thereby helps in increasing sperm count and also helps in bettering the quality of sperms. It also helps in regulating sex hormones for proper functioning. Shilajit effectively increases the fertility of the person and increases the live cells count in reproductive hormones. It increases the core energy responsible for your sexual and spiritual power. It enhances the effect of nitric oxide to produce as erection in males.
- Treating anaemia: It is of great help to anemia patients. Iron is required to make red blood cells. Shilajit is a good source of trace minerals and contains iron. The fulvic acid helps to carry the iron into the body making it bio-available.
- Treating diabetes: Shilajit is very effective in diabetes. Diabetes is due to improper functioning of pancreas, which secretes insulin. Diabetes is the main cause of impairment in vision, clinical blindness, leg amputation and kidney failure. Shilajit counteracts diabetes and regulates the blood sugar level, as it is very helpful in metabolizing glucose in blood stream. Fulvic acid fraction in shilajit is considered responsible for the anti-diabetic activity. FAs showed significant success in preventing and combating free radical damage to pancreatic islet B cells. It significantly increases superoxide dismutase (SOD) activity.
- Cholesterol: Shilajit lowers serum cholesterol, liver cholesterol, serum triglycerides and serum phospholipids in test subjects being fed a high cholesterol diet. Similar effects were also observed in the lipid profile. Treatment with 50, 100 and 200 mg/kg of shilajit produced significant reduction in TCh level, with maximum reduction caused by 100 mg/kg ($P<0.001$). There was dosc-dependent reduction in the TG level. All the three doses of shilajit also produced significant increase in the HDL level with the maximum elevation being produced with the dose of 100 mg/kg ($P<0.001$). Moreover, the combination treatment caused significant improvement in the lipid profile as compared to that of shilajit (100 mg/kg) ($P<0.05$) or metformin ($P<0.05$).
- Brain markers: Recently it has been found that systemic administration of refined extract of shilajit differentially affects cholinergic but not glutamatergic and gamma-aminobutyric acid (GABAergic) markers in rat brain.
- Anti-ageing: After certain age an internal degeneration process sets in the body. For men the energy level, stamina, potency and functioning of the internal organs peaks at the age of 21 years. From then on a wear and tear sets in the internal organs of the body. There is a continuous decrease in electrical potency of the body cells leading to decreasing levels of energy metabolism. For women, youthfulness peaks at the age of 18 years. Starting from 18 years onwards there is a decrease in hormonal levels, lower secretion of body fluids, slowing of metabolism, low levels of stamina and high levels of fat build up. The stress of modern day life, pollution, lack of exercise and orientation towards junk food aggravates the process of aging and degeneration further. Another major reason for many modern day diseases and faster ageing

process is chemical farming. It depletes the soil and the plants of naturally available minerals. Wrinkles on skin, thinning hair, receding hemlines, and tendency to put on more weight in spite of consuming less food, decreasing energy and stamina, decrease in level of sexual life, susceptibility to major degenerative ailments are the signs that mean that ageing process is setting in our body. The number of cancer patients is increasing at an alarming rate every year and heart diseases are taking a heavy toll. Even school children are being diagnosed with type 2 diabetes. And in early thirties joint pains seem to set in. Alzheimer's and Parkinson's are fast becoming household names.

Shilajit effectively arrests and reverses the biological ageing process. Each body cell has certain electric potential. Due to ageing and some other factors like emotional stress, infections, pollution, free radical attack, food habits etc., the electrical potential of the cell decreases. When the electrical potential of the cell becomes zero the cell disintegrates and dies ultimately. Decrease in electrical potency of the cells lead to decrease in energy metabolism of the cells. Energy metabolism is the internal biological reaction in our body cells—deep in the mitochondria and the protoplasm of the cells converting mass into energy. This process slows down progressively with age. This is the reason why more mass is being stored in our body instead of being converted to energy. Biological ageing can be arrested and reversed by restoring the electrical potency of the cells. Body cells need minerals more than anything else to arrest the natural degeneration process and stay young for long. Minerals act as the catalysts for metabolic reactions and biological functions and are required for the assimilation of all vitamins. Even after consuming food rich in minerals and vitamins or taking wide range of mineral supplements, the body cells can never have sufficient minerals. The minerals that are taken as over the counter like the tablets, tonics etc. are in pure inorganic metallic form (also known as the colloidal form). But the calcium in eggs is in ionic form and is absorbed in a much better way than the raw calcium. As the body ages the cell walls become thick and it becomes increasingly difficult for the minerals to be transported into the cells. This is the major reason why ageing and degeneration sets in.

- Antioxidants: Antioxidants can safely neutralize a free radical without becoming a free radical themselves. Shilajit is a powerful nervine tonic and antioxidant that has the added benefit of being able to cross the blood-brain barrier. Shilajit is very effective in treating brain degenerative ailments like Alzheimer's, Parkinson's and dementia and other nervous disorders like epilepsy. Shilajit is known to boost memory and the ability to handle stress.
- Treating obesity: During teenage years in spite of having the rich food and very little physical activity one tends to put on less weight but in mid twenties and above, there is more and more weight gain and loss of stamina and energy levels in spite of strict diet control and stringent physical exercise. This is because energy metabolism process slows down with age and as a result more mass is being stored on the body instead of being converted to energy. This ultimately leads to excess body weight and decrease in energy levels. Shilajit is a very potent excess fat dissolver in the body. It possesses this property because of its light and dry properties. It helps in scraping away of extra fat accumulated in body thus helps in providing slim and healthy looks. Shilajit effects balanced energy metabolism in the body and makes it burn excess fat.
- Anti-ulcerogenic and anti-inflammatory: Shilajit increases the carbohydrate/protein ratio and decreases gastric ulcer index, indicating an increased mucus barrier. Shilajit was found to have

significant anti-inflammatory effect in carrageenan-induced acute pedal oedema, granuloma pouch and adjuvant-induced arthritis in rats.

- Fertility agent: The effects of shilajit on spermatogenesis and ovogenesis were studied using male and female rats. In the male rats, the number of sperms in the testes and epididymides was significantly higher than in the control. A histological examination revealed an apparent increase in the number of seminiferous tubular cell layers in the testes of the treated rats. However, there were no significant differences in the weights of heart, spleen, liver, kidney, brain, testes and epididymides. In the female rats, the effect of shilajit was estimated by the ovulation-inducing activity. It was estimated that shilajit had both a spermatogenic and ovogenic effect in mature rats. Effects of concomitant administration of processed shilajit in Swiss mice have also been evaluated on the development of tolerance to morphine-induced analgesia. Chronic administration of morphine to mice resulted in the development of tolerance to the analgesic effect of morphine. Concomitant administration of processed shilajit with morphine resulted in a significant inhibition of the development of tolerance to morphine-induced analgesia. Processed shilajit did not elicit any significant analgesia in mice; nor did the chronic concomitant administration of processed shilajit alter the morphine-induced analgesia. These findings with processed shilajit indicate its potential as a prospective modifier of analgesic tolerance to morphine.
- Siotone is herbal formulation comprising of *Withania somnifera, Ocimum sanctum, Asparagus racemosus, Tribulus terristris* and shilajit, all of which are classified in Ayurveda as rasayanas which are reputed to promote physical and mental health, improve defence mechanisms of the body and enhance longevity. These attributes are similar to the modern concept of adaptogenic agents, which afford protection of the human physiological system against diverse stressors. Gastric ulceration, adrenal gland and spleen weights, ascorbic acid and corticosterone concentrations of adrenal cortex, and plasma corticosterone levels, are the stress indices. Chronic stress also induced glucose intolerance, suppressed male sexual behaviour, induced behavioural depression and cognitive dysfunction, and immunosuppression. All these stress-induced perturbations werc attenuated by Siotone.
- As a suspending agent for the formulation of antacid preparations: Shilajit produced effects on sedimentation volume similar to those produced by sodium carboxymethyl cellulose (CMC), but at lower concentrations. It induces better flocculation with a moderate increase in viscosity compared to CMC. It did not interfere with the acid-consuming capacity of the suspensions.
- Although some promising results have been achieved by acetylcholinesterase inhibitors, an effective therapeutic intervention in Alzheimer's disease still remains an important goal. Sitoindosides VII–X, and withaferin-A, isolated from aqueous methanol extract from the roots of cultivated varieties of *Withania somnifera* and shilajit are used in Indian medicine to attenuate cerebral functional deficits, including amnesia, in geriatric patients. Administration of shilajit led to reduced acetylcholinesterase staining, restricted to the basal forebrain nuclei including medial septum and the vertical limb of the diagonal band.

SIDE-EFFECTS

There are potential risks of using crude Shilajit. Firstly, the place of origin is an important factor that determines its composition and likely medicinal value. Shilajit samples from different regions of the world can have markedly different physiological properties. Shilajit from India and Nepal decreased the volume of gastric secretion and acid output, whereas samples from Pakistan and Russia did not. Secondly, some samples of shilajit are heavily contaminated with fungal organisms such as *A. niger, A. ochraceous* and *T. roseum*. These are capable of producing lethal mycotoxins. Shilajit is heavily infested at their periphery with a large array of microorganisms, some of which are producers of mycotoxins. Thirdly, native, unprocessed shilajit contains significant amounts of stable free radicals. Two types of free radicals have been detected using electron spin resonance, long lifetime stable radicals (which are stabilized by metal ions present within the sample) and short lifetime radicals. Free radicals can cause non-repairable cell damage. The concentration of free radicals in humic material increases as the pH is raised; above pH 8.0 such radicals are stabilized due to the formation of semiquinone species. Russian shilajit has a pH in this range and, for this reason Russian doctors do not give shilajit to ulcer patients who have a greater risk of developing cancer. Fourthly, amounts of such bioactive compounds have large amounts of contaminants, e.g. high molecular weight polymeric quinones, humins, and inorganic substances. Thus, the potential risk of ingesting shilajit in its native form, or only after rudimentary purification, with no control or defined standards, may be harmful. Fifthly, the hazards of collection of shilajit and the scanty amount generally available in any one locale prompt unscrupulous traders to adulterate it with rock soil, plant debris and quercus gums. It is, therefore, imperative to determine certain standards of shilajit on the basis of bioactivity-directed investigation of its chemical constituents.

CONCLUSIONS

Shilajit has been rightly called a panacea—a universal remedy. It can play a prominent role in maintaining good health. This herbomineral preparation can offer a new and promising approach in the long-term management of large number of diseases because of its multifaceted action. It can prove to be a good complementary and alternative treatment. But this exudate still needs to be explored more. Further studies into mechanisms underlying the relationships between shilajit and its curing properties may improve our understanding.

SUGGESTED READINGS

1. Acharya SB, Frotan MH, Goel RK, Tripathi SK, Das PK. Pharmacological actions of Shilajit. *Indian J Exp Biol* 26: 775–777, 1988.
2. Agarwal SP, Khanna R, Karmarkar R, Anwar MK, Khar RK. Shilajit: a review, *Phytother* Res.; 21(5):410–405, 2007.
3. Flaig W, Beutelspacher H, Rietz E. Chemical composition and physical properties of humic substances. *Soil Components* 1: 1–211. 1975.
4. Ghosal S. The facets and facts of Shilajit. (Spl. lecture) pp. 32–35. Proc. National Symposium on Development of Indigenous Drugs in India during past 25 years. Institute of History of Medicine & Medical Research, Hamdard Nagar, New Delhi. 1988.

hearing and proper function of vital body organs such as heart, liver, pancreas, lungs, kidneys etc. Similarly fibre and water are very important constituents of food.

Everyone wants to feel good and healthy. Optimal health can be achieved with the help of four basic but very important factors.

1. *Positive Mental Attitude*: Studies have shown that a positive attitude can enhance our body's immune system. Our mind plays a significant role to keep our body healthy.
2. *Regular Exercise:* exercise need not be vigorous or painful to be effective. Studies have revealed that walking 20 minutes can produce positive results.
3. *Adequate Rest*: There are 20% to 40% adults who have trouble in sleeping. An average Indian needs 6 to 8 hours of sound sleep at night. Major problems due to inadequate rest are fatigue, difficulty in concentrating, irritability, lustreless skin etc. This is one factor which is neglected for days together.
4. *Balanced Diet:* This is one of the most important factors. Surveys show that 95% people do not know about balanced diet, only five percent know about it and there are less than 2% people who take balanced diet.

   Five groups of diet and its Recommended Daily Allowance (RDA) is given below:

   Group I: Wheat, rice and cereals. These are the major sources of starch, carbohydrates, vitamin B and fibres. Six to eight servings to be taken daily.

   Group II: Fruits and green vegetables are rich sources of several vitamins, minerals carbohydrates and fibers. Four to five servings to be taken daily.

   Group III: Meat, fish, poultry and pulses are sources of proteins, vitamin B_{12}, zinc and iron. Two to three servings to be taken daily.

Table 1. Showing quantity of nutrients to be taken daily

Source	Nutrients	Quantity
Whole grain brown rice and cereals	Carbohydrates, fibre, minerals, protein, phytonutrients.	6–8 servings daily depending upon appetite. One serving is: 1 slice of bread/a bowl of cereal/half a cup of cooked rice or pasta.
Fresh fruits and vegetables. Preferably unprocessed organic products	Vitamins, folic acid, beta-carotene, fibre, magnesium, iron, potassium, calcium, carbohydrate, phytonutrients.	At least 4–5 servings every day. one serving is: 2 tbsp vegetables/2–3 tbsp cooked fruit/ 1 cup salad/fruit/medium glass of juice/1 tbsp of dry fruit.
Meat (fish and poultry) and pulses	Protein, selenium, vitamins B, zinc, magnesium, iron, potassium. Phytonutrients in beans, nuts, seeds, tofu.	2–3 servings a day. One serving is: 50–75 g meat/100 g fish/1–2 eggs/3–4 tbsp cooked beans/50 g nuts.
Dairy products and its alternatives such as soya products	Calcium, protein, vitamin B_2, B_{12}, zinc, vitamins D, phytonutrients in the soya foods.	2–4 servings daily. One serving is: 200 ml milk/soy milk/a cup of yogurt / 30 g hard cheese.
Fats, oils and sugars	Fat and sugar, with some essential fats, vitamins, minerals. Phytonutrients in unprocessed olive oil, canola oil and chocolate.	Eat small amounts. Look for lower-fat alternatives to spreads, salad dressings and fast food. Include Omega-3.

Group IV: Dairy products are rich in proteins, calcium and several vitamins such as A, B_2, B_6, B_{12} and D. Two to four servings to be taken daily.

Group V: Fats, oils and sugars are sources of energy. These are to be used sparingly.

The real challenge is that most of the people never bother about eating a healthy food in the 'recommended' quantities, which in turn leads to the deficiencies in one's nutrition. Even if one takes recommended quantities it is not possible to have proper quality as in one of the reports of Center for Science and Environment (CSE) it is stated that an Indian consumes 111 gms of fruits every day, out of which 42% is nothing but pesticides. Further it adds that Indians consume 100 to 6000 times more pesticides than Europeans' safe limit. Few pesticides, their uses and health problems associated with them are listed in Table 2.

Table 2. Disorders associated with pesticides commonly used in India (partial indicative list only)

	Name of Pesticide	Used for	Health problems associated
1.	DDT	Used against wide variety of insects, including domestic insects and mosquitoes.	Chronic liver damage cirrhosis and hepatitis, endocrine, reproductive disorders, immuno-suppression, cytogenic effects, breast cancer.
2.	Endosulfan	It is used as a broad spectrum and acaricide against insect pests on various crops.	Effects kidneys, foetus, liver, immuno-suppression, instrumental in testicular, prostate cancer, defects in male sex organs, breast cancer.
3.	Aldrin	Used to control termites and wireworms	Lung cancer, liver diseases.
4.	Dieldrin	Used against blowflies, ticks, lice and freely used in cattle and sheep dips, to protect fabrics from moths, beetles carrot and cabbage root flies.	Liver diseases, Parkinson's and Alzheimer's diseases.
5.	Heptachlor	soil-inhibiting pests.	Reproductive disorders, blood dyscrasiasis
6.	Chlordane	Soil pests, white grubs and termites.	Brain cancer, reproductive disorders, blood dyscrasiasis, non-Hodgkin's lymphoma
7.	Lindane	Sucking and biting pest and as smoke for control of pests in grain stores. It is used as soil dressing soil insects	liver damage-cirrhosis and hepatitis, endocrine and reproductive disorders, allergic dermatitis, breast cancer, non Hodgkin's lymphoma, polyneuritis.
8.	Fenitrothion	Chewing and sucking pests—locusts aphids, caterpillars and leaf hoppers, domestic insects and mosquitoes.	Retinal degeneration and myopia. Chronic exposure to fenitrothion can cause frontal lobe impairment and neurologic deficits.
9.	Fenthion	It is a persistent contact insecticide valuable against fruitflies, leaf hoppers, cereal bugs, and weaverbirds in the tropics.	Fenthion may be mutagenic: causing genetic aberrations. It may be a carcinogen.
10.	Malathion	Widely used insecticide and acaricide used for the control of aphids, thrips, red spider mites, leafhoppers and thrips.	Malathion and malaoxon are both carcinogenic and cause leukemia in mammals, allergic reactions, ulcers, behavioral effects, eye damage, brain disorders.

There are few other factors which are hard to avoid in daily life. This is a fact that the foods eaten on daily basis are chosen by masses for various reasons neglecting its nutritional value. One of the major reasons for such choice is busy lifestyle but taste is another factor due to which people choose foods; especially children as well as young generation have special attraction for fast food. Most of these foods carry chemically-altered flavors, which satisfy our taste buds, and these flavors are such that one indulges in overeating despite the fact that these foods have little nutritional value i.e. not enough for healthy life. On the other hand if natural food such as fruits or milk are to be consumed, one stops after certain quantity and do not like to eat more similar types any more, this is due to the fact that one craves for variety of foods which is need of our body. Skipping frequent meals further adds to the nutritional deficiency.

Majority of individuals take these facts very lightly ignoring the fact that human body requires daily intake of 13 vitamins and 11 minerals. These vitamins and minerals are essential for millions of chemical reactions. If these nutrients are missed routinely for long duration, the human body fails to carry out hundreds of reactions which leads to some disorders which are termed as lifestyle disorders such as diabetes, cardiovascular disease, autoimmune disorder, chronic fatigue, neurodegenerative disease, eye problems, asthma and allergies, obesity and cancers etc. Few recent reports are listed below

- World Health Organization estimates that 60 percent of cardiac patients in the world will be Indians by the year 2010.
- Asia is expected to have 190 million diabetic cases by the year 2030, out of which more than half of them will be in India and China
- More than 40 million people in India will suffer from osteoarthritis by year 2020.
- In India 24 percent people will be obese by year 2025 as against 9% in 1995.
- UN report says that more than 500,000 premature deaths are caused by atmospheric pollution annually in India.
- WHO report says that poor diet and lack of exercise are leading cause of non-communicable diseases such as cardiovascular disease, type 2 diabetes and certain cancers, which account for nearly 60% of 56.5 million deaths a year deemed preventable.

The reasons for increase of health-related disorders are very straightforward as almost all the individuals are living highly stressed life, many of us supplement it with smoking and excessive intake of alcohol, over and above living in highly polluted environment cannot be avoided. These are few facts, which are responsible for mani-fold increase in these diseases over past several years that effect our daily life.

What we ate yesterday is our today and what we eat today will be our tomorrow. So we have to rectify our today for our better tomorrow and healthy youth of future. The only way out is supplementation of our diet with vital nutrients i.e. proteins, vitamins, minerals and omega-3 fatty acids.

OXIDATIVE STRESS: THE ENEMY WITHIN US

Cancers, strokes, and cataracts seem to be different from one another as any diseases could be. It's impossible to imagine that they share a single cause. The chain behind these and many other diseases, perhaps behind aging itself, includes a common link: a class of molecules known as *free radicals*.

Oxidative Stress is a term used to describe the steady state level of oxidative damage in a cell caused by the reactive oxygen species. This damage can affect a specific molecule or the organ. Free radicals and peroxides represent a class of molecules that are derived from the metabolism of oxygen and exist inherently in all aerobic organisms. There are many different sources by which the reactive oxygen species are generated. Most reactive oxygen species come from the endogenous sources as by-products of essential metabolic reactions, such as energy generation from mitochondria or the detoxification reactions involving the liver cytochrome P-450 enzyme system. External sources include exposure to cigarette smoke, environmental pollutants such as emission from automobiles and industries, consumption of alcohol in excess, exposure to ionizing radiation and bacterial, fungal or viral infections.

Oxidative stress is very fast becoming the nutritional and medical keyword for the 21st century. Involved in a growing list of diseases, from cataracts to cancer, the health conscious people must take steps to protect themselves against the havoc of free radicals. Despite the growing dangers of free radicals, there are some simple, but very powerful, weapons against it. An avoidance factor that contributes to oxidative stress is a diet of whole, organic, unprocessed foods supplemented with anti-oxidants.

Oxidative Stress is not a disease, in itself, but creates a condition that can lead to a serious life-style disease. Oxidative stress occurs when the available supply of the body's antioxidants is insufficient to handle and neutralize free radicals of different types. The result is massive cell damage that can result in cellular mutations, tissue breakdown and immune weakening etc.

Free radicals are highly unstable molecules that interact quickly and violently with other molecules in our bodies to create abnormal cells. Free radicals are unstable because they have unpaired electrons in their molecular structure. This causes them to react almost instantly with any substance in their vicinity. They are capable of penetrating into the DNA of a cell and damaging its basic information. Now these cells will cause mutation, which replicate without normal controls. They accelerate aging and contribute to the development of many diseases, including cancer, heart disease and other abnormalities.

EXTERNAL SOURCES OF OXIDATIVE STRESS

The pollutants produced by modern technologies generate free radicals in the body. The food most of us buy contains chemicals, such as fertilizers, pesticides and preservatives, which produce free radicals when these are ingested by us directly or indirectly. Prescription drugs often have the same effect on our body.

Processed foods frequently contain high levels of lipid peroxides, which produce free radicals damaging the cardiovascular system. Cigarette smoke generates a very large quantity of free radicals, most of the lung diseases associated with smoking are caused by free radicals. Air pollution and alcohol also generate free radicals in the body.

In addition, exposure to sunlight generates free radicals resulting in ageing of the skin, causing roughness and wrinkles. If the exposure is for longer times, skin cancer may be caused.

FREE RADICAL DEFENCES

Every cell in the body creates its own army—antioxidant enzymes whose specialty is to neutralize the free radicals.

The substances created by plants, which fight free radicals, can help the human body to do the same thing. Thus, as a second line of defence the body uses is many standard vitamins and other nutrients to quench the free radical's thirst for electrons. Mainly used substances are vitamin C, vitamin E, beta-carotene, co-enzyme Q10, grape seed extract and bioflavonoids.

One should try to purchase organic foods to minimize pesticide and hormonal residues.

THE CONSTITUENTS OF OUR NUTRITION

CARBOHYDRATES

The word "carbohydrate" means "carbon plus water." Plants use sunlight (photosynthesis) to convert water and carbon dioxide into carbohydrates and oxygen. These are the main source of energy. If we are not consuming enough carbohydrates our body will use other sources of energy such as proteins, which is not very efficient source of energy.

Sources: Carbohydrates are of two types i.e. simple and complex. Both are composed of units of sugar. The only difference is how many sugar units they contain, and how they link together. Simple carbohydrates are sugars that give instant energy and typically have no nutritional value. These include sweets, candy etc. Complex carbohydrates release energy slowly and often contain fibre. These are healthier forms of carbohydrates. They include bread, rice, potatoes, cereals and pasta.

RDA: Carbohydrates typically should consist of 45–60% of our total caloric intake. Research suggests that adult individuals should consume a minimum of 120–125 grams of carbohydrates per day to satisfy basic needs.

Caution: If we consume excess carbohydrates and participate in little or no physical activity, these excess carbohydrates will be converted and stored in the body as fat – which may lead to weight gain and other health risks.

PROTEIN

Protein is an essential nutrient whose name comes from the Greek word "protos," which means "first". These consist of amino acids, the building blocks of protein, which are essential for cell regulation, growth, and repair.

About half of the protein which we consume daily, is converted into enzymes, the specialized proteins regulate the speed of biological reactions in our body and allow them to perform functions such as digestion of food and metabolize to make new cells and chemical substances. To perform these functions, enzymes often need specific vitamins and minerals too.

To make all the proteins required in our body, we require 22 different amino acids. Nine are considered to be essential, meaning that they are not synthesized by the body and must be obtained from food only. Our bodies can produce the other 13 from fats, carbohydrates, and other amino acids. So, these are termed as non-essential amino acids. They form antibodies to combat invading bacteria and viruses. They are part of the enzyme and hormonal system, build nucleoproteins (RNA and DNA), carry oxygen throughout the body and aid in muscle activity. Following are the nine Essential Amino Acids which if not consumed may lead to different health disorders.

Tryptophan

A natural relaxant helps curing insomnia by inducing normal sleep, reduces anxiety and depression, treatment of migraine headaches, strong immune system, helps reduce the risk of artery and heart spasms, clubbed with lysine it reduces cholesterol levels.

Lysine

It ensures the proper absorption of calcium, forms collagen, helps in production of antibodies, hormones and enzymes. Its deficiency may result in tiredness, irritability, inability to concentrate, growth retardation, hair loss, anemia and reproductive disorders.

Methionine

Being a main supplier of sulfur, which prevents hair, skin and nails disorders, increases the liver's production of lecithin, which helps lower cholesterol levels. It protects the kidneys and reduces liver fat, a natural chemical for heavy metals detoxification, promotes hair growth.

Phenylalaine

It produces norepinephrine in brain, which communicates signals between nerve cells and the brain, keeps us awake and alert, acts as antidepressant and improves memory.

Threonine

It helps preventing fat build-up in the liver and is an important constituent of collagen, it helps digestive and intestinal tracts function more smoothly.

Valine

It promotes mental vigor, muscle coordination and calm emotions.

Leucine and Isoleucine

These provide ingredients for the manufacturing of other essential biochemical components in the body, some of which are used for the production of energy, and help us to be more alert.

The non-essential amino acids are Arginine, Tyrosine, Glycine, Serine, Glutamic Acid, Aspartic Acid, Taurine, Cystine, Histidine, Proline, Alanine.

Sources: Meat, poultry, fish, eggs, cheese, nuts, legumes, and soy.

RDA: An average healthy adult man or woman needs about 1 gram of protein per every kilogram of body weight. Infants, growing children, pregnant women, persons with injuries, and athletes may require more protein on a daily basis.

Caution: Several medical conditions make it difficult for people to digest and process proteins properly. As a result, waste products build up in different parts of the body.

FATS

Fats are essential for good health. They help in energy production, cell building, oxygen transport, blood clotting, and the production of very active hormone-like substances called prostaglandins.

Fats can be either saturated or unsaturated. Our bodies can produce monounsaturated and saturated fats whereas the body cannot produce polyunsaturated fats, or essential fatty acids; they must come from the diet only.

Sources: Monounsaturated fats are found in olive and canola oils. Saturated fats are found in animal products such as butter, cheese, whole milk, ice cream, cream, and fatty meats, as well as some vegetable oils, coconut, palm, and palm kernel oils. Polyunsaturated fats are found in safflower, sunflower, corn, and soybean oils.

RDA: Upper Limits. Keep total fat intake between 20 and 35 percent of calories, with the major consumption of polyunsaturated and monounsaturated fatty acids, such as fish, nuts, and vegetable oils and less than 300 mg/day of cholesterol should be consumed.

VITAMINS

Vitamin A

Vitamin A is a fat-soluble vitamin, which is stored primarily in the liver. Vitamin A is available in two forms: Beta-carotene, which our body converts into vitamin A and Retinol, which is found in animal products. Vitamin A is essential for healthy skin, eyesight, reproduction and growth. It performs a variety of important functions.

Beta-carotene is a powerful antioxidant that neutralizes free radicals. It promotes the growth of strong teeth and bones. Its deficiency may lead to dry skin, dandruff, poor night vision, frequent infections, mouth ulcers.

Sources: Liver, meat, fish, eggs, cod liver oil, carrots, sweet potatoes, tomatoes, yellow fruits and vegetables, leafy vegetables.

Caution: Vitamin A and beta-carotene can be easily destroyed by heat.

Vitamin B

Each vitamin B has its own individual properties and its own specific biological role to play. As a group, these vitamins have so much in common that they are often thought of as a single entity and known as B complex.

Most of the B vitamins are removed when the grains are highly refined and processed.

Vitamin B_1 (Thiamin)

It plays a key role in the body's metabolic cycle for generating energy, aids in the digestion of carbohydrates, essential for the normal functioning of the nervous system, muscles and heart, promotes growth and muscle toning.

Deficiency symptoms: May lead to the loss of appetite, fatigue, paralysis, numbness and pricking in legs, insomnia, loss of weight, water retention in body, depression, constipation, heart and gastrointestinal problems.

Caution: It is easily destroyed by alcohol consumption, caffeine, stress, and smoking.

Vitamin B_2 (Riboflavin)
It is necessary for carbohydrate, fat and protein metabolism, helps in the formation of antibodies and red blood cells, carry out cell respiration, important for the maintenance of good skin, nails, vision and hair, helpful in curing eye fatigue

Deficiency symptoms: Itching and burning eyes, mouth ulcers, cracking of lips, dermatitis, retarded growth, poor digestive system, trembling, sluggishness and oily skin.

Sources: Yeast extract, liver, whole grains, milk, green leafy vegetables.

Caution: Its absorption is decreased by the use of oral contraceptives, excessive exercise and consumption of alcohol.

Vitamin B_3 (Niacinamide)
It improves blood circulation and reduces the cholesterol level in it, reduces high blood pressure, increases energy through proper absorption of food, useful for healthy skin, maintains nervous system, helps to metabolize protein, fat and sugar.

Deficiency symptoms: May result in pellagra, nervousness, headaches, fatigue, gastrointestinal, depression, aches and pains, irritability, loss of appetite, insomnia, muscular weakness, indigestion, bad breath, eczema, poor memory.

Sources: Whole grains, yeast, cheese, fish, eggs, lean meat.

Caution: Is absorption is decreased by the use of oral contraceptives, excessive exercise and stressful lifestyle.

Vitamin B_5 (Pantothenic Acid)
It boosts immune system, gets energy released from carbohydrates, fats and protein, helps in the utilization of vitamins, reduce stress, helps the adrenal glands.

Deficiency symptoms: Painful and burning feet, skin abnormalities, retarded growth, tremors, headache, muscle cramps, overall fatigue.

Sources: Nuts, liver, kidney, whole grains, royal jelly.

Caution: Food processing, preservatives, cooking, freezing easily destroy it and its deficiency is very common. Its absorption is decreased by the use of oral contraceptives, smoking, caffeine and alcohol.

Vitamin B_6 (Pyridoxine)
It is necessary for the synthesis and breakdown of amino acids, helps in fat and carbohydrate metabolism, formation of antibodies and white blood cells, maintains the central nervous system, aids in the removal of excess fluid of premenstrual women, regulates sex hormones, reduces muscle spasms and leg cramps, hand numbness, nausea and stiffness of hands, helps maintain a proper balance of sodium and phosphorus in the body.

Deficiency symptoms: Tingling limbs, nervousness, insomnia, skin eruptions, loss of muscular control, anemia, mouth disorders, muscular weakness, dermatitis, arm and leg cramps, loss of hair, slow learning, and water retention.

Caution: Its absorption is decreased by the use of oral contraceptives, antibiotics, alcohol and smoking.

Vitamin B_{12} (Cobalamin)
It helps in the formation and regeneration of red blood cells, thus helping prevent anemia. Necessary for carbohydrate, fat and protein metabolism, maintains a healthy nervous system, promotes growth in children, increases energy, needed for calcium absorption.

Deficiency symptoms: May lead to pernicious anemia, poor appetite, growth problems in children, tiredness, brain damage, nervousness, neuritis, degeneration of spinal cord, depression, lack of balance.

Sources: Liver, beef, pork, white fish, eggs, yeast extract, milk.

Folic Acid
It is necessary for DNA and RNA synthesis, which is essential for the growth and reproduction of all body cells, essential to the formation of red blood cells by its action on the bone marrow, aids in amino acid metabolism.

Deficiency symptoms: May result in, anemia, poor appetite, premature grey hair, birth defects (abnormal development of spine).

Sources: Peanuts, bananas, liver, wheat gram spinach, broccoli and other leafy vegetables.

Biotin
A member of vitamin B group, sometimes called vitamin H. In association with other B vitamins it converts food into energy. Aids in the utilization of protein, folic acid, pantothenic acid and vitamin B_{12}, promotes healthy hair.

Deficiency symptoms: May lead to extreme exhaustion, drowsiness, muscle pain, loss of appetite, depression.

Sources: Peanuts, almonds, egg yolk, walnuts, chicken.

Inositol and Choline
These members of B-complex family have special ability of breaking down fats and improve the brain function. Necessary for the formation of lecithin, which helps reduce blood cholesterol, helps prevent thinning of hair.

Deficiency symptoms: May result in high blood cholesterol, constipation, eczema, hair loss. high blood pressure, hardening of the arteries.

Sources: Liver, steak, egg yolk, peanuts, citrus fruits, whole meal bread.

PABA (Para Amino Benzoic Acid)

It helps healthy bacteria in producing folic acid, aids in the formation of red blood cells, have sun-screening properties.

Deficiency symptoms: May cause extreme fatigue, eczema, irritability, depressions, nervousness, constipation, headaches, digestive disorders, hair turning prematurely gray.

Sources: Bran flakes, yeast, chicken liver, eggs, lamb liver, rice, whole wheat.

Vitamin C

Vitamin C is also known as ascorbic acid. It is known to be a powerful antioxidant in the body. Vitamin C is a water-soluble essential vitamin that is quickly lost from our body, so daily intake is vital. Prevents scurvy, essential for healthy gums, teeth and bones, helpful in healing wounds, builds resistance to infection, helps in the prevention and treatment of the common cold, strengthens blood vessels, helpful in the absorption of iron, required for the synthesis of collagen. It is also one of the major antioxidant nutrients. Vitamin C will decrease the risk of getting certain cancers by 75%.

Deficiency symptoms: May lead to bleeding gums, swollen or painful joints, slow-healing wounds, loss of appetite, nosebleeds, tooth decay, muscular weakness, anemia.

Sources: Black currants, green pepper, mangoes, oranges, cabbage, tomatoes, and potatoes.

Caution: Water, cooking, heat and light all reduce the levels of the vitamin C available in food sources. Vegetables begin to lose vitamin C as soon as they are cut.

Vitamin D

When the skin is exposed to sunlight our body produces vitamin D. It is also found in foods of animal origin. It improves absorption of calcium and phosphorus in the body, essential for bone and teeth formation, helpful in maintaining a stable nervous system and normal heart action.

Deficiency symptoms: May lead to rickets, tooth decay, delayed healing of fractures, softening of bones, lack of energy, muscular weakness, retention of phosphorus in the kidneys.

Sources: Cod liver oil, salmon, herring, sardines, margarine, tuna, and cheddar cheese.

Vitamin E

Vitamin E is a key antioxidant. It is particularly important for a healthy heart and blood supply. It is also very good for our skin. Fat-soluble vitamin E is stored in our body for a short period of time only, so regular intake is essential.

It supplies oxygen to the blood which is then carried to the heart and other organs, it reduces fatigue, strengthens the capillary walls and prevents the red blood cells from destruction, prevents and dissolves blood clots, has also been used by doctors in helping prevent sterility, muscular dystrophy, calcium deposits in blood walls and heart conditions.

Deficiency symptoms: May lead to a rupture of red blood cells, loss of reproductive powers, lack of sexual vitality, abnormal fat deposits in muscles, degenerative changes in the heart and other muscles, dry skin, beneficial for a healthy immune system.

Sources: Sunflower oil and seeds, almonds, pine nuts, peanut, sweet potato, spinach, avocado and broccoli.

Caution: Heat processing (deep fat frying) and oxidation readily destroy vitamin E food sources. It has an increased ability to be absorbed with adequate intake of vitamin C and selenium.

Vitamin K

Vitamin K is found in food, but the bacteria (*E. coli*) in our intestines also create it. K stands for the Danish word "koagulation," meaning coagulation or clotting. Vitamin K is used by the body to make specialized proteins found in blood plasma. These proteins are mainly responsible for blood clotting. It is also needed to make bone and kidney tissue.

Deficiency symptoms: Vitamin K is manufactured in the intestine, and is widely available in food. Deficiency in this vitamin is very uncommon.

Sources: Broccoli, cabbage, kale, lettuce, spinach and turnip greens, cheese, liver and cereals.

MINERALS

Vitamins can do nothing for us without minerals. Although the body can manufacture a few vitamins, it cannot manufacture a single mineral. All tissues and internal fluids of our body contain varying quantities of minerals. These are constituents of the bones, teeth, soft tissue, muscle, blood, and nerve cells. They are essential for overall mental and physical health.

Minerals act as catalysts for many biological reactions occurring in our body, including muscle response, the transmission of messages through the nervous system, the production of hormones, digestion, and the utilization of nutrients in foods.

Calcium

Calcium is best known for helping to build strong bones and teeth. 99 % of the calcium from food goes straight to the teeth and bones, where it is used for building and maintaining strength. Calcium is very useful throughout the body to help muscle contraction, enable blood to clot, transmit messages through nerves cells, produce energy, keep the heart beating, and maintain the immune system. If we take less calcium, the body takes it from the bones causing bones to become thin and brittle.

Deficiency symptoms: Tooth decay, joint pain, muscle cramps, sleeplessness and osteoporosis

Sources: Milk and its products, cheese, small fish with bones, dark green vegetables, eggs and tofu.

Chromium

Chromium works with insulin in the metabolism of sugar and stabilizes blood sugar levels, reduce cholesterol and triglyceride levels, transport amino acids to the needed place. Research has shown that persons with low levels of chromium in their bodies are more susceptible to having cancer and heart problems and becoming diabetic. It also helps in reducing physiological hunger symptoms and plays a role in breaking down of fats. It is involved in extracting energy from fats and carbohydrates.

Deficiency symptoms: Excessive thirst, glucose intolerance, arteriosclerosis, heart disease, depressed growth, obesity, drowsiness, frequent hunger, hot/cold sweats, tiredness.

Sources: Yeast, egg yolk, meat, cheese, spinach, bananas and whole grains.

Caution: The milling during the processing of grains may remove up to 83% of the chromium.

Copper

It is essential for the absorption and utilization of iron, collagen production, converts iron into hemoglobin, helps proper bone formation and maintenance, is also needed to use vitamin C effectively by our body.

Deficiency symptoms: May result in general weakness, anemia, change in skin color, loss of taste sense.

Sources: Liver, sunflower seeds, mushrooms, shellfish, oysters, peanuts.

Iodine

Derived from Greek word *iodes*, meaning 'violet like'. It is essential in the development and functioning of the thyroid gland. Speech, condition of the hair, skin and teeth are dependent upon a functioning thyroid gland. It regulates the body's production of energy, helps burn excess fat by stimulating the rate of metabolism.

Deficiency symptoms: May result in an enlarged thyroid gland, slow mental reaction, dry skin and hair, weight gain, loss of physical and mental energy.

Sources: Milk, onions, salmon, prawns.

Iron

Iron is essential for life. It is a vital component of hemoglobin, the red pigment in blood that transports oxygen to the cells and removes carbon dioxide from the cells. Iron is one of the top six nutrients that are found to be commonly deficient in people globally.

Deficiency symptoms: White/brittle nails, pale skin, frequent illness, loss of appetite.

Sources: Cereals, liver, dried fruit, parsley, watercress, spinach and other leafy vegetables.

Caution: Boiling vegetables can reduce the iron content by 20%.

Caffeine interferes with the iron absorption in the body. Avoid drinking caffeine along with iron-rich foods or supplements.

Magnesium

Magnesium is essential for every biochemical process that takes place within the body. Magnesium is vital for turning food into energy for the body and help in the release of insulin regulating blood sugar levels. It helps in repair and maintenance of body cells. It controls the movement of calcium in and out of cells and is important for strong bones and teeth.

Sources: Milk and its products, nuts, peanut, white fish, eggs.

Caution: Major part of magnesium in food is lost during processing, milling removes approximately 59% of the magnesium from whole wheat. Cooking foods in water causes magnesium to leach out.

Manganese

It is an antioxidant nutrient, important in the breakdown of amino acids and the production of energy, necessary for the metabolism of vitamin B_1 and vitamin E, activates various enzymes which are important for proper digestion and utilization of foods, helps nourish the nerves and brain, necessary for normal bone development, needed for sex hormone production.

Deficiency symptoms: Paralysis, convulsions, dizziness, hearing problems, digestive problems, blindness and deafness in infants.

Sources: Tea, almonds, coconut, plums, bananas, watercress, wholemeal bread.

RDA: It is not established. 5 mg is considered to be safe.

Potassium

Potassium works with sodium to regulate the body's waste balance and normalize heart rhythms, sends oxygen to the brain, maintains proper alkalinity of body fluids, stimulates the kidneys to eliminate harmful body waste, helps in reducing high blood pressure, promotes healthy skin.

Deficiency symptoms: Poor reflexes, nervous disorders, respiratory failure, cardiac arrest, muscle damage, swollen abdomen, vomiting and diarrhea.

Sources: Tomato, banana, spinach, cauliflower, orange, chicken, red wine.

Selenium

Selenium is a powerful antioxidant. It works best with vitamin E. In other words, both are more powerful when they are taken together. It prevents free radical generation hence decreasing the risk of cancer, heart disease and hardening of blood vessels. Medical surveys show that increased selenium intake decreases the risk of breast, colon, lung and prostate cancer. Selenium also preserves tissue elasticity, slows down the aging and hardening of tissues through oxidation, helps in the treatment and prevention of dandruff.

Deficiency symptoms: Cataract, infertility, dandruff, premature aging, heart disease, skin wrinkles.

Sources: Seafood, meat, grains, sunflower seeds, walnuts, fruits, vegetables, dairy products.

Zinc

It is an antioxidant, necessary for protein synthesis, wound healing, vital for the development of the reproductive organs, prostate functions. It also help in male hormone activity, normal tissue function, digestion and metabolism of phosphorus, plays an important role in protection and repair of DNA.

Deficiency symptoms: Delayed sexual maturity, poor wound healing, white spots on fingernails, retarded growth, stretch marks, fatigue, decreased alertness, frequent infections, skin diseases, loss of sense of taste and smell.

Sources: Wheat gram, pumpkin seeds, egg yolk, oysters.

OMEGA FATTY ACIDS

Omega-3s are termed essential fatty acids (EFAs) because they are very critical for good health. Since our body is unable to produce them, omega-3s must be obtained from food. Studies have shown that the consumption of omega-3 fatty acids may aid in lowering overall blood pressure level, reduce hypertension and improve cholesterol levels, stabilize irregular heartbeat protecting the heart. Researchers believe that alpha-linolenic acid (ALA) may be particularly beneficial in protecting against heart and blood vessel disease, lowers cholesterol and triglyceride levels.

Sources: Omega-3 fatty acids include eicosapentanoic acid (EPA), docosahexanoic acid (DHA), and alpha-lineolenic acid (ALA). EPAs and DHAs are found in oily cold-water fish such as salmon, tuna, mackerel, and also fresh seaweed. ALAs are found primarily in dark green leafy vegetables, flaxseed oils and few vegetable oils.

RDA: There is no recommendation of daily intake for omega-3s, but a healthy diet containing significant amounts of foods rich in this essential fatty acid is recommended. These fatty acids are very safe to consume. However, many experts recommend limiting fish consumption two to three servings a week, because of the fact that many fishes may be polluted with mercury and other contaminants. Fish oil capsules do not usually present this same risk.

Table 3 showing Recommended Daily Amount (RDA) and deficiency of vitamins and minerals.

PHYTONUTRIENTS—THE FUTURE VITAMINS

Phytonutrients is a more recent evolution of the term that means the plant source compounds. Most of these are protective, disease-preventing. Although they are not absolutely essential to human health, yet they are helpful in fighting against various diseases. The vital role of fresh, whole plants in human nutrition is undisputed. Nutritional role of these compounds is becoming more credible every day as research uncovers more of their amazing benefits.

In the past, the phytonutrients found in fruits and vegetables were classified as flavonoids known as vitamin P, cabbage factors were called vitamin U, and ubiquinone was vitamin Q. Tocopherol somehow stayed on the list as vitamin E. These were dropped from vitamin list, as specific deficiency symptoms could not be established. Perhaps soon, phytonutrients may indeed be classified as essential nutrients and they are termed **the Future Vitamins**.

Sources: Plant-based food sources, including brightly-colored fruits and vegetables.

RDA: No specific recommendations exist at this time for most or all phytonutrients. It is recommended to consume five different-colored fruits and vegetables daily to benefit from a variety of different phytonutrients.

Recently research has made possible to group the phytonutrients into classes on the basis of similar protective functions as well as individual physical and chemical characteristics of the molecules. A few very important phytonutrients are discussed below:

Terpenes

It is found in green foods, soy and grains, comprise one of the largest classes of phytonutrients. The most studied terpenes are carotenoids. As proved by the many recent studies on beta-carotene, it acts as antioxidant, protecting lipids, blood and other body fluids from damage by free radical.

Table 3.

Recommended Daily Amount (RDA) and deficiency of vitamins and minerals	**Deficiencies**	**RDA**
Vitamins		
Vitamin A	20% of diets deficient	800 mcg
Vitamin B_1 (thiamine)	Commonly deficient in elderly	1.4 mg
Vitamin B_2 (riboflavin)	Deficient in 30% of elderly people	1.6 mg
Vitamin B_3 (niacin)	Commonly deficient in elderly	18 mg
Pantothenic acid (vitamin B_5)	Average elderly diet contains 60% of RDA	6 mg
Vitamin B_6 (pyridoxine)	71% of male and 90% of female diets deficient	2 mg
Vitamin B_{12}	Serum levels low in 25% of hospital patients	1 mcg
Biotin	Uncommon	150 mcg
Choline and Inositol		600 mg*
Folic acid	Average diet contains 60% of RDA	200 mcg
Para-aminobenzoic acid		400 mcg
Vitamin C	20 to 50% of diets deficient	60 mg
Vitamin D	62% of elderly women's diets deficient	5 mcg
Vitamin E	23% of male and 15% of female diets deficient	10 mg
Vitamin K	Deficiency in pregnant women and newborns common	Not established
Essential fatty acids (EFAs)	Very common	As directed
Minerals		
Calcium	Average diet contains 40 to 50% of RDA	800 mg
Chromium	90% of diets deficient	200 mcg*
Copper	75% of diets deficient	1-2 mg*
Iodine	Uncommon as supplementation of salt with iodine	150 mcg
Iron	Most common mineral deficiency	14 mg
Magnesium	75 to 85% of diets deficient: average diet contains 50 to 60% of RDA	300 mg
Manganese	Unknown, normally noticed in women	5 mg*
Potassium	Commonly deficient in elderly	3000 mg*
Selenium	Average diet contains 50% of RDA	75 mcg*
Zinc	68% of diets deficient	15 mg
Phosphorus		800 mg
Sodium		1600 mg
Optional Supplements		
Coenzyme Q10		30 mg
Garlic		As directed.

* RDA not established but level is considered to be useful/safe

Carotenoids

It consists of bright yellow, orange and red plant pigments found in vegetables such as oranges, tomatoes, parsley, pink grapefruit, spinach and red palm oil. We even find carotenoids lending bright colors to animals, flamingos owe their color to carotenoids, as do shellfish, egg yolks are yellow because of carotenoids that protect the unsaturated fats in the yolk.

There are more than 600 naturally-occurring carotenoids which seem to offer protection against lung, colorectal, breast, uterine and prostate cancers, enhance immune response and protect skin cells against UV radiation.

Limonoids

It is found in citrus fruit peels, appears to be specifically directed to protection of lung tissue, induction of both Phase I and Phase II detoxification enzymes in the liver.

Phytosterols

It is present in most plant species. Although green and yellow vegetables contain significant amounts, the seeds of pumpkins, yams, soy, rice and herbs have very good amount of phytosterols. It competes with dietary cholesterol for uptake in the intestines. It has ability to block the uptake of cholesterol. It also blocks the development of tumors in colon, breast and prostate glands.

Phenols

These are contained in blue, blue-red and violet colorations seen in berries, grapes and purple egg plant. The excellent feature of phenols is their ability to block specific enzymes that cause inflammation. They also modify the prostaglandin pathways and thereby protect platelets from clumping.

Flavonoids

These enhance the effects of ascorbate-vitamin C. Flavonoids were once lumped together as vitamin P. But there were well over 1,500 of them found in grapefruit, rutin, citrus fruits, silybin, milk thistle, ginkgo, buckwheat.

It acts against allergies, inflammation, free radicals, hepatotoxins, platelet aggregation, microbes, ulcers, viruses and tumors. Flavonoids also inhibit specific enzymes.

Anthocyanidins

Known as “flavonals,” they provide cross links that connect and strengthen the intertwined strands of collagen protein hence strengthening soft tissues, tendons, ligaments and bone matrix. Its great tensile strength depends on preservation of its cross links.

It neutralizes free radicals encountered in tissue fluids. This is a very important property and is beneficial for athletes and other sportsmen as heavy exercise generates large amounts of free radicals.

Catechins, Gallic Acids, Isoflavones, Thiols, Glucosinolates, Allylic Sulfides, Indoles, Isoprenoids, Tocotrienols and Tocopherols, Lipoic Acid and Ubiquinone are other important phytonutrients

Since there are thousands of phytonutrients, discussing all these are beyond the scope of this chapter. In fact, we just wanted to tell you the importance of phytonutrients in our life.

LIFESTYLE DISEASES AND NUTRIENTS

In the 1930s and 1940s, some scientific and clinical evidence suggested that there might be beneficial uses of vitamins C, E and B_3 in large doses. Beginning of 1930s, the Shutes in Canada developed a megadose vitamin E therapy for cardiovascular and circulatory complaints, naming it the Shute protocol. Thereafter curing the diseases with right nutrients is emerging as very strong subject of research.

While living in China in the early 1920s, Carl Rehnborg noticed the relation between malnutrition and disease. An important connection between the foods eaten and overall health of people. He observed that in farming communities, where fresh fruits and vegetables were plentiful, people were much healthier than people in the cities, where diets included far more sugar, salt, and fat. He also noticed that there were many more instances of critical diseases, such as scurvy and beriberi. Subsequently, during political unrest in Shanghai, Carl was isolated in a protected enclave. There, he conceived the idea of using nutritional elements missing from the typical diet as a way to improve overall health. Carl supplemented his meager diet with soups from what was available: local herbs, grasses, vegetables, along with rusty nails (for iron), lime stone and ground-up animal bones (for calcium). He shared his soup with a few friends, although the flavor was far from tasty.

Months later, Carl and his friends, who ate the nutrient rich soup emerged from their enclave much healthier than those who had eaten only army rations. Returning to the United States, Carl set up a small laboratory on California's Balboa Island, where, over the next six years he conducted experiments using different plant varieties.

He developed a way to dry these plants while still preserving their important nutrients along with phytonutrients. He created a formulation of alfalfa, watercress, and parsley, which became the basis of his first product. Later on he established a company named Nutrilite.

What I personally feel is that our body is a very complicated system, which is not easy to understand. It has millions of sensors, feedback circuits and rectifying mechanisms. If given the right nutrients it has ability to fight back with any disorder. As most of disorders specially lifestyle disorders are associated with our unhealthy habits of eating, excercise, rest and thinking.

Few important lifestyle diseases are discussed below very briefly as detailed discussions are beyond the scope of this chapter. This is just to give an idea about prevention and cure by nutrients.

Heart disease

World Health Organisation estimates that 60 percent of cardiac patients in the world will be Indians by 2010. That is the biggest challenge in front of us. Whenever one thinks about heart disease the first thing, which strikes the mind, is cholesterol. The recent research has shown that cholesterol is not the major culprit but the main reason is inflammation of blood vessels. As it has been found that in US more than half of the heart attack patients are having normal cholesterol. Instead of concentrating on lowering the level of cholesterol, we should take necessary steps to decrease the cause of inflammation thereby preventing and even reversal of heart disease.

There are two type of cholesterol i.e. High Density Lipoproteins (HDL) and Low Density Lipoproteins (LDL). Normally LDL is considered to be bad cholesterol. In fact this is the component very essential for building good cell membrane, other cell parts and different hormones. So what is bad in it? In fact nothing except that we have changed our lifestyle very drastically and free radicals oxidize LDL and convert it into bad one, which is sticky in nature. In 1989, Dr. Daniel Steinberg published in *New England Journal of Medicine* that if we take enough antioxidants we could avoid LDL becoming bad cholesterol. There is another factor responsible for inflammation. This is something called homocysteine.

It is estimated that an elevated homocysteine level in blood is responsible for approximately 15 percent of all heart attacks and strokes in the world today. One must learn more about this major killer especially, when one comes to know that this can be corrected by supplementing our diet with vitamin B complex.

So we can expect that heart problem can be treated with antioxidants (vitamin C, vitamin E, CoQ10, grape seed extract) and vitamin B complex along with good multivitamin, multimineral supplement and protein consisting of essential amino acids.

Diabetes

Another dominating disease in Indian population is diabetes. It is called slow killer as it gradually induces the secondary complications. Astonishing fact is that many of us do not know if we are diabetic. There are two types of diabetes, type I and II. Type-I diabetes normally occurs in children and is an autoimmune disorder, where pancreas stops making insulin so patient has to take insulin from outside. But most of the cases are of type-II in which person develops insulin resistance. A physician and professor of Stanford University, Dr. Gerald Reavens described a group of problems that have a common cause, the insulin resistance; he termed it as syndrome X.

To maintain proper sugar level in the diabetic patients, our body system needs higher levels of insulin in the blood, which give rise to syndrome X problems i.e. inflammation of artery, increased blood pressure, triglycerides, LDL and HDL, increased tendency to form blood clots and uncontrolled weight gain specially around the middle.

Syndrome X (the insulin resistance): Mostly, we do not get our insulin level checked, so we do not come to know if we are heading towards syndrome X. On the other hand, every one gets routine blood tests, which includes lipid profile. We are familiar with the ratio obtained by dividing total cholesterol by HDL cholesterol. But if we divide triglyceride level by HDL cholesterol we get a ratio, which indicates if we are heading towards syndrome X. If this ratio is greater than two we should understand that we have taken first step towards a cluster of serious diseases and are heading towards diabetes.

Preventing and treating diabetes: Diabetes is a lifestyle disorder, which is developed due to some nutrient deficiencies in our diet. When we start medications and reorganize our diet habits our diet becomes further nutrition-deficient which means we add up more deficiencies to our system aggravating our problem further. Thereby we become totally dependent on medications only. Hence weakening our system further and giving arise to secondary complications. If we start taking a good quality organic vitamin and mineral tablets, sugarless protein with nine essential amino acids along with antioxidants, chromium and magnesium we can start rejuvenating our system and can very effectively fight with this disorder and rectify it.

CONCLUSION

In the present scenario the saying an apple a day keeps your doctor away has lost its meaning, today you have to supplement your apple and rest of your balanced diet with high quality nutritional supplements. Remember that nutritional supplements are about health, not disease. Although we can come out of disease with nutritional supplements, but all of us know that it is easier to maintain health than to regain it.

If you provide all these nutrients to your body at optimal levels, your homocysteine level will decrease, your oxidation stress will decrease resulting in oxidation resistant LDL, protection to lungs, enhanced immune system, decreased risk of heart disease, stroke, cancer, cataract, arthritis, asthma, diabetes and many more.

Remember, this toxic world combined with stressful life make it essential that our immune system and antioxidant defence system be kept at its best all the time.

Optimal health compels us to choose every day to be our best – by selecting the balanced diet and supplements, engaging in exercise, proper rest, by incorporating healthy habits and balance in everything we do.

SUGGESTED READINGS

1. Death by Prescription by Ray D. Strand, Nelson, 2003
2. Lack of Association of Increased C-Reactive Protein and Total Plasma Homocysteine. Auer J.W., Berent R, Elier B. Circulation, 104 (25) 2001.
3. Nutrilite.com
4. One-Third of Pregnant and Lactating Women May Not Be Meeting Their Folate Requirements from Diet Alone Based on Mandated Levels of Folic Acid Fortification. Johann W. Auer, et el, J. Nutr. 136:2827-2833, November 2006.
5. Quest for nutrition, published by Nutrilite.
6. Role of Diacetyl Metabolite in Alcohol Toxicity and Addiction via Electron Transfer and Oxidative Stress, P. Kovacic and A.L. Cooksy. *Arch. Toxicol.* 79, 123-128, 2005.
7. The evolution of free radicals and oxidative stress: McCord J.M., The American Journal of Medicine, Volume 108, Number 8, p. 652-659(8), June 2000.
8. The Wellness Revolution: How to Make a Fortune in the Next Trillion Dollar Industry by Paul Zane Pilzer, Wiley, 2002.
9. Vitamins and Minerals: How to get nutrients your body needs by Sara Rose, Hamlyn, 2003.
10. What your doctor does not know about nutritional supplements may be killing you by Ray D. Strand, Nelson Books, 2002.

11

Genetically Engineered Foods and Health

Anil Kumar, Richa Goyal and Anant Singh

INTRODUCTION

The world population has topped to 6.5 billion and is predicted to double in the next 50 years. Worldwide, over 58 million hectares of land is used for the commercial production of genetically modified (GM) crops. The area under agriculture is greatest in United States i.e. 66% of total global area, followed by Argentina (23%) and Canada (6%). These countries are the major producers of GM products, but other Asian and European countries such as China, Australia, Mexico, India and Spain, France and Germany respectively are also following suit. The main crops planted per million hectares are soybean (62%), maize (21%) and cotton (12%).

Ensuring an adequate food supply for a booming population is going to be a major challenge in the years to come, particularly for the developing countries like India, where resources are limited or inadequate. Apart from increasing population, the nutritional requirements of people are also on the rise. During recent years, there has been a considerable advancement in modern biotechnological tools and techniques for the production of foods, feeds and drugs. Genetically modified (GM) products are one of the examples of modern biotechnology advancement in areas of Science and Technology. Genetically modified products are those foods or organism in which the hereditary genetic traits have been modified by adding other genetic materials (isolate single genes of known function from one organism and transfer a copy of that gene to a new host to introduce desirable characteristics). Genetically modified foods have been developed to solve the world's hunger and malnutrition problems. With the emergence of transgenic technologies, new ways to improve the agronomic performance of crops for food, feed, and processing applications have been devised. In addition, ability to express foreign genes using transgenic technologies has opened up options for producing large quantities of commercially important industrial or pharmaceutical products in plants. The gene of interest which is transferred is expected to impart desirable characteristics for improving better health (medicine, vaccines), better products (modified crops, fruits, vegetables) and better environment. Although, their safety and toxicities need to be evaluated. Genetically modified foods include tomatoes, squash, yeast, corn, potatoes, and soybeans (which are used in 60 % of all processed foods, such as bread, pasta, candies, ice cream, pies, biscuits, margarine, meat products and vegetarian meat substitutes). Research is undergoing in its full momentum to develop banana, tomato with vaccine against cholera and rabies respectively. Cures even for diabetes and cancer have been attempted with encouraging results through the production of immune proteins in plants. Besides these global developments, cultivation of GM crops for both food and industrial purposes are most important in the Indian context. GM foods promise to meet the ever-increasing needs in a number of ways such as increased production and productivity, improved food quality, possible reduction in herbicide application, possible reduction in the use of chemical substances, nutrition, pharmaceutical, phytomedication pest resistance, herbicide tolerance, disease resistance, and cold tolerance etc. On the other hand, GM products are also proposed to have their own limitations and consequences. Despite this high adoption rate and future promises, there is a multitude of concerns about the impact of genetically modified (GM) crops on the environment. Potential contamination of the environment and food chains has prompted detailed consideration of how such crops and the molecules that they produce can be effectively isolated and contained. One of the reasonable steps after creating a transgenic plant is to evaluate its potential benefits and risks to the environment and these should be compared to those generated by traditional agricultural practices. The precautionary approach in risk management of GM plants may make it necessary to monitor significant wild and weed populations that might be affected by transgene escape. Effective risk assessment and monitoring mechanisms are the basic prerequisites of any legal framework to adequately address the risks and watch out for new risks. Several agencies in different countries monitor the release of GM organisms or frame guidelines for the appropriate application of recombinant organisms in agro-industries so as to assure the safe use of recombinant organisms and to achieve sound overall development. It is also desirable to establish an internationally harmonized framework for the safe handling of recombinant DNA organisms within a few years. Guidelines for marketing of GM products are not stringent and their safety studies are still inadequate. Therefore the present article will highlight GM products and their related aspects such as their advantages, limitations, regulation and their impact on health and environment.

PRINCIPLES OF GENETICS

Genetic engineering (GE) is a new technology that involves the direct transplantation or artificial manipulation of the gene or genetic material of a plant, animal, or bacteria, or the transfer of foreign genes into the DNA of an organism, which results in the creation of a transgenic organism or genetic food products. These genetic manipulations modify the genetic code of the recipient permanently. GM products contain altered genes that do not exist in nature, and possess new qualities or characteristics that are passed onto their offspring through heredity. Unlike traditional selective breeding techniques used from centuries, genetic engineering allows researchers or farmers to cross species boundaries. Never before was it possible to transfer animal genes or synthetic genes to plants. Various steps involved in developing a genetically modified organism (GMO) are described as (Fig. 1).

Identify gene of interest: The first step is to identify a particular characteristic from any organism (for example, a plant, animal or microorganism) and determine which gene or genes in that organism

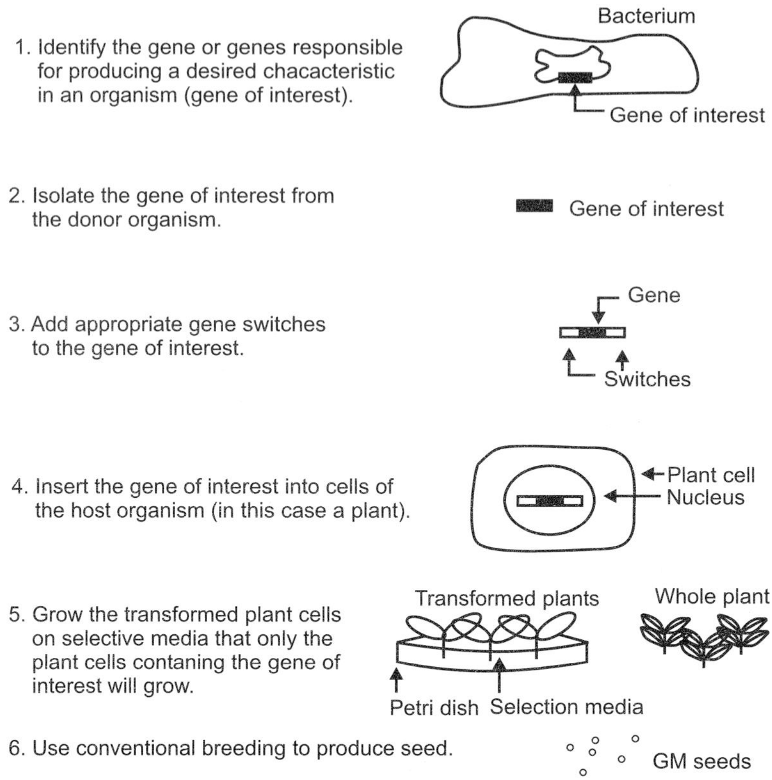

Fig. 1. Principles of Genetic Engineering.

are responsible for producing that characteristic. Such a gene which is responsible for the desired characteristics is termed as "gene of interest".

Isolate the gene of interest from the donor organism: For developing a GMO, appropriate gene switches are added to the gene of interest, to allow it to function in the cells of the recipient organism.

Modifying the gene of interest developing a genetically modified organism: Molecular biology techniques are employed to isolate and copy the required gene. (If the gene of interest was found in a bacterium it will have 'on switch' (and other switches) that function only in bacterial cells. For a gene to function in a different type of cell (for example, in a plant cell), the bacterial switches need to be removed and different switches are added which allow it to function in a plant cell. These switches may function in all plant cells or in specific plant tissues (for example, in root tissue only), so that the new trait will only be switched on in the roots.)

Insert the gene of interest into cells of host organism: In this step, gene of interest, together with its gene switches, is inserted into the cells of the host organism, using molecular biology techniques. Cells with the new gene inserted into their genome are referred to as 'transformed' cells. They contain all the usual genes of the recipient organism as well as the new gene or genes.

Growing of transformed cells in nutrient media: In the case of GM plants, Step 5 is to grow transformed cells into whole plants in the laboratory; this is done using selective nutrient-rich media. The plants are examined to ensure that they have the desired physical characteristic conferred by the new gene.

Using conventional breeding to produce seed: The genetically modified plants are bred with conventional plants of the same variety to produce seed for further testing and possibly for future commercial use. The entire process from the initial gene selection to commercial production can take upto ten years or more.

INDIAN GM PRODUCTS

Genetically modified (GM) products are as summarized in Table 1, will soon become a reality in the country as the field trials for several such products are being carried out and toxicity studies involving animals are in progress. In all possibilities, *GM cotton* would be the first product available for commercial use. Field trials in India suggest that GM crops have more benefits when compared to the conventional crops. GM cotton would be the first product available for commerical use. Results of field trials in India show that GM cotton is insect resistant, high yielding and requires less pesticide when compared to non-bioengineered crops. It is also likely to be the first GM food as its oil would be available for consumption. Fibre of the GM cotton would be used for making fabric and seeds would be used for extracting oil. Products are likely to be commercialized this year after getting approval from Genetic Engineering Approval Committee (GEAC). However, **GM mustard**, which has been produced by Belgium-based Pro Agro Plant Genetic System (Pro Agro PGS) still requires field trials. Preliminary tests on animals showed that GM mustard is safe and free from any acute toxicity. However in spite of controversial introduction of GM cotton (known as Bt cotton *Bacillus thuringiensis*), the Bt craze has "caught up" with Indian farmers very quickly.

Another product in the pipeline is **GM potato**, which has been developed at Jawaharlal Nehru University (JNU) in New Delhi and has increased protein content. JNU in collaboration with Indian Council of Agriculture Research are conducting field trials on GM potato, while animal studies are under way. The commercial growing of a genetically modified potato, which contains nutrients lacking in the diets of may of the poorest, is expected to be approved in India within few months. Experts say malnutrition affects huge number of people in India. The present proposed GM potato contains a third more proteins than normal, including essential high-quality nutrients, and has been created by adding a gene from the protein-rich amaranth plant. But critics describe the plan as risky, naive and a propaganda tool to promote the merits of GM food in India. GM potato is in its final stage of regulatory approval. It is further expected that GM potato would be used against malnutrition, which causes blindness due to vitamin A deficiency. This technological product has the advantages of improved yield, better quality food and larger quantities.

At the Indian Agricultural Research Institute and the Hyderabad-based ICRISAT (International Crops Research Institute for Semi-Arid Tropics) work is going on for modifying rice and wheat where the effort is to genetically engineer vitamin-enriched (say, Vit B_{12} and Vit A) rice and wheat, said sources. However, so far the work in the country on GMF is only experimental and contained,

Table 1. Genetically modified products and their features and status

GM product	Beneficial features	Status
GM apple	Stay fresh for long time	Used in national and international markets
GM cotton	Expected to resist insects, reduced pesticide use and increased yield compared with non-bioengineer crops	Under field trials
GM mustard	Seed would be used for extracting oil, which is consumed	Under field trials
GM potato	1. Increased protein content 2. Expected to provide more nutrients that lack in the diets 3. Expected that GM potato would be used against malnutrition, which causes blindness due to vitamin A deficiency	GM potato is in its final stage of regulatory approval in India
GM tomato	Increased shelf life	Under field trials
GM rice and wheat	Expected to be vitamin enriched (Vit B_{12} & Vit A)	Under field trials
GM tomato, brinjal, cauliflower, cabbage and paddy	Shall kill worms, which attack cotton	Under field trials
GM organism	Resistant against diseases, environment	Being used in both agriculture as well as experimental studies

added the source. This should be read in the background of the hot global debate on GMFs owing to the personal nature of food coupled with apprehension of newer technologies and varied results.

Besides, other products such as GM tomato, brinjal, cauliflower, cabbage and paddy are to have an increased shelf-life under investigational phase at Indian Agricultural Research Institute (IARI). According to Dr Kailash Bansal (Principal Scientist at IARI's Department of Biotechnology), GM tomato, which was developed it by a team led by Prof V L Chopra, has proven to increase shelf life. Further she claimed that field trial had shown favorable results and further trials are yet to begin. However, toxicity studies have not begun. Besides field trials, GM crops which are under investigations at IARI include GM brinjal, rice and cabbage, all of which had Bt gene (responsible for killing worms which attack cotton). Bansal said a salt and drought tolerant GM mustard had been developed by introducing an imported gene. If the experimental works going on at the Indian institutes is fruitful and the government finds the products safe, the country will witness the launch of GMF in near future. Future GM products are projected to include crops with pharmaceutical benefits, vitamin-enhanced grains, and those with amino acids and other nutritional features that are better suited to human needs than current varieties. Crops with increased drought and extreme-temperature tolerance, and resistance to a variety of pests and diseases also are envisioned to be developed.

ADVANTAGES

GM products are expected to provide following advantages:

- **Herbicide tolerance:** Crops such as soya beans, maize, oilseed and cotton are made resistant to a company's broad-spectrum herbicide (e.g. Monsanto's glyphosate, AgrEvo's glufosinate ammonium or Rhone-Poulenc's bromoxynil). Use of herbicide tolerant crops will improve the effectiveness of weed control. This may further reduce populations of a number of, already rare, weed species and reduce food for the wildlife that depends on weeds within cropped areas.
- **Insect resistance and environment protection:** Crops modified with genes of bacterial origin which produce proteins toxic to insects but harmless to plants and humans would afford insect resistance. Farmers can also protect the environment by using less crop spray.
- **Altered ripening and freshness:** They can remain fresh longer. They can be allowed to ripen in the sun before picking resulting in better taste. Fruit (e.g. tomatoes) can be modified to allow it to ripen without softening.
- **Altered fertility:** To produce hybrid seed by conventional breeding or to cause the crop to die before it can pollinate.
- **Harvesting:** Producers have the advantage that all the tomatoes can be harvested simultaneously

As the range of plants being modified has expanded extensively, some of the main targets currently being developed that would provide advantage to the consumer include:

- Resistance to bacteria, viruses or fungi.
- Improving product quality (e.g. changing oil profiles, amino acid composition, carbohydrate metabolism, carotenoid content).
- Changing the agronomic properties of plants (e.g. improving growth rates, tolerance to cold, drought, stress, changing nitrogen metabolism, maturation rates, yield, etc.).

SAFETY CONCERN AND DANGER

In principle, an absolute safety of food, whether produced conventionally or from GM plants, cannot be guaranteed. It is common knowledge that conventionally produced food can be the cause of allergies for predisposed persons. Food of plant origin can also contain toxic or carcinogenic substances. Nature has provided plants with a large arsenal of defence substances as a protection against damage from feeding insects or from bacterial and fungal infections. Moreover, fungal toxins, a number of which are strongly carcinogenic, may contaminate plant products.

Although GM products have enormous potential to improve agriculture production yield, quality of food, nutritional values of food and health, but uncertainties also exist because there are limited scientific evidences regarding the health of consumers. Safety of GM products is not fully validated till now as these products are introduced in the market without adequate safety and toxicities testing. Besides methodologies used for assessing the risks are not robust enough or sensitive enough. Governing their marketing control and its rule and regulations are also not much stringent as compared to drugs and cosmetics. Molecular and genetic effects of the technology are also unpredictable in nature.

SAFETY EVALUATIONS

As a starting point, the safety assessment of a GM food involves a careful assessment of the following informations:

- The amounts of the GM food that people are likely to consume, including both average and extreme consumption.
- A detailed description of what the food is and how it is produced.
- A history of any possible adverse health effects linked to the organism being modified.
- A detailed description of the genetic modification process.
- An evaluation of any possible nutritional effects of the modified food.
- An evaluation of any toxicological effects of the modified food.
- An evaluation of any adverse microbiological effects of the modified food.
- An evaluation of any data on people eating the modified foods under controlled conditions.

ENVIRONMENTAL CONCERN

Public and scientists have expressed widespread concerns about the effect of GM foods on the environment, lack of consumer benefits, ethical issues and the perception that a few large multinational corporations will be the prime beneficiaries and would dictate world markets. Society is recognizing side effects such as nuclear pollution, global warming and the toxic effect of pesticides and herbicides. Genetic engineered product poses the greatest danger of any technology yet introduced. Many of the damaging effects of genetic engineering products are irreversible and uncontrollable. A number of concerns have been expressed (Table 2) about genetic modification. These include direct effects on the ecosystem and indirect effects, which might arise as a result of changing farming practices. Some of these concerns are outlined below.

Possible direct effects on ecosystem: Predators eating the GM crop might be damaged, something that could have serious consequences in the case of beneficial predators, such as ladybirds, which eat agricultural pests.

Possible indirect effects: Genetic modification may permit changes to the locations, altitudes and climates within which crops can be grown with consequent destruction of uncultivated areas that are currently important for wildlife. Much of the biodiversity that is associated with traditional agriculture depends on the combination of habitats resulting from particular cropping patterns. There is concern that GM agriculture may result in changes to these cropping patterns and the possible loss of wildlife habitats.

The precautionary approach is essential if we have to protect ourselves and other generations to come. Environmental activists, religious organizations, public interest groups, professional associations and other scientists and government officials have all raised concerns about GM foods, and criticized agribusiness for pursuing profit without concern for potential hazards, and the government for failing to exercise adequate regulatory oversight. It seems that everyone has a strong opinion about GM foods. Most concerns about GM foods include unintended harm to other organisms, reduced effectiveness of pesticides, gene transfer to non-target species.

Other reasons for objections to GM crops come from persons concerned about environmental implications. For many years, U.S. agriculture has focused on "integrated crop management" where pest management practices are used selectively, depending on the need. Many times, for example, chemical control of corn borers is not used because it is not needed. In some years, spraying for corn borers may be done in some but not all fields and only in geographic areas where insect infestation is severe. The Bt corn, in contrast, is visualized as being used every year, with concern that it might increase the risk that a Bt-resistant borer would develop. Major concerns about the risks of this new technology are:

1. Genetically engineered food crops may harm wildlife and change habitats. Unanticipated allergic responses to novel substances in foods, the spread of pest resistance or herbicide tolerance to wild plants, inadvertent toxicity to benign wildlife, and increasing control of agriculture by biotechnology corporations.
2. Uncharacterized genetic material and gene products may be added to foods.
3. Genes from the genetically modified crop could be transfered to the pests. The pests then become resistant to the crop spray and the crop spraying becomes useless.
4. Plant can pollinate weeds.

Effects of GMOs on the environment vary depending on the receiving organism, the type of genes introduced through genetic engineering, environmental conditions and the type of application. The potential long-term effects are especially difficult to assess because genes, as opposed to chemicals, may reproduce in the natural environment. Potential effects can be expected from gene dispersal (e.g. out crossing), the establishment of wild populations of crop plants and the displacement of natural populations. Besides, a toxic effect that is intended for a pest organism might be harmful to other organisms in the ecosystem. Through changes in agricultural practices, indirect effects (e.g. on the application of pesticides, the development of resistance) are also possible. For example, the ways and means of applying a herbicide has an influence on the weeds and thus the insect populations in a field. By feeding on these insects, birds and vertebrates may then, although sometimes with a delay, be indirectly affected.

On the other side, a number of environmental benefits are potentially associated with genetic modification, these include:

1. Genetic modification for herbicide tolerance, and the replacement of a cocktail of selective herbicides with a single application of a broad-spectrum herbicide may have the following environmental benefits: reduced persistence in the environment of the chemical used, reductions in the total amount of chemical applied, reductions in the number of passages over the land by a machine and therefore reductions in soil compaction and chemical run-off.
2. Increased yields from existing agricultural land may reduce the demand for more agricultural land. This could lessen destruction of semi-natural habitats and increase potential for agricultural land to be allowed to revert to its seminatural state.
3. The introduction of insect and disease resistant crops would reduce the use of chemicals to control these problems.
4. Reductions in fossil fuel usage as a result of reduced passages over crops and reduced transport of chemicals.
5. It has also been suggested that careful management of the genetic modification to take them into account could overcome many of the concerns of environmental groups.

Table 2. Effect of GM products on health and environment

Impacts on Environment	Consequence
• **Nuclear pollution, global warming** • Toxic effect of pesticides and herbicides • Unintended harm to other organisms • Reduced effectiveness of pesticides • Incorporation of uncharacterized genetic material and gene products to foods • Gene transfer to non-target species • GM food crops may harm wildlife and change habitats • Unanticipated allergic responses to novel substances • Spread of pest resistance or herbicide tolerance to wild plants **Direct** • Damaged predators eating the GM crop **Indirect** • Changes to the locations, altitudes and climates • GM agriculture may result in changes to these cropping patterns and the possible loss of wildlife habitats	• Damaging effects of genetic engineering products are irreversible and uncontrollable • Serious consequences in the case of beneficial predators, such as ladybirds, which eat agricultural pests • Destruction of uncultivated areas that are currently important for wildlife

HUMAN HEALTH CONCERN

GM food may not be safe to eat. A recent study from CSIRO, Australia showed that when a gene for pest resistance from beans was genetically engineered into peas, the GM peas provoked inflammation of the lung tissue and allergic reactions in mice. A secret study conducted by the US food giant Monsanto came to light a few months ago. The study showed that rats fed on a diet of GM corn resulted in severe organ damage. The rats had developed lesions in important organs like kidneys and liver, and blood cells were so affected that the immune system collapsed. Some years ago it was found that rats fed with GM potatoes showed similar organ malformations and altered immune response. There are numerous other reports of stomach lesions in rats, false pregnancies in cows, uncontrolled cell growth and damage to animal immune system following feeding studies conducted with GM foods. Experts think that because of genetic engineering and the shuffling of foreign genes into natural crops, the genetically engineered plants may start producing new kinds of products in their cells. These new products could be poisonous, produce allergies or damage tissues to create lesions or destroy the immune system. When the first genetically altered tomato "Flavr Savr" was fed to rodents in the labs in 1994, they refused to eat the GM tomatoes. Data revealed that many of the rats that ate the GM tomato developed lesions in their stomachs. Seven of the forty rats that were fed with GM tomatoes died unexpectedly within two weeks. Bt Cotton is a genetically modified cotton plant. It contains the Bt gene. The Bt gene is taken from a soil bacterium *Bacillus thuringiensis*, it triggers production of a poison within the plant that kills the insect pest attacking the plant; the Bt plant in other words, is a pesticide producing plant.

All foods, including those from GM plants, pose the same types of inherent risks to human health. They can cause allergic or toxic reactions, or they can block the absorption of nutrients.

Some foods from GM plants have contained allergens, toxins and antinutrients. The power of genetic modification techniques raises possible hazards to human health and environment, giving way to economic problems. This includes:

- Allergenicity, unknown effects on human health.
- Loss of biodiversity, economic concerns and aggravation of prosperity gap.
- Possibly new toxicants may be added.
- Nutritional quality of engineered food may be diminished.
- New substances may significantly alter the composition of food.
- New proteins that cause allergic reactions may enter the food supply.
- Genetically engineered food may cause unexpected effects.
- Many studies have shown that DNA does not always fully break down in the digestive system. Gut bacteria can take up genes and fragments of DNA, which could potentially lead to the spread of antibiotic resistance.
- Crude and unpredictable nature of genetic engineering techniques could lead to "the biosynthesis of molecules that are toxic, allergenic or carcinogenic. GM (genetically modified) food is not a safe option, given our current lack of understanding of the consequences of recombinant technology".
- Long-term animal feeding studies required testing for possible toxic and reproductive effects.

LABELING OF GM FOODS

The safety of a GM food is assessed before it can be sold as food; therefore, the purpose of labeling is simply to provide information to consumers, allowing them to purchase or avoid GM foods depending on their own views and beliefs. The GM food labeling regulations, which are among the most stringent in the world, represent a balance between the needs of consumers and what governments can realistically enforce. If a food, food ingredient, additive or processing aid contains novel DNA or protein that has come from an approved GM food, it must be labeled with the words 'Genetically Modified'. Labeling regulations also cover GM foods that differ in composition to the conventional counterpart; in this case, the labeling must make clear to consumers the altered characteristics of the food. There are some exemptions to the labeling requirements for GM foods. Foods that do not need to be labeled as 'genetically modified' include highly refined foods that contain no DNA or protein (for example, oil made from GM soy beans), and foods in which GM ingredients are present accidentally and make up less than 1% of the final food.

REGULATIONS

The following are the existing guidelines in India for transgenic crops:

1. The EPA Act 1986 and Rules1989 of Ministry of Environment and Forests, namely which deals with rules and procedures for handling GMOs and hazardous organisms. The Genetic Engineering Approval Committee (GEAC) established by the Ministry acts as a statutory body for review and approval from environmental angle of activities involving large scale use of GMOs and their products in R&D, industrial production, environmental release and

field application. Government. Issues for action includes review and control, and monitoring of large-scale use of GMOs in R&D and industrial production, environmental release and experimental field trials.

2. The Review Committee on Genetic Manipulation (RCGM) under the **Department of Biotechnology** (DBT), Ministry of Science and Technology, monitors the safety related aspects of ongoing research projects involving GMOs. It brings out manuals of guidelines specifying procedures for regulatory process, activities involving GMOs in research, use and application from environmental safety angle (Recombinant DNA Safety Guidelines 1992 and Revised Guidelines for Research in Transgenic Plants 1998).
3. In addition there is a provision of State Biotechnology Coordination Committee (SBCC) and District Level Committee (DLC).

LIMITATIONS OF THE REGULATORY SYSTEM

1. Lack of adequate standards for risk assessment.
2. The procedure and information required for a full environment risk.
3. Assessment and appropriate safety and emergency responses not laid down.
4. Instructions and conditions for use on labeling and packaging of products.
5. Containing GMOs yet to be specified.
6. Detailed safeguards as embodied in the Cartagena Protocol yet to be incorporated.
7. Lack of infrastructure for risk assessment.
8. Shortage of skilled personnel from laboratory researchers to extension service officers.

Patent—The US Government allows crops to be patented but the Indian law does not allow patents on plants or plant parts like cells or genes. An Indian law called the Protection of Plant Varieties and Farmers Rights Act, grants Plant Breeder's Rights to the developer of a new plant variety, including a GM variety. The Indian law acknowledges both Farmer's and Breeder's Rights.

CONCLUSION

Genetically modified foods have the potential to solve many of the world's hunger and malnutrition problems, and to help protect and preserve the environment by increasing yield and reducing reliance upon chemical pesticides and herbicides. Yet there are many challenges ahead for governments, especially in the areas of safety testing, regulation, international policy and food labeling. Many people feel that genetic engineering is the inevitable wave of the future and that we cannot afford to ignore a technology that has such enormous potential benefits. However, we must proceed with caution to avoid causing unintended harm to human health and the environment as a result of our enthusiasm for this powerful technology.

SUGGESTED READINGS

1. Singh OV, Ghai S, Paul D, Jain RK. Genetically modified crops: success, safety assessment, and public concern. *Appl Microbial Biotechnol.* 71(5): 598–607, 2006.

2. Schubbert R, Lettmann C and Doerfler, W. Ingested foreign (phage M13) DNA survives transiently in the gastrointestinal tract and enters the blood stream of mice. *Molecules, Genes and Genetics 242*, 495–504, 1994.
3. Schubbert R, Hohlweg U, Renz D and Doerfler W. On the fate of orally ingested foreign DNA in mice: chromosomal association and placental transmission in the fetus. *Molecules, Genes and Genetics* 259, 569–576, 1998.
4. Schubert, D. A different perspective on GM food," *Nature Biotechnology*, 20, 969, 2002.
5. The Safety of Genetically Modified Foods Produced through Biotechnology. *Toxicological Sciences*, 71, 2–8, 2003.
6. GM Foods. Safety assessment of genetically modified foods. Food Standards Australia New Zealand, 2005.
7. Jefferson V. The ethical dilemma of genetically modified food. *J Environ Health.* 69(1), 33–4, 2006.
8. Bhalla PL. Genetic engineering of wheat—current challenges and opportunities. *Trends Biotechnol* .24 (7): 305–11, 2006.

12

Bioaerosol: Impact on Human Health

Geeta Shukla and D.V. Rai

INTRODUCTION

Bioaerosols are defined as aerosols or particulate matter of microbial, plant or animal origin. Bioaerosols or organic dust (Synonymous of Bioaerosol) may comprise of either pathogenic or non-pathogenic live or dead matter e.g. bacteria, fungi, algae, viruses, protozoa, high molecular weight allergens (HMW), peptidoglycans, β $(1 \rightarrow 3)$ glucans, enzymes, pollen, plant fibres, gases, vapors, excreta or fragments of insects, skin scales or hair of mammals, or other components, residues or products of microorganisms e.g. bacterial endotoxins or fungal mycotoxins that are present in air. Bioaerosols are present everywhere. Microorganisms have long played an important role in human ecology. Normally bioaerosols pose no threat to humans but some of them can have a wide range of adverse effects on human health resulting into infectious diseases, acute toxic effects, allergies and cancer. Respiratory symptoms and lung function impairment are the most common bioaerosol-associated health effects. With advancement in industrial technology and activities, humans have modified the environment via cultivation, landscaping and building structures, which in turn disturbed the ecological balance. The distributions of moisture and nutrients has been altered to a point where it is quite common to encounter reservoirs of bacteria, fungi, algae, protozoa, insects of mites and cockroaches. Humans confront components of microorganisms continuously through airborne dispersion, ingestion or direct contact. However, their influence on individual depends on genetic polymorphisms. Predisposed individuals may not experience the severity of reactions immediately after the exposure to particular bioaerosol because of prior sensitization. Once an individual is sensitized, e.g. asthmatic attack might be delayed hours following the exposure event due to preformed antibodies. However there are many infectious diseases caused by bacteria, fungi, protozoa and viruses that require no prior sensitization before illness develops.

GENERAL CHARACTERISTICS

Bioaerosols have specific characteristics as follows:

1. Single spores, pollen grains, bacterial cells, or viruses.

Table 1. Common bioaerosols, sources and diseases.

Bioaerosol Types	**Common Sources**	**Diseases**
Pollens, spores, plant parts	Plants, trees, grasses, ferns, harvesting, cutting, shiploading	Hay fever, allergic rhinoconjunctivitis, asthma upper airway irritation
Fungi	Plants material, skin, leather, oils; bird, bat and animal droppings; feathers, soil nutrients, glues, wool	Asthma, allergic diseases, Infection, Toxicosis, Tumours
Bacteria	Humans, birds and animals (e.g., saliva, blood, dental secretions, skin, vomit, urine, faeces), water sprays and surf, humidifiers, hot tubes, pools, drinking water, cooling towers	Endotoxicosis, Tuberculosis Pneumonia, respiratory and wound infections, legionellosis, Q and Pontiac fever
Other allergen sources	Mite excreta, insect parts (cockroach, spiders, moths, midge)	Asthma, Dermatitis
Athropods, Vertebrates	Dander and saliva from cats, dogs, rabbits, mice and rats, bird serum, farm animal dander	Hypersensitivity Pneumonitic
Viruses*	Infected humans, animal excreta, insect vectors, protozoa**	Respiratory infections, colds, measles, mumps, hepatitis A, influenza, chicken-pox, Hanta virus
Protozoa	Dust, air, food, water	Toxoplasmosis, primary meningoencephalitis, granulomatous amoebic encephalitis (GAE)

* Viruses are included in table for completeness but are not covered in this chapter.
** Protozoa in the form of free-living amoebae can be direct-acting pathogens or allergens; they can also interact with bacteria (e.g., Legionella growth within amoebae).
Source: Burge (1995)

2. Aggregates of several spores or cells.
3. Fragments of spores and cells or their metabolic products.
4. Biological material carried by other, non-biological particles.
5. Cells or particle size

Many bioaerosol particles, such as fungal spores and pollen, are designed by nature for transmission to other areas and to stay alive during the transmission. Therefore, this type of hardy bioaerosol is resistant to environmental stresses, such as shear forces acting on them during sampling. Vegetative cells of bacteria are easily damaged and their viability may be compromised by both environmental and sampling stresses. Some bioaerosol particles are bits and pieces of biological origin that are shed by an ecosystem. Bacteria emitted from aerated wastewater are residues of evaporated droplets. Many of the bacteria shed by humans, for example, are residues of the skin. They probably stay viable while airborne because they are adapted to the dry conditions of their environment and protected by their original substrate, i.e., the skin scale. Otherwise, most vegetative cells of bacteria are prone to damage as a result of becoming airborne (Table 1).

Microbial cells in an aerosol may be viable or nonviable. The definition of the viability of a cell is not explicit, but generally, viable cells are able to reproduce or they have metabolic activity. Nonviable organisms are unable to reproduce as they are dead. In addition, the nutritional requirements of many environmental microorganisms are not known and not all microbes can be grown on laboratory media. From a natural soil or water sample, for example, it is typically possible to culture less than 1% of the microbes present, and this may be true for airborne microbes as well. Bioaerosols that are not whole cells, such as endotoxins, mycotoxins, or various allergens, may also be present. Therefore, depending upon the detection method, the results are commonly expressed as colony-forming units (cfu). The results of chemical analyses, e.g., for endotoxins, are expressed as ng/m^3. The density of microbial cells is somewhat variable, depending upon the degree of cell hydration, the reserve materials and lipid content of the cell. They consist mainly of water, about 70% of their weight. The rest of the cell material consists of macromolecules, such as nucleic acids, proteins, lipids, carbohydrates and their combinations. The density of microbial cells has been reported to be 1.07–1.09 g/cm^3, 0.9–1.3 g/cm^3, 1.09–1.24 g/cm^3 or 1.5 g/cm^3. Bioaerosol particles cover a wide size range. Bacteria and fungal spores cover a size range from about 0.2 up to 100 nm. Pollen, algae, protozoa, and dander are several tens to hundreds of micrometers in diameter. When microbial cells or spores are carried by other materials or present as aggregates, their migration and deposition depend on the overall size of the whole unit (Table 1).

Although most of the bioaerosols are harmless constituents of normal environments, some bioaerosol particles may be infectious agents, allergens, irritant or may carry toxic metabolites. To be infectious, an organism must be viable, but to cause allergenic or toxic effects viability is not a prerequisite. Thus, dead cells as well as cell residues may affect human health.

Cell or particle size is an important factor in determining risks associated with microbial contamination. In general, the smaller the particle, the greater the risk. This relationship is due to the fact that smaller cells and spores become trapped within lung tissue and are not easily expelled. As previously mentioned, cell fragments can be many times smaller than the intact cell or spore and their smaller size poses even greater health risks.

Can bioaerosols cause health problems?

Bioaerosols pose great threat to human beings and can cause severe health problems. Some, like viruses and bacteria (Fig. 2), cause infections (like a cold or pneumonia). Others cause allergies. Both allergic responses and infections may be serious or even fatal. An allergic reaction occurs when a substance provokes formation of IgE antibodies in a susceptible person. Substances, which will cause an allergic reaction in some people are known as *antigens* or *allergens*. Bioaerosols may cause allergic reactions on the skin or in the respiratory tract (Fig. 1). Rashes, hay fever, asthma (tightness in the chest, difficulty in breathing), and running nose are common allergic reactions. A few people develop a severe allergic reaction in the lung, which can destroy lung tissue. This is called *hypersensitivity pneumonitis*. It is not an infection, but repeated episodes can lead to infections of the lung, such as bacterial pneumonia. Hypersensitivity pneumonitis can be triggered by exposure to very small amounts of the allergen, once a person is sensitive to it. Symptoms can range from tightness in the chest, cough, and difficulty in breathing, to low-grade fever, muscle aches, and headaches.

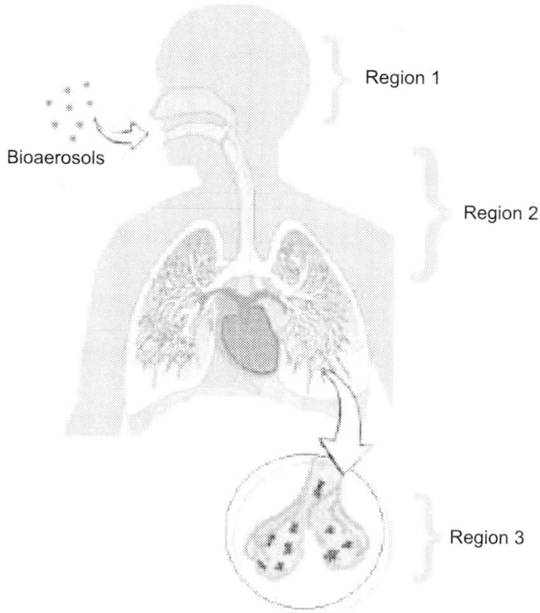

Fig. 1. (1) Nasopharynx or Head, Deposition of spores > 20 μm; (2) Larynx, Trachea, Bronchi, Deposition of spores < 20 μm; (3) Bronchioles and Alveoli, Deposition of spores < 1–4 μm.

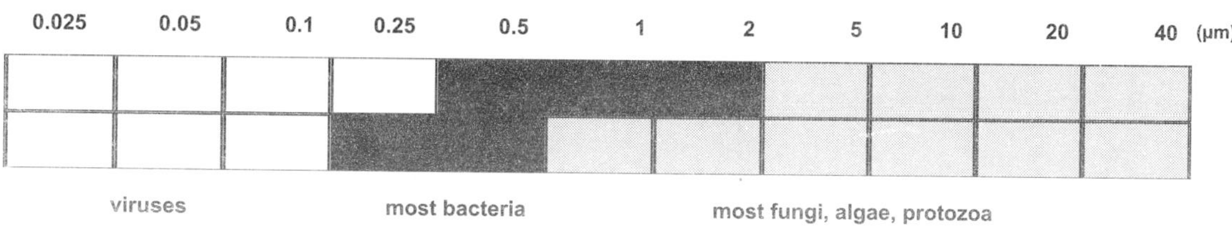

Fig. 2. Cell size in microns.

SOURCES OF BIOAEROSOLS

Bioaerosols may originate from almost any natural or man-made surface and each source may give rise to an entirely unique assemblage of bioaerosols. Many indoor bioaerosols originate outdoors. The surfaces of living and dead plants are probably the most important sources of airborne fungal spores and bacteria. Therefore, the mechanical movement of plants, as in many farming activities, or soil, e.g., farming or construction, generates bioaerosols together with other dust. Actinomycetes apparently become airborne from soil. All natural waters as well as anthropogenic waters, such as sewage lagoons and cooling water systems, contain large numbers of microorganisms, e.g. gram-negative bacteria. Actinomycetes and algae are common constituents of water ecosystems. Therefore, water or liquid droplets resulting from rain, splashes, or bubbling processes may contain bioaerosols, which may remain airborne after the water evaporates.

Strong sources of bioaerosols may exist in many work situations, when organic material is handled, such as plants, hay, straw, wood chips, cereal grains, tobacco, cotton, organic waste, or wastewater. In agricultural and horticultural occupations, exposure to fungal and actinomycete spores may be severe. In non-industrial situations, specific bioaerosol sources may develop due to microbial growth in a building's heating, ventilation and air-conditioning systems or in the structure itself. Generally, the prerequisite for microbial growth is excessive and accessible moisture and temperature. Standing water is always a good reservoir for microbial growth and, therefore, a potential source of microbial aerosols when disturbed. In addition to water, microorganisms only need minute quantity of nutrients, available in the water or building materials, such as cellulose, wood, or concrete. Therefore, a source of spores or other bioaerosol material may develop wherever water is leaking or condensing inside a building.

In non-industrial indoor environments, the most important source of airborne bacteria is the presence of humans. Air with a high concentration of human bacteria is not necessarily a health hazard, but indicates the presence of many humans, their physical activity, or inefficient ventilation. In such cases, pathogenic bacteria or viruses may also occur in air, although their presence may be difficult to verify by conventional bioaerosol sampling techniques. Therefore, the concentration in air of normal human skin bacteria is often used as an indicator of indoor air quality. Bioaerosol concentrations are different in different environments and these concentrations are not necessarily representative of normal or extreme levels as given in Table 2.

Table 2. Concentration of bioaerosols in different environments

Category	Activity Type	Bacteria (CFU/m^3)	Fungi (CFU/m^3)
Agricultural	animal facilities	10^3–10^5	10^2–10^8
	composting	10^3–10^6	10^2–10^7
	harvesting, storage	10^2–10^3	10^3–10^9
Air systems	HVAC	10–10^3	10^2–10^7
		10–10^4	10–10^3
Indoor surfaces	ceilings & walls	10–10^3	10–10^4
	carpet	10^3–10^6	10^2–10^5
	house plants	10–10^4	10^2–10^5
	operating room	10–10^2	10–10^2
Industrial	saw mill	10–10^3	10^4–10^8
	food processing	10–10^3	10^2–10^4
	manufacturing	10^2–10^6	10^2–10^6
Water treatment	aeration tanks	10^2–10^3	10–10^2
	activated sludge	10^2–10^6	10–10^3

Types of Bioaerosols

Bacteria

Bacteria are generally found in all soils, waters, plants, and animals. In air, bacteria may occur as vegetative cells or endospores. They may be carried by other parties, such as water droplet residues, plant materials, or the skin fragments of animals. Bacteria are single-cell microorganisms with a size range of 0.2–30 nm. Their shape varies from spherical to rod-shaped, spiral or filamentous. Many spherical bacteria occur as pairs, chains, tetrads or clusters, e.g., *Micrococcus, Staphylococcus* and *Streptococcus.* The rod-shaped bacteria may occur as single cell or in chains, e.g., *Lactobacillus.* Bacteria can be divided into two groups based on the ability of the cell wall to retain crystal violet. Gram-positive bacteria, such as *Bacillus* and *Staphylococcus,* retain the dye while gram-negative bacteria, such as *Pseudomonas* and *Legionella,* do not retain it and acquire the counter stain and thus appearing red or pink in colour. Bacterial cells that are actively metabolizing and dividing into new cells are called vegetative cells. Endospores are formed within the vegetative cells of certain bacterial genera, e.g., *Bacillus* and *Thermoactinomyces.* Bacterial endospores are metabolically dormant cells, resistant to cold, heat, radiation, and other environmental stresses. The size of bacterial spore ranges from 0.5 to 3 nm. They are easily carried away by air current because gravitational settling is fairly insignificant for particles of this size range. Bacterial endotoxins are lipopolysaccharides specific to the cell wall of gram-negative bacteria. Endotoxins are heat-resistant, chemically stable and maintain their biological activity even after the death of a bacterial cell. Endotoxins cause acute toxic effects that include fever, malaise and decrements in pulmonary function. They can be abundant in agricultural environments, some industries and in humidified indoor air, and may contribute to symptoms of humidifier fever.

Pathogenic bacteria are known to cause disease in humans, animals, and plants. Pathogens are often specific and cause disease only to a certain species of animal or plant. Most animal and plant pathogens are different from human pathogens. Knowledge about human pathogens is the major focus of clinical microbiology. Environmental or saprophytic bacteria are found everywhere but vary in their nutritional and temperature requirements. Only a fraction of environmental bacteria appear to have been identified and characterized so far. Some of the environmental bacteria are opportunistic pathogens, i.e., they may attack an individual having a weakened immunological response e.g. *Legionella* species are opportunistic pathogens. *Legionella* species are common in natural waters but, are difficult to culture in the laboratory. For reasons presently not known, some *Legionella* species find an ecological niche in man-made warm water systems, where they may multiply and, if aerosolized, cause serious disease in exposed people (Table 1).

Actinomycetes

Actinomycetes are a group of soil bacteria that resemble fungi in their growth morphology. Their spores may contribute significantly to occupational exposure in agricultural work situations. In occupational and environmental hygiene, important genera of actinomycetes are *Streptomyces,* which give soil its characteristic odor, and the thermophilic genera are *Theromoactinomyces* and *Faenia.* These organisms cause hypersensitivity pneumonitis. Actinomycete spores may be found in office or residential buildings which have excessive microbial growth due to moisture accumulation within the structure or inside the heating, ventilation and air-conditioning (HVAC) system. In air, bacteria may occur alone or may be carried by other particles. Bacteria tend to grow in colonies in their natural habitats, such as water and soil, and on different surfaces such as biofilms. Therefore, whenever they become aerosolized, they often occur as aggregates or microcolonies attached to other materials. For example, the skin scales of mammals, which are abundant in indoor air, usually contain colonies of bacteria, e.g., *Micrococcus* and *Staphylococcus* (Table 1).

Fungi

The fungi are also omnipresent microorganisms. They are responsible for most of the aerobic decay of natural organic material. The terms mold and mildew refer to visible fungal growth on surfaces. Fungi can be unicellular, e.g., yeasts, but are usually multicellular, forming long chains of cells called hyphae which, in mass, are called mycelium. Fungi are classified into groups based on their methods of spore production. Fungal spores are usually the primary dissemination for the organism and are well adapted to airborne transport. The size range of fungal spores, 0.5–3.0 nm or sometimes even larger, allows for their transport by winds to long distances. They are often resistant to various environmental stresses such as dryness, cold, heat, and ultraviolet radiation. Fungi are well-known allergens that play an important role in the development of hypersensitivity pneumonitis. The common genera are *Pencillium* and *Aspergillus*, occurring at very high levels in work environments e.g. composting facilities, farms etc. Some fungi can utilize living plant materials and cause crop diseases of great economic importance. A few fungi, e.g., *Histoplasma, Blastomyces,* and *Coccidiodes,* readily invade the living animal tissue and cause infectious disease. Others are opportunistic pathogens, e.g., *Aspergillus* and *Cryptococcus,* and cause infection only in people with impaired immunity. However, most fungi are saprophytic, i.e., they utilize and grow on any non-living organic material, provided adequate moisture is present. The moisture levels required for fungal growth are often quite low. Sufficient moisture can be absorbed from air by some organic materials, e.g., human skin scales, if the relative humidity exceeds 70%.

Most fungal aerosols can cause allergic reactions and diseases, such as asthma, allergic rhinitis, or, in some cases, hypersensitivity pneumonitis. Studies of these allergens usually focus on fungal spores, although fragments of mycelium and metabolites can also become airborne. Although a few genera, such as *Cladosporium, Alternaria, Basidiospores* and *Ascospores*, dominate the outdoor aerosols over most of the world yet there is some geographical variation. In areas of seasonal variation, the levels of fungal spores are highest in summer and fall to lowest in winter. Massive exposure to fungal spores may occur in farming and food handling occupations as well as in some industries. In office and residential indoor environments, the outdoor air is an important source of fungal spores. The detection of airborne spores resulting from growth occurring on indoor substrates can be difficult in the presence of normal background levels of outdoor bioaerosol. Mycotoxins are toxic chemicals produced by some fungi. Several species of the genus *Aspergillus* produce carcinogenic toxins. These toxins have been studied primarily with respect to ingestion. There is some indication that aerosol exposure is a hazard as well. Another category of mycotoxins, the trichothecenes, are produced by fungi such as *Fusarium* and *Stachybotrys*. These can cause serious acute effects, including headaches, dizziness, and immunosuppression, and have been isolated from the air.

Protozoa

Protozoa are the unicellular eukaryotic cells measuring 10–50 nm, consist of plasma membrane, cytoplasm, locomotory organelles (pseudopodia, flagella and cilia), contractiles, vacuoles and nucleus. Single protozoal cell performs all important functions of life e.g. reproduction (binary fission, conjugation), digestion, respiration, excretion etc. Man harbours more than 50 species belonging to more than 20 genera ranging from non-pathogenic to pathogenic protozoa e.g. malaria, leishmanianses, sleeping sickness, filariasis, caryptosporidium and toxoplasma gondii are increasingly being implicated as major pathogens in people infected with HIV and immunousuppessed individuals. Another important class of protozoa are opportunistic amoebae which are free-living amoebae belonging to genera *Naegleria* and *Acanthamoeba*. These opportunistic amoebae are ubiquitous in nature and found commonly in soil and water where they feed on bacteria. As opportunitst they may produce serious infection of central nervous system and eye. Man acquires infection either by nasal contamination during swimming in fresh water lakes, ponds or swimming pools or by inhalation of dust containing infective amoeboid forms. These invade the nasal mucosa and travel along the olfactory nerves to brain leading to a fatal infection known as primary amoebic meningoencephalitis (PAM) mostly affecting healthy children and young adults with the history of swimming in fresh water. Patient usually dies within 72 hrs after the onset of the symptoms.

Acanthamoeba causes a chronic disease known as granulomatous amoebic encephalitis (GAE) in immunosuppressed patients like AIDS and those undergoing immunosuppressive therapy and keratitis in healthy individuals. One acquires infection by inhalation of aerosol or dust containing trophozoites and cyst. One may also acquire infection by direct invasion through broken or ulcerated skin or eye. Keratitis occurs by direct contact of the cornea with amoebae, which may be introduced via minor corneal trauma with exposure to contaminated water or contact lenses. Both *Naegleria* and *Acanthamoeba* can be demonstrated by microscopic identification of live or Giemsa-stained amoebae in cerebrospinal fluid. Diagnosis of acanthamoeba keratitis can be made by identifying trophozoites or cysts in corneal scrapings. These can be cultivated on non-nutrient agar (1.5%) spread with a lawn of washed *Eseherichia coli* or *Enterobacter aerogenes* and incubated at 37°C. The amoebae will grow on the moist agar surface and will use the bacteria as food, producing plaques as they eat up the bacteria. As colonies grow and expand, cysts are formed; thus strains can be maintained by transfer of either trophic or cystic form. At present no satisfactory treatment is available for both PAM or GAE.

Algae

These are ubiquitous, many of which are unicellular complex organisms capable of photosynthesis and the synthesis of a complex compounds that constitute the cell. Algae may live as aquatic in terrestrial and subterranean algae. They differ from green plants in having simple reproductive structures for sexual reproduction e.g. unicellular algae may act as gametes, also produce flagelleted spores or non-motile spores in sporangia (sac-like structure within which asexual spores are produced). Algal cell wells are important source of agar algenic acid and carageenan which is used as solidifying agent or in making solution viscous. Algae is used as food (a rich source of protein) for humans and domestic animals in Japan and China. Beside these pathogenic algae causes systemic and subcutaneous infections in humans but are also parasitic on higher plants, liver works, duckweeds. Some of these algae produce extracellular toxins lethal to fish and other animals.

Viruses

Viruses differ from other microorganisms because they are intracellular parasites and can reproduce only inside a host cell and never grow on non-living substrates. Viruses infect either bacteria, plants, animals, or humans and they are mostly very specific within these groups. Viruses that replicate only within bacterial cells are called bacteriophages. Viruses are the smallest of all microorganisms, 0.02–0.3 nm. They consist of only one type of nucleic acid, either RNA or DNA, and are, therefore, not able to generate genetic information without a complete host cell. Viruses are surrounded by a protein layer called a capsid and can be transmitted through air in the absence of the host cell. Although individually small, viruses travel in air as droplet nuclei carried by other materials such as respiratory secretions. Thus, their particle sizes may vary greatly and depend on many factors, including relative humidity. Viruses can cause various infectious diseases, e.g., influenza, measles, polio and chicken-pox, and the viral sources are almost always other infected humans. There is epidemiological evidence for the transmission of viral infections inside and between buildings.

Pollen

Pollen grains are produced by plants to transmit the "male" genetic material to "female" flower structures. Many ornamental plants produce pollen that is transported by insects. However, many trees, grasses, and weeds produce pollen that is adapted for airborne dispersal. To be successfully transmitted, this kind of pollen is usually produced in large amounts. For example, one shoot of hemp may produce over 10^9 pollen grains. Airborne pollen types are resistant to environmental stresses such as desiccation, temperature, and light; hence, they tend to resist sampling stresses. Pollen from different plants varies in size, surface structure and, to some extent, shape. Size range is approximately 10–100 mm, with many grain types being between 25 and 50 mm. Many kinds of pollen contain important allergens. Patterns of prevalence for airborne pollen vary with geography and climate. For example, ragweed pollen is considered to be one of the most important allergens in North America, while birch pollen is considered the most important in Northern Europe.

Volatile organic compounds (VOC)

These are volatile organic compounds which have high enough vapour pressure under normal conditions to enter the atmosphere. A wide range of carbon-based molecules, such as aldehydes, ketones, hydrocarbons are VOC. These are normally produced from the paint thinners, dry cleaning solvents and some constituents of petroleum fuels. Exposure to VOC can cause eye allergy, choke upper respiratory tract, lead to nasal congestion, rash, headache and nausea. Benzene has been branded as a carcinogen by International Agency for Research on Cancer. According to WHO estimates, lifetime exposure to even $1 mg/m^3$ of benzene increases the risk of leukemia in four in one million people. Breathing very high levels of benzene can result in death while high levels can cause drowsiness, dizziness, rapid heart rate, headaches, tremors, confusion and unconsciousness. Long-term exposure to benzene affects bone marrow, decreases red blood cells leading to anemia, can also affect the immune system.

Indoor bioaerosols

Beside these there are indoor bioaerosols too that have delitrious effect on public health. These indoor bioaerosols are released by the microorganisms via air-conditioners, ventilation, doors, windows, cracks in the walls, potable drinking water or via foot wears/clothes. The growth of indoor microorganisms depends on water, humidity, temperature, nutrients (e.g. wood, dust, paper and paint), oxygen and light. Most common indoor bioaerosols are bacteria and fungi. Fungi e.g. *Penicillium* and *Aspergillus* are spore formers and some produce mycotoxins (poisonous substance) or volatile organic compounds. Most common diseases caused by indoor bioaerosols are hypersensitivity and infectious diseases.

13

Energy Sources and Health Effects

D.V. Rai, L.M. Aggarwal and Y.P. Verma

INTRODUCTION

Issues of energy sources and quality of environment have remained an open concern to the human kind. The developing countries need a massive and growing supply of energy for economic development which is crucial both for their well-being and for harmonious international relations. The magnitude of their requirements can be gauged from the fact that the developing countries are today consuming only about 600 kg of coal equivalent per capita of commercial energy as against the world average of more than 2500 kg per capita. At the same time, the continuous increase in the price of oil and problems of its availability are leading many countries, both developing and industrialized, to switch to other energy sources such as coal, liquefied natural gas or nuclear power etc. The challenge, therefore, is to ensure that the increasing requirements of energy are met from non-oil sources without damage to the environment and human health.

A wide variety of options exist for energy sources in the future. These options will have different implications on health. Therefore, in terms of health, future energy systems should anticipate rather than merely react to change. To evaluate any option, the entire energy system should be analyzed, with due attention to its final use. Energy, after all, is the means of doing work. Energy productivity can be greatly improved, and such improvements may lead to the use of much less energy. In addition, a combination of fossil fuels, nuclear energy and an array of renewable energy, sources may fill the gap left by the declining use of oil and gas.

UNITS OF ENERGY

i) **Calorie:** Energy required to raise the temperature of one gram of water by 1°C.

ii) **British Thermal Unit (BTU):** Energy required to raise the temperature of one pound of water by 1°F.

All these units are related with each other through conversion factors given in the following table (Table 1).

Table 1. Energy Conversion Rates

	Value in joules	Value in kilowatt-hours	Value in BTU
1 Joule	1	2.78×10^7	0.95×10^3
1 Kilowatt-hour (kWh)	3.6×10^6	1	3.41×10^3
1 Calorie	4186	1.16×10^3	3.97
1 BTU	1055	2.93×10^4	1

Every source of energy has a definite amount of energy hidden in it. In Table 2 the energy contents of various fuels are given in different energy units.

Table 2. Energy Content of Various Fuels

Fuel	Amount	Energy		
		Joules	Kilowatt-hours (kWh)	BTU
Coal	1 ton (2000 1b)	2.8×10^{10}	7800	2.7×10^7
Oil	1 barrel (42 gal)	6.1×10^9	1700	5.8×10^6
Natural gas	1 cubic foot (ft)	1.0×10^6	0.29	0.95×10^3
Wood	1 ton	1.0×10^{10}	2900	0.95×10^7
TNT (explosion)	1 ton	4.4×10^9	1200	4.2×10^8
Gasoline	1 gallon	1.3×10^8	36	1.2×10^5
Uranium (fission)	1 gram	8.2×10^{10}	2.3×10^4	7.8×10^7
Deuterium (fusion)	1 gram	2.4×10^{11}	6.6×10^4	2.3×10^8

I. FOSSIL FUELS

The ecological consequences of fossil fuel production, transportation and use, such as oil spills, acid rain, damage to forests and vegetation and the greenhouse effect of increasing carbon dioxide emissions into the atmosphere are the areas of considerable study. Decisions about fuel use now cannot be linked with such effects; hence the following discussion will be limited to the description and documentation of the possible effects of the use of fossil fuels on health. It has been known as the risks to health of the production of energy, including exploration, extraction, transport, storage and burning of fossil fuels and the disposal of waste products. It is appropriate here to summarize the effects on health, without necessarily specifying pollutants, and to note how likely these effects are to occur. The effects were arbitrarily described either as definite effects that are known, demonstrated or at least convincingly identified, or as potential effects that are possible, suspected or inferred. The effects were also categorized according to the elapsed time between a person's exposure to the fuel cycle and the appearance of any effects. Many other considerations that may influence the effects of fossil fuel use on health are included where they are known.

The arbitrary scale of the time between exposure and the appearance of effect has categorized acute effects as those that occur from within minutes to a few days following exposure. Intermediate effects require from a few days to two decades for manifestation. Lastly, long-term effects are those that usually appear more than 20 years after exposure. Naturally, the longer the latent period between exposure and effect, the greater will be uncertainty about their relationship. Furthermore, long-term human experiments are unethical and long-term evaluations of exposure when they occur naturally are costly and technically difficult.

Motor vehicles account for about 20-30% of the fossil fuel energy budget of any country or region. The associated risk of motor vehicle accidents accounts for a sizeable number of deaths or injuries.

Effects on Human Health

Fossil fuels play an important role in the overall growth of a country. The ill-effects caused by the use of fossil fuels cannot be ignored and hence we need to look for alternative sources causing less or no harm to environment and human health. But it is not possible to switch over to other sources at once replacing all the fossil fuels. Table 3 gives an idea on some of the health effects caused by use of fossil fuel or the effects they can cause in longer run.

Fires and explosions can occur in procuring and storing fossil fuels. Work accidents occur in coal mining, offshore oil exploration and in fossil fuel transport.

Eye irritation is primarily the result of photochemical pollution. This has been widespread in the urban southwestern United States and can occur in some places in Europe. Eye irritation is associated with high levels of ozone, but the specific causal agents are not well established and it can be caused by high local concentrations of other forms of air pollution.

Respiratory tract irritation and impairment is the most prevalent effect of fossil fuel energy use on the health of the general public. Since stationary power sources or variations in household energy use and household ventilation contribute to these effects, it is impossible to determine the proportion caused by the use of fossil fuels, although the surveillance of children and adults for these conditions may be valuable.

Aggravation of respiratory and cardiovascular conditions can occur among the part of the population afflicted with chronic respiratory conditions or with cardiac or circulatory conditions. It may increase some children's respiratory difficulties and it has been correlated with high levels of ambient sulfur dioxide (SO_2) and airborne particulate concentrations, as in the correlation between high SO_2 concentrations and on the problems that people with asthma or cardiovascular disease experience when exercising.

Increase in respiratory problems in children and adults can occur. Scientists have concluded, from experiments on animals, that respiratory defence against bacterial infection may be impaired by exposure to oxides of nitrogen or SO_2.

Exacerbation of cardiovascular conditions can occur. Exposure to carbon monoxide affects the transportation of oxygen by the blood and can limit the mobility and probability of survival of people with cardiovascular disease.

Table 3. The effects of exposure to the fossil fuel cycle.

Health effects of fossil fuel cycle	Definite	Potential
Acute		
1. Motor vehicle accidents	+	
2. Fires and explosions	+	
3. Work accidents	+	
4. Eye irritation	+	
5. Respiratory tract irritation and impairment	+	
6. Aggravation of respiratory and cardiovascular conditions	+	
7. Neuropsychological impairment	+	
8. Odour and noise	+	
9. Psychological effects	+	
10. Skin irritation	+	
11. Specific acute toxicity	+	
12. General effects on mortality		+
Intermediate		
13. Increase in respiratory conditions in children and adults	+	
14. Increase in cardiovascular conditions		+
15. Effects of organic pollutants in the human body		+
16. Haematological effects		+
17. Effects on reproduction, fetal development and infant growth		+
Long-term		
18. Carcinogenic effects		+
19. Mutagenic effects		+
20. Consequences for health of ecological damage		+

Neuropsychological impairment has also been documented and is correlated with exposure to high levels of carbon monoxide.

The odour of SO_2 is offensive at levels found in heavy pollution. Ozone, while more hazardous, perhaps does not smell as bad. Diesel exhausts have long been known to be offensive, as is the odour of the mercaptans used to odorize natural gas. The noise made by motor vehicles and aircraft can also be a nuisance.

Psychological or behavioural effects can affect mood, attitude towards work and recreation. While assumed to be unequivocally present these are difficult to measure and may also result from accompanying factors such as odour, noise, soiling and visibility impairment.

Skin irritation is an important problem for people exposed to hydrocarbons at work and has been noted in association with occupational exposures to arsenic.

Specific acute toxicity is likely to be confined to occupational exposure and is related to high concentrations of hydrocarbon vapours and trace in miners.

General effects on mortality include some increases in the rates of mortality of people briefly exposed to high levels of SO_2 and particulate matter, as well as people who are exposed to lower levels of these every day. It is not clear whether, or to what extent, past conditions will repeat themselves in the future.

Organic pollutants such as benzene may cause interference with blood cell production. Other medium-term effects may be important, particularly for people exposed to these pollutants at work.

Inorganic pollutants such as trace metals, silica and fly ash are most likely to affect people exposed to them at work. These people should be studied to determine the long-term effects of ozone, carbon monoxide, lead and oxides of nitrogen. The effects of exposure to smoke are being re-evaluated on the basis of the size of the particles in it. When fossil fuels are used correctly, suspended matter is not likely to affect health.

Haematological effects have occurred in some people exposed to hydrocarbons at work. While occupational exposure to lead is known to have haematological consequences, the effects of lead additives in fuel were excluded from this analysis.

Effects on reproduction, fetal development and infant growth can be possible. Conclusions derived from certain animal toxicity studies are causing concern, although evidence of effects on people is scant and unconvincing. If this concern is justified, the exposure of parents to fossil fuels at work should be studied.

Potential carcinogenic effects are the most serious and the most uncertain of the effects of pollutants. Carcinogenic agents, such as polynuclear aromatic hydrocarbons and traces of radioactive material, are almost certainly present in some fossil fuels or their combustion products. These can affect workers as well as the community. Uranium and iron miners have been at risk of lung cancer from inhaling radioactive particles. Coal gas retort workers have had excess rates of chronic respiratory disease and lung cancer mortality. However, improved standards of industrial hygiene make it unlikely that carcinogenic effects will continue at the same level in the future. Concentrations of radioactivity in coal from some locations can be quite high, and under certain conditions might approach the limits of acceptable occupational exposure to radiation.

The excess rates of respiratory cancer in urban areas seem to show that pollution has a carcinogenic effect on the general community. The large role of cigarette smoking in overall cancer rates makes it difficult to quantify the role of pollution. Migration from rural to urban areas also complicates the analysis. Past exposure to fossil combustion products might have been responsible for a small percentage of the excess rates of lung cancer in the population, but present techniques of fossil fuel use are much less likely to lead to risks of cancer in the community from non-radioactive exposure.

Careful dosimetry will be required for any meaningful estimates of the effects on health of radioactive products of fossil energy use. Their role in causing cancer may be minimal.

The determination of potential *mutagenic effects* is subject to many of the same problems as carcinogenic effects. The documentation of potential human mutagenicity is inherently even more difficult than documentation of carcinogenicity.

The *consequences for health of ecological damage*, or secondary health effects resulting from ecological effects, are causing increasing concern. The consequences of climatic changes, such as the greenhouse effect of increasing temperature, are clear. Deforestation leads to the loss of topsoil, potential water pollution and damage to recreational facilities. These may have health consequences of unknown magnitude and uncertain location. Finally, problems related to acid precipitation, studied systematically first in Scandinavia and more recently in Canada and the United States, may lead to loss of fish as a food source and affect lake ecology, and eutrophication may affect water quality where lakes are a major water source. However, the certainty, severity and location of such potential health effects remain to be under study.

Some airborne pollutants, such as carbon monoxide (CO), carbon dioxide (CO_2) and oxides of nitrogen (NO_x), result from the burning of all fossil fuels and others are specific to certain fuels. In general, the manner of fuel use is vital to the emission of pollutants; black or tarry smoke, for example, results from incomplete combustion with an inadequate air supply. In addition, effects on health depend entirely on levels of exposure. These depend on the effectiveness of the dispersion of the pollutants, the activity patterns of the pollution and, for indoor sources, ventilation. Exposure can be minimized by the use of control measures at the source; these can often be more effectively applied in large, central fuel-burning installations than small domestic ones.

Domestic heating and cooking

The use of coal for domestic heating and cooking in the past created major problems of urban air pollution, with demonstrated harm to health. Suspect pollutants include smoke, sulfur compounds and perhaps oxides of nitrogen. Smoke can be avoided to a large extent through the use of hard coal or manufactured solid smokeless fuels, however, the sulfur problem remains. Because of poor dispersion from low chimneys, domestic use produces higher local concentrations of sulfur in the air than major industrial use.

Light oils and natural gas can be used more effectively than coal for instances, as with unflued gas cookers or water heaters, there may be risks of indoor pollution by NO_x or CO, and the use of any fuel-burning appliance with blocked fuels in the home can result occasionally in carbon monoxide poisoning from the unburnt fuel. It is reported that sharp reductions in deaths caused by carbon monoxide were seen after the change from coal to natural gas.

Generation of power from central stations

While central power stations generally emit little smoke, health problems are associated with the SO_2 from burning coal or heavy oil and with coal fly ash. Each can be controlled to some extent but the real key to minimizing health risk is often effective dispersion of pollutants. This assumes that their effects are non-stochastic and not linearly proportional to dose. In this way the exposure of local people to SO_2 in particular can be reduced to a low level but problems from the long-distance transport of pollutants will then arise, with associated environmental effects such as acid rain.

Transport

Current problems, and those of the near future, are associated mainly with the use of internal combustion engines using gasoline and the use of diesel oil in commercial vehicles, cars and railway locomotives. The range of pollutants and types of problems created are rather different from those of stationary sources and the exposure of people in Europe to primary pollutants such as CO, NO_x and particulates depends heavily on how close people are to busy streets and how much time they spend in them. The most substantial and clearly defined effects of road traffic on health are not those arising from pollution but the injuries and deaths resulting from motor vehicle accidents.

Potential problems of the development of new coal technology

In *coal conversion* a considerable fraction of the fuel is used to generate steam and electricity and this use of coal is associated with the same problems as the burning of coal. The use of the converted part of the coal will release only a small amount of effluents. Problems may arise from the final use of the conversion products. Using synthetic petroleum from coal as a primary source may cause certain problems, especially in cases where the direct or indirect synthesis of the petroleum produces polycyclic aromatic hydrocarbons that can affect the health of workers as well as the general public. The disposal of waste products may pose a substantial problem.

Fluidized bed combustion, through the addition of limestone, produces much less SO_2, as well as lower levels of oxides of nitrogen effluents, because of the lower burning temperature. The fly ash produced may be softer and more absorbent than that of conventional coal combustion; therefore, its toxic elements may be more soluble than would be normally expected.

All the effects of coal use on health have many causes. They are generally related to a variety of environmental factors, including lifestyle, smoking habits, diet, aspects of sexual behaviour, drug use, socioeconomic conditions, indoor and general air pollution (not all resulting from fuel use), exposure at the workplace, natural and medical radiation and water pollution. The strength of the association of these factors with cancer and respiratory and cardiovascular diseases ranges from strong to very weak. The magnitude of such associations varies considerably as well. For example, the associations between tobacco smoking or radiation and cancer, or between sulfur oxides and smoking and respiratory disease are well established, but the percentages of disease associated with each factor differ widely. In addition, some factors and agents may interact with one another in a synergistic or antagonistic way. For this reason, considering each factor as a separate agent can give no more than an approximation of real conditions.

Some pollutants occur in gaseous, particulate or aqueous emissions. For their effects on health, the way in which they enter the body may matter less than the dose or concentrations of pollutants, singly or in combination, in organs or systems of the body. While many agents may have similar effects on health, especially since many pollutants in the air affect the lungs, problems of long-term exposure to carcinogens or mutagens are also likely to involve many types of exposure. The pollutants most likely to travel *multiple routes of exposure* are carbon monoxide from vehicles and combustion, including cigarettes and trace contaminants in coal and oil or their combustion products, such as heavy metals, radon and polynuclear aromatic substances.

II. NUCLEAR ENERGY

The generation of electricity through the use of nuclear energy exposes both workers and the general public to ionizing radiation. This exposure can do immediate as well as delayed harm to health, such as cancer and genetic damage. While measured risks to health from the operation of nuclear power reactors have been very low, both the authorities and the public fear the potential for harm to their health and the environment. The following is a consensus opinion of the risks associated with this energy source.

Consideration of the potential radiological and non-radiological damage to health must include all the steps in the fuel cycle: mining, milling, reactor operation, decommissioning, storage, transport, reprocessing of spent fuel, and the disposal of radioactive waste. Analyses must consider normal operations as well as the potential for accidents. Instruments and techniques exist to quantify and monitor the potential and actual dose of radiation to people at all stages of the fuel cycle.

More is known about the toxicity of ionizing radiation than the consequences for health and the environment of other energy sources. Damage to health has been estimated and radiation risk coefficients for carcinogenic and genetic effects have been calculated and published in previous reports at national and international level. The biological effect of exposure to radiation can be both stochastic and nonstochastic. In the nuclear fuel cycle, an increased incidence of lung cancer has been observed in uranium ore miners and particularly in those who smoke cigarettes. However, the available scientific data do not seem to bear out public concern about the effects of exposure to low level of radiation.

In the absence of evidence of harm to health from the doses of radiation that the general public incurs at present from nuclear energy, the international radiation protection agencies' philosophy of avoiding unnecessary exposure to radiation is particularly relevant. People should not be exposed to radiation unless the benefit clearly outweighs the risk. The costs of additional risk protection should be optimized, and limits should be placed on both individual and population doses.

People are most afraid of the terrible consequences of accidents at nuclear plants because they might result in deaths from cancer and genetic damage in future generations. Such accidents, however, are not very common. The normal operation of the nuclear power plants now operating exposes the population to considerably less radiation than other man-made sources of ionizing radiation, such as medical radiation, or natural radiation. The disposal of high-level radioactive waste is the method under active consideration today.

III. RENEWABLE SOURCES

Energy sources other than fossil fuels and nuclear power provide relatively little energy in several countries with the exception of hydroelectricity and, to a much smaller extent, wood burning. This is a consequence of both economic and technical factors. Although technical progress, such as improvements in direct conversion of light to electricity, may lead to a more substantial contribution from these other sources, the use of solar, ocean, tidal, geothermal, wind and biomass energy is impeded by one or more of the following shortcomings.

Energy is harvested at relatively low temperatures, leading to poor efficiency in conversion to electricity and correspondingly to large amounts of waste heat. The energy flux density can be quite low, requiring large investments per kilowatt installed, and large land areas.

These limitations influence potential effects on health directly and indirectly. The general public perceives alternative energy sources as relatively safe and environmentally benign. This perception is enhanced by the fact that the use of these sources would entail a shift to decentralized energy systems and therefore to more local control of decisions about the production and use of energy and safety and health standards. This does not necessarily apply, however, to large-scale hydroelectric power. Unfortunately, authorities in decentralized systems can often exercise less control over energy use, so health and safety problems may actually be exacerbated. An example is the use of wood combustion for space heating.

Alternative sources

Hydropower has long been considered a relatively clean, safe, cheap and renewable source of energy. In many countries this perception continues and hydropower is used. In many industrialized nations, however, most of the best sites are either already developed or are unsuitable because their use would have unacceptable ecological effects. These effects might include the flooding of unique scenic or historic areas. Consequently, in industrialized countries pumped storage seems to be the only major option for large-scale hydropower development. In some areas the development of small-scale hydropower plants (Fig. 1) may have a marginal positive effect.

In most countries the development of hydropower can have effects on health and the environment that must be considered. Amongst these are possibilities of:

- loss of life due to dam failure and lives saved because of the prevention of annual flooding;
- the spread of waterborne diseases such as schistosomiasis in some areas near the Mediterranean;
- loss of fisheries due to the change in the thermal gradient;
- salinity changes;
- blockage of the upstream and downstream movement of anadromous fish;
- loss of rich farming areas along rivers and loss of nutrients from the deposition of silt on flood plains;
- possible increase in seismic activity;

Fig. 1. Hydel power.

— increased water loss by evaporation;
— loss of farm area downstream of dams due to increased erosion;
— flooding from poor water management; and
— groundwater sinking.

A review of the entire fuel cycle should consider the health effects of the large-scale use of concrete, iron and steel in the construction of large facilities and the associated diversion of large amounts of capital from other, possibly beneficial, uses. The latter factor is common to all capital-intensive energy systems.

Ocean and tidal current power (Fig. 2) is used little in Europe, although it has been used at LaRance, France. Moreover, the prospects for increased energy generation seem remote. For ocean thermal energy, the small temperature gradients require large heat transfer surfaces, with various attendant difficulties. The environmental consequences of developing this energy appear small, but are still largely unknown. Some people are interested in the use of ocean currents, but this is in a preliminary state of development. While the concept of using tidal energy remains attractive, its exploitation requires large expenditures to develop an always-intermittent source of energy.

Fig. 2. Turbine for sea tidal power generation.

Wind power has been used mostly to generate electricity, but smaller applications have been employed to pump water and desalinate seawater. Wind energy can be expected to provide about 2-3% of electricity generation and depends on several factors. A problem associated with wind power, as with tidal power, is the irregular nature of the supply and the accompanying necessity for energy storage.

Large-scale wind generators (Fig. 3) can directly affect the environment by influencing the local climate over a distance measured at about ten times the diameter of the propeller. Furthermore, the generators are noisy. Indirect effects arise from the need for storage and back-up systems and the technology used for storage. The uncertainties about the development and effects of large-scale wind generators include their acceptability to the public, particularly in densely populated areas. The rate and potential effects of the breakdown of large propellers are largely unknown.

Fig. 3. Windmill.

Small-scale wind generators that generate electricity require storage systems, such as batteries, that may have significant consequences for health. As with large-scale generators, noise and local climate changes may have significant effects. When small-scale generators are used for mechanical energy, such as pumping, it may be useful to consider the net positive effects. That is, it is important to consider that the source of energy being displaced might be a less desirable fuel, such as diesel oil.

Solar power is generally produced from small local sources or large central stations, on land or satellites. Unlike fossil fuel technology, solar technology does not make significant emissions to the environment during operations, and unlike nuclear technology, it does not produce hazardous waste products during operation. Solar water heating, especially in relatively low-density areas of housing in the southern part of the European region, can usefully contribute to comfort and hygiene at modest cost and with virtually no risk to health. The largest fraction of potential effects on health of the installation, operation and discontinuing of solar technology are likely to be associated with the massive extraction of materials and construction required to build solar energy systems. Land-based solar voltaic technology requires large collection areas per unit of installed capacity. For example, a photovoltaic solar cell plant (Fig. 4) would require painting black about more than 150 square kilometers plus land for storage and retrieval equal to a 1000 MW nuclear plant. In most of Europe this requirement can be difficult to meet, because of the climate, latitude and availability of land.

Satellite-based solar energy stations may require less land and materials but, depending on the specific system, are likely to expose large groups of people to low levels of microwave radiation. All solar technology will require the large-scale fabrication of solar conversion materials, and the manufacture and ultimate disposal of this material can increase the exposure of both workers and the general population. Silicon, cadmium, germanium, arsenic, complex organic coating compounds and other materials may be used in large quantities. More information on their effects on health and their transport in the environment will be required.

Fig. 4. Solar cell.

Biomass energy is created by activities that range from the direct burning of wood or the gasification of agricultural residues to the recovery of biogas containing methane from municipal refuse landfills (Fig. 5). Techniques will have to be developed to improve the production and harvesting of biomass, as well as the production of fermentation alcohol to supplement gasoline. Its effects on health vary. The careless and inappropriate use of stoves to heat homes can cause fires. Even when properly used, stoves generate carbon monoxide and mutagenic material in the smoke, as well as possibly dangerous volatile and condensable organic compounds. While nitrogen emissions are at somewhat lower levels than those of most other fuels, there is a serious concern about their escape into the living area. Wood ash does not appear to be toxic, and while wood combustion does not appear to generate large quantities of oxides of sulfur or heavy metals, wide use may have serious effects on health. Alcohol used as a transport fuel may cause formaldehyde emissions.

Biomass production requires extensive cultivation and harvesting, with some associated hazards. However, biomass that is currently considered waste can be used and produced on otherwise unproductive land. There are associated concerns about the large amount of irrigation water needed, and associated potential for soil leaching. The wide distribution of small generating units using biomass can lead to accidents and difficulties in maintenance and quality control. On the other hand, the low intensity and altitude of emissions can create a local rather than an international control problem. It is difficult to apply and monitor the effectiveness of control technology for widely spread biomass units of varying construction. However, effective control techniques should be feasible for larger industrial, commercial or community units used to produce heat or electric power. Also, the use of catalysts to remove aldehydes and other organics from motor vehicle exhaust should be feasible.

The wide distribution of varied types of sources and the lack of adequate research to identify other, as yet unanticipated, variables or to develop safe technologies for transforming biomass into useful energy may be major problems. Improving the efficiency of combustion in domestic stoves, for example, would lower health and environmental hazards. The various energy sources that would be replaced by the use of biomass technology must be considered, so that the net energy effect is

Fig. 5: Biomass plant

properly understood. For example, although formaldehyde emissions may result from the use of alcohol-based fuels, the gasoline or diesel that is displaced may create even more harmful products.

Geothermal energy has been derived by a limited number of methods. The most common has been the direct use of natural hot fluids from deep geothermal layers. Other techniques, based on the artificial pumping of water from the surface down through layers of hot rocks, are being developed.

The direct use of underground high-temperature fluids usually includes the extraction and frequently the dispersion of toxic substances, vapours and sometimes gases that may escape containment. These include sulfur and boron compounds, as well as radon-222. Trace elements in various concentrations may also be released and at least one potentially useful geothermal field had high concentrations of arsenic that seriously complicated its exploitation.

Geothermal energy could affect people's health by exposing them to toxic or potentially toxic elements, including natural radionuclides as well as non-nuclear agents. Each source will probably have its own spectrum of pollutants and, while they may be easily identified, information on their potential effects on health is scanty, particularly for long-term, low-level exposure.

Particular attention should be paid to concentrations of trace elements since small increases in their concentrations in the environment and food chain could be harmful. Because of the uncertainties about this technology there is no consensus on acceptable levels of environmental contamination from geothermal effluents. Fig 6 below depicts a geothermal-based energy station.

IV. ENERGY CONSERVATION

The effects on health of the production and use of energy are generally assumed to have a direct relationship with the amount of energy consumed by society. Therefore, the conservation of energy should lead to a reduction of energy consumption and a parallel reduction of the effects on health and the environment of energy use. Conservation of energy should not mean reducing or eliminating needed services or goods, but their production with less consumption of energy resources, that is,

Fig. 6: Geothermal energy.

higher energy productivity. Choosing the appropriate technology for each purpose can reduce effects on health and the environment, although the wrong choice can save energy at the expense of health.

Large improvements can be made in the efficiency of energy use, and recent studies in developed countries have concluded that energy savings of up to 50% can be achieved, even with a moderate growth of up to 40% in the consumption of goods and services. These goals, however, are not likely to be achieved soon. Success will mean the implementation of the best and most economical available technology. Savings and improvements can be made in several areas discussed below.

a. Energy Systems

The efficient use of energy storage systems should be promoted. Even if no energy is saved, benefits to health may result.

b. Transport

Designs should be introduced for energy-efficient vehicles with lightweight, reduced friction and aerodynamic improvements, as well as engines that use fuel more efficiently. To encourage the acceptance of lightweight vehicles, however, their safety must continue to be improved. Further, the improvement and expansion of mass transport systems in large cities would lead to reductions in travel mileage and lower emissions per unit of transport.

c. Domestic and Commercial Uses

Improvements in space heating in northern Europe, as well as cooling in southern Europe, can affect important savings. Energy efficiency should be an integral part of the designs of new homes. Domestic energy consumption can be significantly reduced when proper attention is given to climatological factors that affect thermal insulation designs, weather-proofing, the choice of appliances and heating and air-conditioning systems. The full use of energy conservation measures in northern climates could reduce space-heating requirements by a factor of 5 in existing buildings. Improvements to existing buildings could result in energy savings of 50%. Modern insulating materials may, however, cause health problems for people who work with them.

The costs to health of the implementation of conservation measures are difficult to determine. The improved sealing of homes, for example, reduces ventilation rates and can thereby significantly increase concentrations of radon and radon daughters in the air, and expose people to more radiation. In areas where the natural radon concentration in buildings is already high, unacceptable risks to health might accompany the implementation of severe conservation measures. Alternatively, heat losses from ventilation can sometimes be minimized through the use of forced ventilation, combined with heat recovery. Improvements in the design of electric home appliances that use large amounts of energy can now result in considerable energy savings.

d. Industrial Uses

Energy surveys and improvements of existing facilities can lead to savings of up to 40% by the detection of unnecessary losses from deficiencies in design or maintenance. New industrial facilities should be energy-efficient. Furthermore, energy productivity can be improved when new technology is developed and introduced and when production shifts towards the production of services. Scheduling for a more economical spread of peak energy use and adding cogeneration can conserve more energy.

Improved efficiency in the production of energy would result from the following measures:

- — economic incentives, either inherent or from government policies;
- — regulations prescribing energy efficiency standards that are both technically achievable and economically reasonable;
- — internalization of the cost of health and environmental detriments, so the polluter pays for the damage caused;
- — taxes on emissions and pollution; and
- — determining whether policies on energy prices really can lead to more conservation.

CONCLUSION

Each nation is concerned about the environmental and health effects of the internal and international use and generation of energy. While each country in the world may have different energy resources and different approaches to energy problems, groups of countries could benefit from cooperation in handling these concerns, particularly in counteracting international effects. Each country is served when all create joint solutions and recognize that energy pollution respects no national boundaries. While the costs of controlling and preventing releases of pollution may be large, correcting damage to health and the environment from unrestricted releases would be far more expensive. Today's polluter may well receive contamination from another nation tomorrow. The development and use of energy systems makes no country immune to the problems of importing and exporting pollutants. Pollutants from fuel cycles are cooled, transported and diluted in the air, surface water and groundwater. Acid rain from fossil fuel combustion and the thermal and chemical pollution of river water used for agriculture and drinking are well-known international problems. Examples of international effects include:

— the effects on health and the environment from the transport of combustion products through air and water;
— the potential threat to health from radiation released by accidents at nuclear power plants;
— the increased risk of respiratory illness from chronic exposure to pollutants from fossil fuel combustion;
— the harm to health and the environment from leaks and accidents in oil and gas pipelines;
— the health effects of high-voltage transmission lines; and
— the effects of accidents in the transport of nuclear fuel, oil, coal and gas.

Therefore, the question before the researchers is what's the optimum solution? Renewable energy sources are not the final solution as they also pose a threat to the environment. Because every form of renewable energy involves vast infrastructures such as concrete, steel, and access to road. Hence renewable sources are renewable but they are not green. According to studies if we want to minimize new structures and rape of the nature, probably nuclear energy is the best option.

SUGGESTED READINGS

1. WHO Regional Publication, European Series No. 19, Health impact of different energy sources. 1986.
2. Energy—Today and Tomorrow by Dr. Maheshwar Dayal, Publication Division, Ministry of Information and Broadcasting, Govt. of India, New Delhi, 1989.
3. Environmental Biophysics, edited by D.V. Rai, Panjab Univ. Publication, Chandigarh, 2006.
4. Divergent Dimensions of Environment, edited by R.C. Sobti and Vipin Sobti, Radvision, Mohali, 2002.

14

Medical Geology

Naresh Tuli

INTRODUCTION

Medical geology is defined as the science that studies the links between geological materials and processes and the occurrence of human and/or animal diseases. At first instance, geology may seem far removed from human and animal health. However, rocks and minerals and the soils derived from them contain all the naturally occurring major and trace elements that are critical to our well-being. Most of these elements are taken into the human body through food, water and air. This interaction with natural material is generally harmless, perhaps even beneficial as it supplies essential nutrients. However, the interaction with minerals and trace elements can have detrimental effects and in most severe cases can be fatal. The overall scenario reflects the geology of a given environment directly influencing the chemical make-up and health of the life within it. For instance, people living on impoverished soils may lack essential trace elements, giving rise to a variety of diseases. On the other hand, excessive intake of some chemicals such as arsenic and fluorine can cause serious diseases.

The elements are not distributed equally in the earth crust (Table 1) and problems to life can occur when there is a deficiency or even abundance of elements (Tables 2 and 3). The inability of environment to have the correct mineral balance leads to serious health problems. Exposures of certain trace elements such as arsenic, mercury, cadmium, selenium, lead and uranium etc. have been found harmful. These elements enter the food chain and affect the human health.

The damage to an environment originates from a combination of natural and anthropogenic input. The pollutants in the environment include radioactive elements, organic and metalloids and their chemical species. The heavy metals and metalloids are considered to be potentially toxic to life forms as reported by U.S. Environmental Protection Agency and the World Health Organization. The physical, chemical and biological conditions in an environment such as temperature, state of matter, acidity (pH), bacterial activity and biological oxygen demand (BOD) influence the mobilization, deposition, distribution and concentration of toxic metals which effect the health of organisms.

Table 1. Average abundance of some elements in the earth's crust and rocks (all values in parts per million, ppm).

Element	Earth's crust	Ultrabasic	Basalt	Granite	Shale	Limestone
As	1.80	1.000	2.00	1.50	1.5	2.50
Cd	0.20	—	0.20	0.20	0.2	0.10
Co	25.00	150.000	50.00	1.00	20.0	4.00
Cr	100.00	2000.000	200.00	4.00	100.0	10.00
Cu	55.00	10.000	100.00	10.00	50.0	15.00
Pb	12.50	0.100	15.00	20.00	20.0	8.00
Se	0.05	—	0.05	0.05	0.6	0.08
U	2.70	0.001	0.60	4.80	4.0	2.00
W	1.50	0.500	1.00	2.00	2.0	0.50
Zn	70.00	50.000	100.00	40.00	100.0	25.00

Table 2. Diseases associated with deficiency of the elements

	Element	Associated Diseases	Causative Factor	Permissible Limit
1.	Selenium	Keshan Disease (China) Bone Disease, Arthritis, Cancer, Cardiovascular failure.	Deficiency of selenium in food chain	0.01 mg/lt.
2.	Zinc	Dwarfness (Iran & Egypt) Infertility, delay in filling up of the sores, loss in resistance, impaired taste and smell etc.	Deficiency in agriculture produce due to low zinc content in soil	0.5 mg/lt.
3.	Copper	Anemia Skeletal defects	Heart ailment, deficiency of blood, disorders of nervous system, defects in bones.	0.05 mg/lt.
4.	Chromium	Disturbances in the glucose metabolism, kidney problem, diabetic problem, cholesterol disorders	Released through industrial effluents of tanneries	0.01 mg/lt.
5.	Magnesium	Depression, diseases of nervous system, malformations of skeleton, convulsions, dysfunction of gonads, urolithiasis.	Deficiency in food intake	30 mg/lt.
6.	Iodine	Goitre	Excessive calcium in water sinks iodine and does not allow to go in drinking water for human use. Elevated places from m.s.l. gives rise to iodine deficiency.	0.3 mg/day
7.	Fluorine	Dental caries	Deficiency of fluorine (<0.5 mg/lt.)	0.05 to 1.2 mg/lt.
8.	Molybdenum	Mouth and esophageal cancer	Deficiency in food intake	2.8 mg/day
9.	Coblat	Anaemia, white liver disease	Deficiency in food intake	<670 mg/day
10.	Iron	Anaemia	Deficiency in food intake	0.3 mg/lt.
11.	Sodium	Coma	Deficiency in food intake	0.9 mg/day

Table 3. Diseases associated with excessive presence of elements

	Element	Source	Associated Diseases	Causative Factors
1.	Mercury	Dust in atmosphere, volcanic eruption, forest fire, vegetational loss, salt spray, geyser thermal liquid gaseous release from mantle surface of earth.	Maniamate (Japan, 1953) Inactiveness of organs, haemorrhage of brain cells, blindness, deafness, faintness, death, neurological disorders.	0.001 mg/lt.
2.	Cadmium	Flying dust, volcanic eruption, forest fire, spray of sea water. Electroplating, Cement industry, Fertilizers, Sewage, storm pigment, gasoline burn	Itai-Itai (Japan, 1950) joint pain, softening of bones, skeletal deformity, prostate gland cancer, lung cancer, renal heart disease.	0.01 mg/lt
3.	Fluorine	Leaching of rocks, volcanic ash, bentonite, flying dust, fertilizer seepage, industrial effluent.	Yellowing of teeth, skeletal deformations–osteo and skeletal fluorosis	1.5 mg/lt
4.	Lead	Coal burning, gasoline burning, paint smelter, exhaust release from automobiles	Lung cancer neurological disorders	0.05 mg/lt
5.	Arsenic	Arsenic is absorbed on clay, oxides and (OH) iron and manganese	Liver disease, cancer of skin and lungs, ulcer, blood deficiency	0.05 mg/lt $(ASO_4)_3$ is toxic

The natural sources of metals in the environment lie within the rocks and processes by which they are disintegrated after lithification. Soils can develop from the weathering of any class of rock (Fig.1). The natural chemistry of water on land is influenced mainly by the chemical composition of

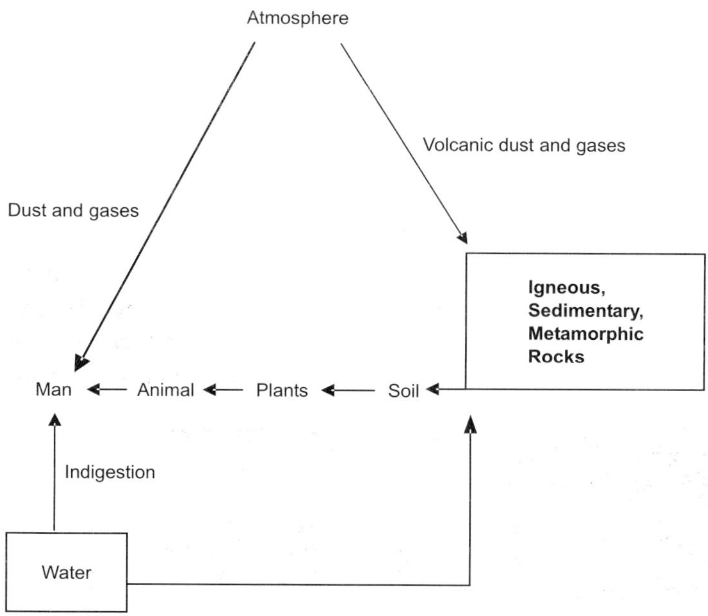

Fig. 1.

the rocks (soils) and the water flows on or flows through it. The waters take on the chemistry of rock forming minerals and any sulfide minerals present in the rocks. Thus, water will have a different pH depending upon the rock type through which it passes.

The atmosphere carries toxic metals as gases, aerosols and particulates. These metals originate from geological phenomena i.e. volcanic emissions, thermal springs, and volatilization from deposits and soils.

The heavy metals present beyond permissible limit of Bureau of Indian Standards develops toxicity in land, water and air. Table 4 shows the presence of elements in some rocks, minerals and ores.

Table 4.

Rocks/ores/minerals	Elements
Shales and bituminous shales	V, U, As,Sb, Mo, Cu, Ni, Co, Cd, Ag, Au, Pt, Se, Al.
Iron ores	Fe, V, As, Sb, Se, Fe.
Manganese ores	Mn, Ba, Li, K, Ti, W,Co, Ni, Cu, Zn, Pb.
Limestone/dolomite	Ca, Mg, Fe, Ba, Sr, Pb, Mn.
Olivine	Fe, Cl, Ni, Co, Mn, Zn, Cu, Mo.
Hornblende	Fe, Al, Ni, Co,Mn, Se, V, Zn, Cu, Ga.
Augite	Al, Ni, Co, Mn, Se, V, An, Cu, Ga.
Biotite	Mn, Fe, Al, Rb, Ba, Ni, Co, Li, Mn, V, Zn, Cu, Ga.
Apataite	Rare Earths, Pb, Sr.
Anorthite	Al, Sr, Cu, Mn, Ga
Andesite	Al, Sr, Cu, Mn, Ga
Oligoclase	Al, Cu, Ga

(*Source:* Metchell, 1995)

Prominent health hazards related to geological causes may be dealt under the following heads.

Fluorosis

The disease fluorosis is due to an element known as fluorine, a negative charged ion having low dissociation energy. As a result, it is highly reactive and has an affinity to form compounds called fluoride by reacting with other elements. The link between the fluoride geochemistry of water in an area and the incidence of dental fluorosis, a tooth damaging condition is well established. While the WHO standards (1991) permit only 1.5mg/l as a safe limits for human consumption, the new Indian

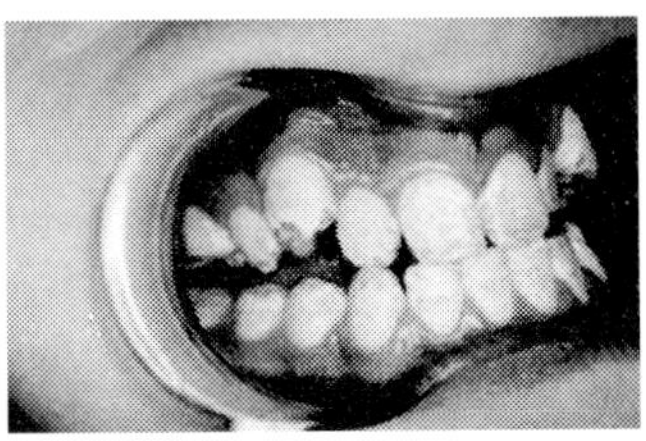

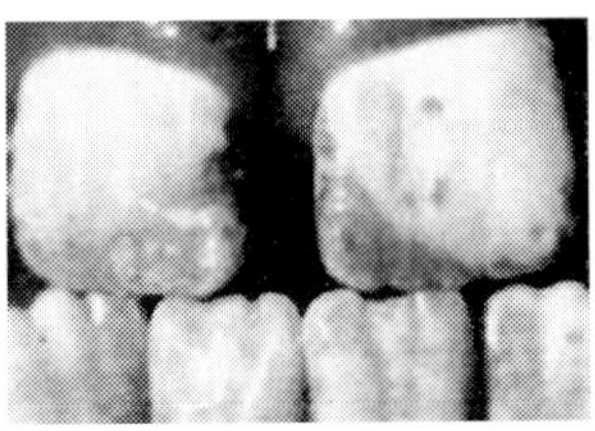

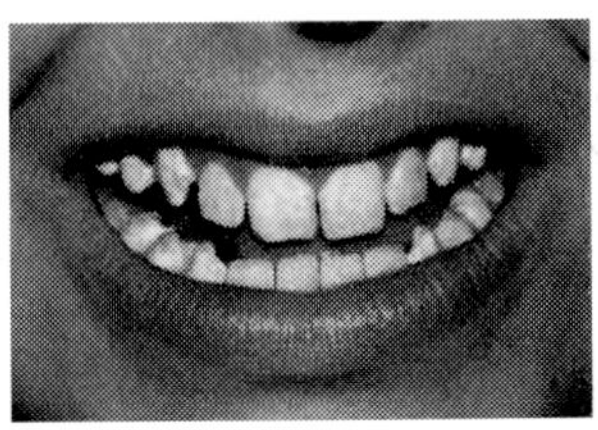

standards permit only 1 mg/l. When such fluoride-rich water is consumed, the teeth develop a dark brown color and a mottling. The deficiency of fluoride also gives rise to dental cavities.

When water flows through rocks it leaches out fluoride from these rocks. The source of fluorine is fluoride-rich rocks containing minerals, such as fluorospar (CaF_2), cryolite – (Na_3AlFPO_6) and fluorapatite – $Ca_3(PO)_2Ca(FCl)_2$.

The highest fluoride contamination is reported from Muktsar, Punjab (42.5 mg/litre) followed by in parts of New Delhi (32.5 mg/litre).

Flourosis is widespread and acute in the state of Rajasthan. The countries that are worst affected by this disease are Pakistan, Bangladesh, Argentina, United State of America, Japan, New Zealand etc. This disease has affected the 17 states of India, which are 50–100% district affected Andhra Pradesh, Tamil Nadu, Uttar Pradesh, Gujarat, Rajasthan 30–50 % district affected—Bihar, Haryana, Karnataka, Maharashtra, Madhya Pradesh, Punjab, Orissa, W.B.; < 30% district affected—J&K, Delhi, Kerala.

The remedial method for control and prevention of fluorosis can be classified into (i) chemical (ii) biochemical (iii) nutritional and (iv) hydrological. The main objective of these methods is to increase the concentration of calcium in water or human body, which would help to lower the fluorine content by chemical principles like absorption and ion exchange etc.

To prevent this disease, some researchers prescribe vitamins C and D and salt of calcium, magnesium or aluminium. It is found highly favorable in children.

To escape from the disease, prevention of fluorosis is the only effective measure. For hydric fluorosis, change or modification in the water supply is the only practical solution. Digging deep bore well has been shown to provide lower fluoride contents. Defluoridation of water can also be performed by adding activated alumina, or by alkalinization and chlorination followed by adding aluminium salts.

Further there is a technique called nanofiltration, a membrane-based process which removes small molecules and ions including fluoride and can sterilize water.

Goitre

Iodine is an essential trace element in soil, water, vegetables and cereals. It has been estimated that nearly 30% of the world population is at risk for some form of iodine deficiency disorder (IDD). The ideal iodine intake for an adult is 0.14 mg per day. Insufficient intake of iodine is the world's most common cause of mental retardation and brain damage.

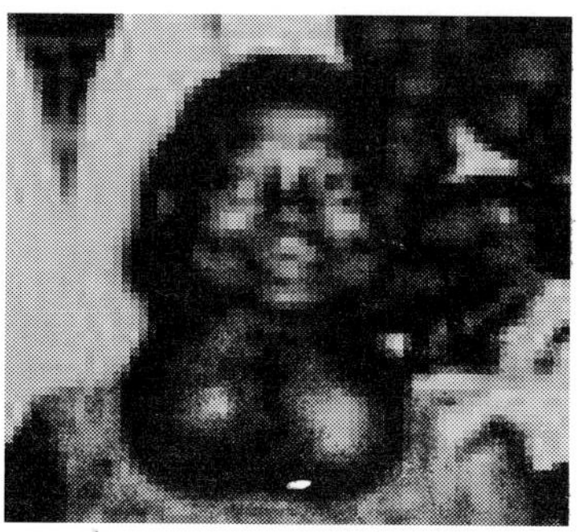

Goitre is the most visible sign of iodine deficiency. It is a swelling in the neck caused by an enlarged thyroid gland. The thyroid increases in size when there is not enough iodine available to it. Person can become dull and less active. Its deficiency also results in high rates of infant mortality. In pregnancy, this deficiency leads to mental and physical defects in the baby.

The sea is a major source of iodine, so there exists a relationship between the incidence of iodine deficiency disease in a region and its distance from the sea. In general, the further away from the sea, the less iodine is available. Other factors such as atmospheric circulation, may play a role in iodine availability, as does topography. In mountainous regions, iodine abundance is quite low.

Goitre is more pronounced in limestone terrains and in the areas yielding high calcium. Geomorphologically, steep gradient in the Himalayan terrain promotes quick movement of the surface water. With the result, water finds little time to transfer its iodine to the soil and in turn to the ground water.

In India, goitre is reported in foothill zone of Himalayas, Bhabbar and Tarai belt in U.P. and in alluvial plains of West Bengal. Large number of people in China, Sri Lanka, Greece, Egypt are affected by goitre.

Goitre Treatment

The remedial measure of the problem is salt iodization. Wherever iodized salt is eaten in the daily diet, IDD has become a thing of the past. The various home remedies for goitre treatment are as follows:

Goitre treatment using Iodine

Iodine is, undoubtedly, most helpful in many cases, but it should be given in its organic form. All foods containing iodine should be taken liberally. These are lettuce, turnips, carrots, garlic, onions, oats, pineapples, whole rice, tomatoes, watercress, strawberries, guavas, citrus fruits, egg yolks, and sea foods.

Goitre treatment using Watercress

Goitre can be treated with watercress which is one of the best sources of the element iodine. It is valuable in correcting the functioning of the thyroid gland. Its regular use, therefore, is highly beneficial in the prevention and treatment of goitre. A paste made of this vegetable can also be applied beneficially over the affected parts. It helps to reduce swelling.

Goitre treatment using Swamp Cabbage

Exophthalmic goitre can be treated using the leaves of swamp cabbage. A teaspoon of the juice with tea almonds should be given once or twice daily as a medicine. The leaves are also useful in myxodema, a condition that result's from lack of thyroid hormones results.

Goitre treatment using Kachnar

Kachnar, botanically known as *Bauhinia variegata*, has been used as a folk medicine for treatment of goitre in India. About 30 ml of the decoction of the bark should be given twice a day in the morning on an empty stomach.

Goitre treatment using Dandelion

Dandelion has been found valuable in treating goitre. The leaves of this salad vegetable should be smeared with ghee, warmed and bandaged over the swollen parts for about two weeks. Any existing discomfort will be relieved by this remedy.

Goitre treatment using Flex

The seeds of flex are also useful in reducing swellings in goitre. The seeds should be ground in water, heated and applied over the affected parts. The leaves of the tree should then be bandaged over them.

Diet for Goitre

Fruit juices

To begin with, the patient should take juices of fruits such as oranges, apples, pineapples, and grapes every two or three hours for five days.

Fruits and Milk Diet

After the juice fast, the patient may spend the next three days on fruits and milk, taking three meals a day of juicy fruits with a glass of milk, at five hourly intervals.

Food to avoid during Goitre

Certain foods and fluids are extremely injurious for goitre patients and should be avoided by them. These include white flour products, white sugar, flesh foods, fried or greasy foods, preserves, condiments, tea, coffee, and alcohol.

Other Goitre Treatment

Rest and bowel cleansing with lukewarm water

For the first five days of the treatment the bowels should be cleansed daily with lukewarm water. The patient should take plenty of rest and spend a day in bed every week for the first two months of the treatment. More and more physical exercise should be taken after the symptoms subside.

Silicosis

It is an incurable lung disease caused be inhalation of dust containing free crystalline silica. It is irreversible and, moreover, the disease continues even when exposure stops.

Slilica dust is released during operations in which rocks, sand, concrete and some ores are crushed and broken. Work in mines, quarries, foundries and construction sites, in the manufacture of glass, ceramics, and in masonry workshops is risky.

Sandblasting is one of the high risk operations for silicosis. Any abrasive blasting, even if the abrasive does not contain silica, may pose a silicosis hazard when it is used to remove materials that contain silica, such as remains of sand moulds from metal castings. Some operations, like dry sweeping, the clearing of sand or concrete, or the cleaning of masonry with pressurized air can generate large dust clouds. Thus even in open air these activities can be hazardous.

Free crystalline silica, SiO_2 is one of the most common minerals in the earth's crust. It is found in sand, many rocks such as granite, sandstone, flint and slate, and in some coal and metallic ores. The three most common forms are quartz, tridymite and cristobalite. Inhaled crystalline silica (in the form of quartz crystobalite) from occupational sources is classified by the International Agency for Research on Cancer (IARC) as a Group 1 human lung carcinogen.

Silicosis is a preventable disease. The key to silicosis prevention is to prevent dust from being in the air. A water hose to wet dust down at the point of generation. When sawing concrete or masonry use saws that provide water to the blade. During rock drilling use water through the drill stem to reduce the amount of dust in the air. Use local exhaust ventilation to prevent dust from being released into the air.

Skin Lesions

Longtime use of drinking water and food with high concentration of arsenic (750 ppb) can lead to health hazards in the form of pigmentation, keratosis, cancer of bladder, kidney and lung.

The arsenic is naturally occurring and is not derived from any anthropogenic source. Its primary source has been mainly the rock formations of the Himalayan Mountain System. The arsenic content could be related to poly-metallic sulphides, biotite, chlorite and other feromangnetite minerals, altered products of which contain limited concentration of arsenic in the Himalayas and also in the peninsular India.

Bangladesh and India are the most arsenic effected countries. With the exception of vaccination, control measures are usually not effective.

HEALTH PROBLEMS DUE TO ANTHROPOGENIC INPUTS

With urbano-industralization, some important heavy metals have been released from industrial units and are the source of health hazards if discharged into river water. Table 5 shows average quantities of some important heavy metals released by industrial unit (Klein et al. 1976).

The presence of toxic elements in the fertilizers and pesticides is also polluting the environmental health of the area (Table 6). The use of fertilizers has revealed that toxic elements are the least in the organic manure as compared to fertilizers.

There are certain minerals which are essential for our health but due to its deficiency the body shows certain symptoms (Table 7).

These problems can be preventive by intake of balanced diet supplemented with fruits, vegetables, milk and milk products.

CONCLUSIONS

The toxicity related diseases in humans and animals are associated to anomalous concentration of trace elements in rocks and minerals and comprise the domain of medical geology. The environmental cycles disturb the distribution of elemental balances and compositions that lead to serious health problems. The environments surrounding the biotas, which are determinants of temperature, acidity,

Table 5

Industry	Average quantity in mg/lt (ppm)				
	Zn	Cu	Ni	Cr	Cd
Bakery	0.25	0.15	0.43	0.33	0.002
Brewery	0.47	0.41	0.04	0.06	0.005
Fat rendering	**3.90**	0.22	0.28	0.21	0.006
Fish processing	1.60	0.24	0.14	0.23	0.014
Fur dressing and dyeing	1.73	**7.04**	**0.74**	**20.14**	**0.115**
Ice-cream	0.78	2.70	0.11	0.05	0.031
Laundry	1.75	1.70	0.10	1.22	0.134
Meat processing	0.46	0.15	0.07	0.15	0.011
Miscellaneous chemicals	0.80	0.16	0.10	0.28	0.027
Miscellaneous foods	1.10	0.35	0.11	0.15	0.006
Soft drinks	2.99	2.04	0.22	0.18	0.003
Textile, dyeing	0.50	0.04	0.25	0.82	0.030

Table 6.

Source of Manure	Pb (in ppm)	Cd (in ppm)	Hg (in ppm)	As (in ppm)
Phosphate fertilizers	4 – 1000	0.01 – 190	0.01 – 2	<1 – 1200
Nitrogen fertilizers	2 – 120	<0.05 – 0.1	0.3 – 3	2 – 120
Lime	20–1250	<0.05–0.1	—	0.1–24
Sewage	2–7000	<1–56	<1–56	2–30
Organic manure	0.4–16	<0.1–0.8	<0.01	0.2->1.25
Irrigated water	<20	<0.05	—	<10
Pesticides	11–26	—	0.6–6	3.30

(*Source:* Geoenvironment, Awasthnarayna, 1995)

Table 7.

Major Mineral	Best source	Deficiency symptoms
Calcium	Milk and milk products	Back and leg pain
Phosphorus	Meat, poultry, fish and eggs	Poor appetite
Potassium	Vegetables, fruits, milk and yoghurt	Lethargy, loss of appetite
Iron	Meat, poultry, fish, leafy vegetable	Lethargy, abnormal heart rhythms
Iodine	Sea food and iodised salt	Obesity and lethargy

bacterial activity, are frequently disrupted through behavior of organic material, radioactive elements and chemical species.

The geochemical recycling is a major cause of commencement of toxicity affecting biotic health, that results from water flowage through soil and atmosphere toxicity of aerosols and particulates. Fluorosis, a tooth-damaging condition, is most prevalent in southern Asia, Argentina, USA, Japan and Germany. Goitre, a syndrome resulting from iodine deficiency, results in mental retardation and general defects in foetuses. India and Bangladesh are prone to skin lesions, which result from food and drinking water with high concentration of arsenic. A major anthropogenic relation with medical geology arises from the introduction of heavy metal products of industry and ultimate release into river waters. Fertilizers, sewage, organic waste and pesticides are major sources of metallic toxicity. Major and much weeded minerals for human health are there that bear calcium, phosphorus, potassium, iron and iodine. Closely relatable health ailments could be timely prevented by intake of balanced dietary elements and their supplements.

SUGGESTED READINGS

1. Anonymous Contribution in Environmental Geology – GSI Spl. Publ. No. 43, 243–245, 1996.
2. Bureau of Indian Standard Drinking water specification IS-10500: Bureau of Indian Standards, New Delhi, 1991.
3. Cargo D.N and Mellary. B.F. Man and his geologic environment, second edition, 507 and 519. 1997.
4. James, R.A. and Ramesh R., In: Subramanian, V. (ed.) Environmental hazards, Capital Publishing Company, New Delhi, India, 160–166, 2002.
5. Komov, I.L., Medical Geology, Newsletter 4.5–6, 2001.
6. O'Dell, B.L. and Sunde, R.A., (Editors) Handbook of Nutritionally Essential Mineral Elements. Moreal Depper, Inc., New York, 1997.
7. Pathak, B.D. Occurrence of fluoride in ground water in Uttar Pradesh (Proc. of Symp. on Fluoride, 237), 1997.
8. Smith, F.A. and Hodge H.C. Fluoride Toxicity, Indian University Press, Bloonington, 1959.
9. Spinner, H.C.W., Minerals and human health. Chaper 11 In: Environmental Mineralogy, Edited by D.J. Vaughan and R.A. Wogelius, European Mineralogican Uniton Notes in Mineralogy, Editors University Press, Budapest, Hungary, 383–412, 2000.
10. World Health Organisation. Fluoride and Oral Health. WHO Technical Report series 846, 1994.

15

Biomedical Geography: Disabled Population

Krishna Mohan and Sandeep Thind

INTRODUCTION

Disability is not a new concept however the increasing global concern for it is certainly a recent phenomenon. Surprisingly, there are no precise figures for the total disabled population in the world. Disability is multi-faceted and neither definitions nor statistics available are comparable. The present paper takes into account the following questions: Is the disability a disease or a result of some disease? Has the concept of disability in India undergone change since its introduction in 1872 census? What is the pattern of distribution of different types of disabled population in India? Does it vary by urban and rural settlement, or by sex? Is the existing pattern a biophysical response?

What is disability? Perhaps neither you nor me are able to give an all time acceptable definition. In fact, every one is disabled during their lifetime in some form or the other however the focus of this paper is limited to physical and mental disabilities which were recorded at the time of 2001 census in India.

Based on this census, the paper attempts to answer the following questions: How are different types of disabled population distributed geographically? Are particular types of disabilities found in some specific areas? Is there any association between the nature of disability and the epidemiological transition? Does it unfold the biomedical status or prevailing regional disparity in access to health services?

METHODOLOGY

The chapter has been organized in three sections. The first section deals with the literature review followed by the definitions as used in India over time. The second section relates to the distribution of disabled population and third covers various types of disabled population and their spatial patterns. The sub-section of each section discusses a particular type of the disability differentials in the rural-urban settlements and by sex. It reflects the biophysical relationship of disabled population in the context of geographic diversity in India. Finally a summary of conclusions is presented.

I. REVIEW OF LITERATURE

Poverty is the major factor associated with disability. There is definite relationship between disability and level of development. In the State of Maryland, 300 per cent more disability was recorded amongst the very poor (Sussman, 1969). The incidence of blindness in Laupula Province of Zambia by 1935 and 1936 was associated with some geographical phenomena using dot map. And this was geographical approach to one aspect of disability (McGlashan, 1972).

Describing the scope of medical geography in India Akhtar (1985) said "in a society where rich and poor, affluent localities and slums, developed regions without resources, backward regions and without resources, exist together and governed by a single political authority, inequality in the provision of medical care is bound to occur. Thus ample scope for the study of the health care geography of various regions exists."

Dutt and Dutta (1985) described that disease occurrence or non-occurrence are dependent upon a systematic integration of three factors: causative, preventive and curative. These factors are keys to an understanding of disease dynamics in India. Along with this he also described the level of economic development has a far reaching effect on health and diseases. Parasitic, infectious and poverty related diseases will continue to be significant in South and South-east Asia as long as this area has low levels of economic development.

Jones and Moon (1987) claimed that mental illness is the leading cause of disability and ultimately death, in most of western countries. They also described that one woman in eight, and one man in twelve, will, at some time during their life, enter psychiatric care. They apprehended that low rates of mental illness would be found in stable, integrated family rather than the society which is characterized by social and residential mobility. The U.S. changed from a colonial to an industrial society during the nineteenth century, so a parallel rise in detected 'madness' occurred.

It is also described that present studies on disease ecology are studied under the political epidemiology. Differentiating the fields of epidemiologist and a medical geographer they said that both are complementary to each other. Differentiation between these two were made by quoting Learmonth "epidemiology is the study of the behavior of disease in population and medical geography is the study of patterns of the similarities and differences in the occurrence of diseases between areas" (Akhtar, 1992).

Mani D. (1988) described that whatever little attention the physically handicapped have received so far by the government as well by the society indicates at best piecemeal and lukewarm approach to solve gigantic and urgent problem of a vast mass of population.

Cataract has a high rate of incidence in tropical countries like India. Nearly 5 million people have become blind in India because of cataract. Trachoma is the most widespread infectious disease, and 400 to 500 million persons in the world are victims of this disease, and almost 2 million of them are completely blind. Deficiencies of protein and vitamin A lead to blindness. In India, about 14000 children lose their vision annually primarily for this reason (Mehta, 1983). Before the age of 25 and after the age of 29 a women is more likely to produce a mentally retarded child (Thomas, 1961).

CONCEPT AND DEFINITION OF DISABILITY

The enumeration of people in census records of India under heads like infirmities from 1872–1931, physically handicapped in 1981 and disabled in 2001 confirms the ambiguity existing in the concept of disability. It is further corroborated by the fact that there was no single category used with similar connotations in any census of India ever since and whenever there was an enumeration of broadly called disabled population. In simple words, the census data are neither comparable nor dependable. However, they are indicative of our concern (Table 1 and 2)

Table 1. India: Census-wise Coverage of Disabled Population, 1872–2001

Census Years	Categories	Terminology
1872	Blind, Deaf and Dumb, Insane, Idiots or Lepers.	Infirmity
1881	Unsound Mind, Deaf and Mutes from birth, Blind, Lepers.	Infirmity
1891	Blind, Insane, Deaf-Mute, Leper.	Infirmity
1901	Blind of both eyes, Insane, Deaf-Mute and Dumb from birth, Suffering from Corrosive Leprosy.	Infirmity
1911	Insane, Totally Blind, Corrosive Leprosy, Deaf and Dumb.	Infirmity
1921	Insane, Totally Blind, Leper, Deaf-Mute.	Infirmity
1931	Insane, Totally Blind, Leper, Deaf-Mute.	Infirmity
1981	Totally Blind, Totally Dumb, Totally Crippled.	Physically Handicapped
2001	Disabilities in seeing, speech, hearing, movement, and mental.	Disabled

Source: Census of India, 1891; A General Report.
Census of India, 1901; Tables, Vol. 1, Part 2.
Census of India, 1911; Vol. 1, Part 1; Report.
Census of India, 1921; Tables, Vol. 1, Part 2.
Census of India, 1931; Vol. 1, Part 1, Report.
Census of India, 1981, Series 1, Part VII. B, The Physically Handicapped Reports and Tables.
Census of India, 2001, Disabled Population in India.

Globally, different terms are in use such as disabled, differently-abled, physically handicapped, infirmities, impairment, crippled, handicapped etc. Though they are used in varied contexts however opinions converge on the point that they are not like most of the human beings on the earth.

II. DISABLED POPULATION

The disabled population of 21.9 million, recorded in 2001 census, in India outnumber the total population of Australia (20.1 million). One in every 50 persons in India was disabled in 2001. In fact, 2.13 per cent of the total population in India was reported as disabled in some form.

The UN estimated share of population in India living with disability was 8–10 per cent. The census of India 1981 recorded 0.2 per cent of the total population of India under the category physically handicapped. It means the proportion of disabled population, as estimated or recorded, vary widely

Table 2. India: Disabled Population, 1881-2001

Census Years	Total Disabled Population
1881	937063
1891	856252
1901	670817
1911	833644
1921	860099
1931	1095678
1981	1118948
2001	21906769

Source: Census of India, 1891; A General Report.
Census of India, 1901; Tables, Vol. 1, Part 2.
Census of India, 1911; Vol. 1, Part 1; Report.
Census of India, 1921; Tables, Vol. 1, Part 2.
Census of India, 1931; Vol. 1, Part 1, Report.
Census of India, 1981, Series 1,Part VII. B, The Physically Handicapped Reports and Tables.
Census of India, 2001, Disabled Population in India.

from 0.2 per cent in census 1981 to 8–10 per cent by UNO. It therefore is always subjective and varies by objectives.

There were five different types of disability recorded in India in 2001 census. These were disability related to seeing, speech, hearing, movement, and mental. These together were 21.9 million persons and shared 2.13 per cent of the total population of India. The absolute number of disabled population in all categories varied from the highest of 3.45 million in Uttar Pradesh to the minimum of about 16 thousand in Goa among states. The proportion of disabled population to total population varied from the highest of 3.77 per cent in Sikkim to the lowest of 1.17 per cent in Goa (Table 5).

Spatially, more than half of the total disabled population was concentrated in six states (53.9 per cent). These were Uttar Pradesh, Bihar, West Bengal, Tamil Nadu, Maharashtra, and Rajasthan (Table 3). The total population share of these states in India was 53 per cent. The notable feature was that the share in disabled population of India for Bihar, West Bengal, Tamil Nadu and Rajasthan in comparison to their share in total population was more. Among these states the gap was most pronounced in Tamil Nadu and Rajasthan. Contrary to it, the share of disabled population as compared to total population of Uttar Pradesh and Maharashtra was less.

Rural-Urban Distribution

Nearly three fourth of the total disabled population in India were recorded in rural areas. Majority of the disabled in all the states was found in rural areas. In the rural areas, the highest proportion of 92.82 per cent disabled to total disabled population was recorded in Himachal Pradesh and the lowest of 54.35 per cent in Goa. In fact, 221 persons out of every 10,000 persons living in rural India were recorded as disabled. Only exceptions were two union territories i.e. Lakshadweep and Pondicherry where majority of the disabled to total disabled population was recorded in urban areas.

Table 3. India: Disabled Population as Per Cent to Total Disabled Population, 2001

Per cent Disabled	Sates	Total
> 10	Uttar Pradesh (15.8)	15.8
5–10	Bihar (8.6), West Bengal (8.4), Tamil Nadu (7.5), Maharashtra (7.2), Rajasthan (6.4), Madhya Pradesh (6.4), Andhra Pradesh (6.2)	50.8
1–5	Gujarat (4.8), Orissa (4.7), Karnataka (4.3), Kerala (3.9), Assam (2.4), Haryana (2.1), Jharkhand (1.9), Punjab (1.9), Chhattisgarh (1.9) Jammu & Kashmir (1.4), Delhi (1.1)	30.5
< 1	Others	2.9

Source: Census of India, 2001
Note: Figures in parentheses denote percentage of disabled population to total disabled population in India.

Male-Female Distribution

The proportion of male (2.37) disabled population was higher than the female (1.87). Out of every hundred disabled persons 58 were males. Among the states, it varied from the maximum of 67 males in Arunachal Pradesh to the minimum of 48 males in Tamil Nadu against every hundred disabled persons. In fact, it was only Tamil Nadu where the number of female disabled population outnumbered the males.

III. TYPES OF DISABILITY

A. Disability in Seeing

The recording of data on visual disability is as old as the history of recording the disabled population i.e. 1872. The visual disability was recorded under the category blind (1872, 1881,1891), blind of both eyes (1901), totally blind (1911, 1921,1931 &1981) and as disability in seeing (2001).

There are three broad causes of blindness: ocular diseases and anomalies, general and systemic diseases, and injuries and accidents. The ocular diseases and anomalies may be prenatal which are related to drug abuse, poisoning, infectious diseases, syphilis or tumor in mother. It may be due to bypthalmia, albinism, retinoblastoma, retinnal detachment, diabetes, and cataract of all kinds.

The proportion of disabled in seeing to total disabled population among all the states and union territories except Goa formed the largest category among five types of disability recorded in 2001 (Table 4 and 6). In case of Goa, it was the second largest after disability in movement.

Almost every second disabled recorded in 2001 census was disabled in seeing in India. Of the total disabled population, 10.6 million were recorded with seeing disability. One in every hundred persons in the country was suffering from seeing disability (Table 6). Half of the total persons with disability in seeing were concentrated in the states of Uttar Pradesh (17.42 per cent), Bihar (9.46 per cent) Tamil Nadu (9.07 per cent), West Bengal (8.11 per cent) and Rajasthan (7.09 per cent). Of the total disabled in the country, the share of disabled in seeing was 48.5 per cent. It varied from the

Table 4. India: Disabled Population, 2001

Disabled Population by Type of Disability	Total Disabled	As per cent to Total Population	As per cent to Total Disabled Population
In Seeing	10634881	1.03	48.5
In Movement	6105477	0.59	27.9
Mental	2263821	0.22	10.3
In Speech	1640868	0.16	7.5
In Hearing	1261722	0.12	5.8
Total	**21906769**	**2.13**	**100**

Source: Census of India, 2001

highest of 69 percent in Arunachal Pradesh to the minimum of 27.9 per cent in Goa. Disability in seeing was the most dominant among five types of disability recorded in 2001. More than 50 per cent of the total disabled population in the states of Arunachal Pradesh (69.3 per cent), Jammu & Kashmir (69 per cent), Tamil Nadu (58.7 per cent), Uttar Pradesh (53.6 per cent), Rajasthan (53.4 per cent), Bihar (53.3 per cent), Assam (53.2 per cent), Sikkim (53 per cent) and Orissa (50.3 per cent) was recorded with seeing disability. Majority of the disabled population in union territories of Daman and Diu, Dadra and Nagar Haveli, Chandigarh and Delhi were disabled in seeing.

Rural-Urban Distribution

Of the total disabled population in seeing, nearly 54 per cent were in rural India. In other words the incidence of disability in seeing is more a rural phenomenon. The rural urban difference in the prevalence of disability in seeing is proved by the fact that all the states and union territories of the country except Kerala (49.68), Tamil Nadu (39.70) and Lakshadweep (49.59) have more than 50 percent of seeing disabled in rural areas. 106 persons in every 10000 of total rural population is in fact suffering from disability in seeing whereas only 97 persons out of every 10000 were having seeing disability in urban areas.

Male-Female Distribution

The prevalence of disabled in seeing is more among the males as compared to females. Of the total disabled in seeing, 54 per cent were males. Among the states and union territories, majority of total disabled in seeing was males except Tamil Nadu (41.20 per cent) and Lakshadweep (48.62 per cent). In simple words, the number of females disabled in seeing outnumber the males in Tamil Nadu and Lakshadweep.

B. Disablity in Movement

The disability in movement constitutes the second largest number of disabled persons in India. It may be due to osteogenesis imperfecta, a congential disorder in which bones are usually brittle and fragile. It has an incidence of one in 20000. Cerebral palsy, muscular dystrophy, and ataxia are other reasons for it.

Table 5. India: Disabled Population, 2001

States/Union Teritories	Disabled Population as per cent to			Urban Disabled Population as per cent to Total Disabled Population
	Total Population	Total Male Population	Total Female Population	
Sikkim	3.77	3.95	3.55	8.82
Arunachal Pradesh	3.03	3.82	2.15	17.59
Jammu & Kashmir	2.98	3.20	2.74	24.10
Orissa	2.78	3.05	2.49	14.06
Kerala	2.70	2.96	2.46	24.75
Tamil Nadu	2.63	2.52	2.74	42.46
Himachal Pradesh	2.57	2.93	2.19	7.18
Rajasthan	2.50	2.86	2.11	21.43
Madhya Pradesh	2.33	2.62	2.02	24.71
West Bengal	2.30	2.55	2.04	26.69
Uttaranchal	2.29	2.62	1.96	20.51
Bihar	2.27	2.62	1.9	10.34
Haryana	2.15	2.41	1.85	25.33
Uttar Pradesh	2.08	2.37	1.75	21.65
Gujarat	2.06	2.29	1.81	31.19
Chhattisgarh	2.02	2.21	1.82	17.84
Assam	1.99	2.16	1.81	11.73
Tripura	1.84	2.04	1.64	17.20
Mizoram	1.80	1.91	1.69	39.47
Andhra Pradesh	1.79	2.01	1.57	23.05
Karnataka	1.78	2.00	1.55	29.71
Punjab	1.74	1.95	1.51	30.03
Jharkhand	1.66	1.90	1.41	22.18
Maharashtra	1.62	1.85	1.37	34.99
Nagaland	1.33	1.39	1.27	11.24
Manipur	1.31	1.41	1.21	26.15
Meghalaya	1.24	1.30	1.18	21.05
Goa	1.17	1.29	1.04	45.65
Lakshadweep	2.77	2.89	2.63	89.34
Pondicherry	2.65	3.03	2.28	94.31
Daman & Diu	2.00	1.92	2.12	40.71
Andaman & Nicobar Islands	1.98	2.19	1.73	27.99
Chandigarh	1.73	1.88	1.52	41.06
Dadra & Nagar Haveli	1.84	1.91	1.74	19.24
Delhi	1.70	1.90	1.46	63.10
India	**2.13**	**2.37**	**1.87**	**25.19**

Source: Census of India, 2001

Note: Data arranged in descending order by Column II.

Table 6. India: Disabled Population as per cent to Total Population Classified by Types, 2001

States/Union Territories	Disabled Population as percentage of Total Population				
	i	ii	iii	iv	v
	Seeing	**Speech**	**Hearing**	**Movement**	**Mental**
Arunachal Pradesh	2.10	0.22	0.28	0.32	0.11
Jammu & Kashmir	2.06	0.17	0.14	0.37	0.25
Sikkim	2.00	0.59	0.63	0.40	0.15
Tamil Nadu	1.54	0.20	0.12	0.57	0.20
Orissa	1.40	0.19	0.23	0.68	0.28
Rajasthan	1.33	0.13	0.13	0.71	0.19
Bihar	1.21	0.16	0.09	0.62	0.20
Uttar Pradesh	1.11	0.15	0.08	0.56	0.17
West Bengal	1.08	0.21	0.16	0.51	0.34
Assam	1.06	0.21	0.19	0.35	0.18
Himachal Pradesh	1.06	0.21	0.25	0.77	0.28
Madhya Pradesh	1.05	0.13	0.14	0.82	0.19
Kerala	1.05	0.21	0.25	0.75	0.44
Uttaranchal	1.01	0.20	0.19	0.67	0.23
Gujarat	0.98	0.13	0.14	0.61	0.20
Haryana	0.95	0.12	0.13	0.72	0.23
Tripura	0.86	0.16	0.18	0.44	0.21
Karnataka	0.83	0.17	0.09	0.50	0.18
Chhattisgarh	0.77	0.15	0.16	0.73	0.21
Andhra Pradesh	0.76	0.18	0.10	0.55	0.20
Mizoram	0.70	0.23	0.27	0.28	0.32
Punjab	0.70	0.09	0.07	0.61	0.26
Jharkhand	0.69	0.15	0.10	0.51	0.21
Maharashtra	0.60	0.12	0.10	0.59	0.22
Meghalaya	0.58	0.15	0.16	0.22	0.14
Manipur	0.54	0.13	0.14	0.29	0.22
Nagaland	0.50	0.22	0.26	0.21	0.13
Goa	0.33	0.14	0.07	0.36	0.27
Daman & Diu	1.20	0.12	0.08	0.44	0.17
Pondicherry	1.09	0.19	0.23	0.91	0.23
Dadra & Nagar Haveli	1.06	0.13	0.15	0.36	0.12
Lakshadweep	0.99	0.34	0.24	0.83	0.36
Chandigarh	0.94	0.10	0.07	0.43	0.20
Andaman & Nicobar Islands	0.93	0.18	0.15	0.53	0.19
Delhi	0.87	0.11	0.06	0.47	0.19
India	**1.03**	**0.16**	**0.12**	**0.59**	**0.22**

Source: Census of India, 2001
Data arranged in descending order by column (i)

Table 7. India: Disabled Population Classified by Type of Disability, 2001

States/Union Territories	Disabled as Per cent to Total Disabled Population in				
	Seeing	Movement	Mental	Speech	Hearing
Andhra Pradesh	42.6	30.5	11.4	10.2	5.4
Arunachal Pradesh	69.3	10.4	3.8	7.3	9.2
Assam	53.2	17.3	9.0	10.7	9.8
Bihar	53.3	27.1	8.8	6.9	3.9
Chhattisgarh	38.1	36.1	10.4	7.2	8.1
Goa	27.9	31.2	22.7	11.9	6.3
Gujarat	47.3	29.7	9.9	6.4	6.7
Haryana	44.3	33.3	10.9	5.5	6.1
Himachal Pradesh	41.1	29.8	11.1	8.2	9.8
Jammu & Kashmir	69.0	12.5	8.2	5.6	4.7
Jharkhand	41.5	30.8	12.5	8.9	6.3
Karnataka	46.9	28.3	9.8	9.6	5.3
Kerala	38.9	27.6	16.5	7.8	9.3
Madhya Pradesh	45.2	35.2	8.2	5.4	6.1
Maharashtra	37.0	36.3	13.6	7.2	5.9
Manipur	41.3	21.8	16.6	9.8	10.6
Meghalaya	46.5	17.8	11.1	11.9	12.7
Mizoram	39.1	15.5	17.8	12.5	15.1
Nagaland	37.6	16.1	9.9	16.6	19.8
Orissa	50.3	24.6	10.1	6.7	8.2
Punjab	40.2	35.3	15.0	5.4	4.1
Rajasthan	53.4	28.4	7.7	5.2	5.3
Sikkim	53.0	10.7	3.9	15.6	16.9
Tamil Nadu	58.7	21.5	7.8	7.6	4.4
Tripura	46.7	23.7	11.3	8.7	9.7
Uttar Pradesh	53.6	26.9	8.3	7.4	3.7
Uttaranchal	44.0	29.0	10.2	8.6	8.2
West Bengal	46.7	22.3	14.7	9.2	7.1
Andaman & Nicobar Islands	47.1	26.5	9.5	9.2	7.7
Chandigarh	54.2	24.6	11.6	5.7	3.9
Dadra & Nagar Haveli	58.0	19.6	6.8	7.3	8.3
Daman & Diu	59.9	21.8	8.6	6.0	3.8
Delhi	51.2	27.5	11.0	6.6	3.7
Lakshadweep	35.9	30.1	12.9	12.3	8.8
Pondicherry	41.2	34.1	8.8	7.0	8.8
India	**48.5**	**27.9**	**10.3**	**7.5**	**5.8**

Source: Census of India, 2001

There were 6.10 million persons recorded as disabled in movement. Of the total disabled in the country, nearly 28 per cent were disabled in movement. Among the states, the proportion of disabled in movement to total disabled was recorded the maximum in Maharashtra (36.3 per cent) and the minimum in Arunachal Pradesh (10.4 per cent). The number of disabled in movement in Maharashtra and Chhattisgarh were almost equal to the number of disabled in seeing. Among the five types of disability recorded, the disability in movement formed the largest category in Goa.

There were six out of every 1000 persons in India recorded as disabled in movement. It varied from the highest in Pondicherry (9) to the minimum in Nagaland (2).

Rural-Urban Distribution

Of the total disabled persons in movement, 64 per cent were recorded in rural areas. The highest proportion of disabled in movement to total disabled population among the states was recorded in Bihar and the minimum in Nagaland (Table 8).

There were 5 persons disabled in movement out of every 1000 persons in urban areas. In rural areas it was 6 persons against every 1000 persons. Among the states, Madhya Pradesh recorded the highest and the Nagaland the lowest disabled in movement both in rural and urban areas.

Male-Female Distribution

Like any other disability, the males suffer more in disability of movement also. Nearly 64 per cent of the total disabled in movement were males. The proportion of male disabled in movement in total disabled in movement was the highest in Uttar Pradesh (66.81 per cent) and the lowest in Nagaland (54.77 per cent).

The disabled population in movement in Pondicherry, Madhya Pradesh, Lakshadweep, Himachal Pradesh, Kerala, Haryana, and Rajasthan revealed that one in every 100 males was disabled in movement. The female disabled in movement was lower than males everywhere.

C. Mental Disability

A very high proportion of mental disability in total disabled population was recorded in Goa, Mizoram, Manipur, Kerala, Punjab, West Bengal, and Maharashtra. The high proportion of mental disability in total disability in the states like Goa, Kerala and Punjab need further investigation since they are our model states. Such a situation is certainly an indicator of major epidemiological transition.

There are many causes of this problem. Most important are epiloia, galactosaemia, Hunter's syndrome, cretinism, microcephaly, Down's syndrome, Edward's syndrome, etc.

The mental disability in Assam, Sikkim and Arunachal Pradesh is the least important cause of disability. There were 2.26 million persons suffering from mental disability. It is the third major disability of all the disabilities recorded in census 2001. In fact 2 persons from every 1000 were having mental disability. The prevalence of mental disability varies from the highest of 4 persons from every 1000 in Kerala to the lowest of 1 person from every 1000 in Arunachal Pradesh (Table 6). Almost 54 per cent of the total mental disabled in the country were concentrated in the five states of Uttar Pradesh (12.65 per cent), West Bengal (11.95 per cent), Maharashtra (9.42 per cent), Bihar (7.30 per cent), Andhra Pradesh (6.86 per cent) and Kerela (6.26 per cent). In terms of proportion of

Table 8. India: Disabled Population in Rural Areas Classified by Type of Disability, 2001

State/Union Territories	Rural Disabled Population as percentage of Total Disabled Population among Different Categories				
	i	ii	iii	iv	v
	Seeing	**Speech**	**Hearing**	**Movement**	**Mental**
Arunachal Pradesh	68.91	58.53	54.01	60.61	53.28
Rajasthan	56.88	64.23	52.75	62.88	64.41
Tripura	56.85	56.29	49.51	60.42	54.42
Sikkim	56.26	51.91	55.85	60.49	53.01
Uttar Pradesh	55.97	59.26	57.25	67.07	65.91
Mizoram	55.41	53.41	53.82	55.99	49.24
Nagaland	55.35	52.18	53.86	53.96	56.09
Bihar	55.22	59.09	58.70	67.86	64.99
Jharkhand	55.09	56.34	55.38	63.04	59.93
Gujarat	54.84	61.84	51.33	61.77	61.08
Haryana	54.83	63.40	54.91	65.96	66.57
Jammu & Kashmir	54.80	57.82	55.15	61.30	59.98
Assam	54.34	55.68	53.30	60.94	57.78
Madhya Pradesh	54.31	60.23	56.98	62.68	61.26
Karnataka	54.12	55.54	50.64	63.41	55.06
Andhra Pradesh	54.10	54.55	48.86	61.33	54.84
West Bengal	54.06	55.72	54.05	64.75	57.40
Punjab	53.95	58.89	52.89	64.74	64.44
Manipur	53.77	53.22	50.59	56.29	56.65
Maharashtra	53.64	56.11	56.36	65.71	56.69
Himachal Pradesh	53.61	58.86	55.29	62.64	61.11
Meghalaya	53.03	51.64	49.53	56.68	52.73
Orissa	52.91	54.48	54.22	60.49	55.74
Uttaranchal	52.90	59.04	55.11	62.69	62.72
Chhattisgarh	51.90	54.60	54.02	57.97	53.57
Goa	51.40	54.77	50.41	64.33	54.92
Kerala	49.68	55.48	46.09	60.28	52.49
Tamil Nadu	39.70	54.50	48.32	62.10	55.91
Chandigarh	65.06	68.69	52.46	68.16	72.39
Daman & Diu	62.88	54.87	40.00	61.26	55.07
Delhi	58.21	61.84	53.38	66.64	68.70
Andaman & Nicobar Islands	57.56	60.98	57.88	67.80	52.52
Dadra & Nagar Haveli	56.67	55.02	52.10	58.40	55.41
Pondicherry	52.26	55.29	47.84	65.13	55.71
Lakshadweep	49.59	60.00	51.02	57.39	45.24
India	**53.63**	**57.40**	**53.68**	**63.92**	**59.57**

Source: Census of India, 2001.

Note: Data arranged in decending order by coloumn (i)

persons with mental disability to total disabled persons, Goa with 22.72 per cent was the highest. In simple words, of the every 100 disabled persons in Goa 23 were mentally disabled. Whereas it was the lowest in Arunachal Pradesh where only 4 persons from every 100 disabled persons were suffering from mental disability (Table 7).

Rural-Urban Distribution

As per census 2001, of the total population effected with mental disability 60 per cent were in the rural areas. Among the states majority of mentally disabled were concentrated in rural areas. But in case of union territories such as Delhi (94.08 per cent), Chandigarh (90 .04 per cent) and Pondicherry (67.80) mental disability was more urban phenomenon. But in urban areas more persons were suffering from mental disability when it is compared with total urban population where 23 persons for every 10000 persons were suffering from mental disability whereas in rural areas 21 persons for every 10000 persons were suffering from mental disability.

Male-Female Distribution

Mental disability in India is more prevalent in males than the females. Almost 60 per cent of the total mental disabled are males. The proportion of male mental disabled to total mental disabled varies from the maximum in Haryana (65.95 per cent) to the minimum in Kerala (52.92 percent) (Table 9). Out of every 1000 males three were suffering from mental disability whereas only two from every 1000 females were suffering from mental disability. Mentally disabled females in the union territory of Lakshadweep outnumbered males. (Table 9)

D. Disability in Speech

According to 2001 census, 1.64 million persons in India were suffering from speech disability. Among the states, it varied from the maximum of 256 thousand persons in Uttar Pradesh to the minimum of 1868 in Goa. However the proportion of disabled population in speech to total population varied from the maximum of 0.59 per cent in Sikkim to the minimum of 0.12 percent in Haryana and Maharashtra. Of the total disabled in the country only 7.5 per cent were suffering from disability in speech. And out of these 7.5 percent disabled in speech half were concentrated in five states of the country namely Uttar Pradesh (15.60), West Bengal (10.36 percent), Andhra Pradesh (8.47 percent), Bihar (7.95 percent) and Tamil Nadu (7.59 percent).

Rural-Urban Distribution

In India, according to the census of 2001 approximately 76 percent of the total disabled in speech were in the rural areas. In all the states of the country majority of disabled population in speech were in rural areas. But, among union territories Pondicherry (39.49), Chandigarh (11.22), and Delhi (6.68) have majority of disabled in speech in urban areas. The maximum proportion of rural disabled in speech to total disabled in speech was recorded in Sikkim where (95 per cent) of total disabled in speech were in rural areas. Whereas the lowest was in Goa (55.03 per cent).

Table 9. India: Disabled Male Population Classified by Type of Disability, 2001

States/Union Territories/India	Male Disabled as percentage of Total Disabled among Different categories				
	i	ii	iii	iv	v
	Seeing	**Speech**	**Hearing**	**Movement**	**Mental**
Arunachal Pradesh	70.55	58.46	54.33	61.23	53.61
Rajasthan	57.11	63.64	52.99	63.24	64.48
Tripura	56.82	56.00	49.25	60.82	55.10
Sikkim	56.53	52.11	55.94	60.13	53.69
Nagaland	56.45	51.96	53.75	54.77	56.20
Uttar Pradesh	56.28	59.16	57.01	66.81	65.58
Mizoram	56.03	54.04	54.03	57.27	50.75
Jharkhand	55.93	56.74	55.80	63.48	60.82
Jammu & Kashmir	55.60	57.49	54.34	61.40	60.41
Haryana	55.40	62.77	54.71	65.22	65.95
Bihar	55.36	59.16	58.73	67.69	64.99
Gujarat	55.33	61.19	50.38	61.62	61.64
Maharashtra	55.16	56.44	56.05	65.74	57.74
Andhra Pradesh	54.80	54.71	48.92	61.69	55.91
Karnataka	54.76	55.66	51.05	63.33	55.63
Assam	54.65	55.79	53.17	61.31	58.26
Punjab	54.52	59.29	53.18	64.23	63.94
Madhya Pradesh	54.47	60.14	56.78	63.14	61.58
West Bengal	54.40	55.67	54.08	64.82	57.77
Himachal Pradesh	54.30	58.86	55.19	62.65	60.99
Uttaranchal	54.20	59.36	55.09	63.05	62.44
Meghalaya	53.58	51.03	48.94	56.43	53.44
Manipur	53.48	54.24	50.63	56.78	56.47
Orissa	53.33	54.79	54.33	61.02	56.34
Goa	52.72	55.51	47.10	62.77	55.42
Chhattisgarh	52.49	54.87	53.81	58.67	54.41
Kerala	50.01	55.44	45.73	59.89	52.94
Tamil Nadu	41.20	54.77	48.04	61.87	56.84
Chandigarh	59.86	62.47	58.32	65.31	60.70
Delhi	59.10	60.76	55.54	65.81	63.56
Dadra & Nagar Haveli	57.67	56.27	52.23	60.00	57.09
Andaman & Nicobar Islands	57.42	58.74	55.96	67.97	53.81
Daman & Diu	56.32	56.08	37.50	60.00	52.92
Pondicherry	55.42	53.96	45.98	62.77	56.61
Lakshadweep	48.92	61.35	51.70	60.20	45.83
India	**53.90**	**57.41**	**53.40**	**63.92**	**59.84**

Source: Census of India, 2001.
Note: Data arranged in descending order by column (i)

Male-Female Distribution

Males outnumbered females in disability in speech (57 percent) as well. Disability in speech in India varied from the maximum of 63.64 percent in Rajasthan to the minimum of 51.03 percent in Meghalaya (Table 9).

E. Disability in Hearing

There were 1.26 million persons recorded as disabled in hearing in 2001 census. It was 0.12 per cent of the total population of India. In simple words, 12 out of every 10000 persons were suffering from disability in hearing. Of the total disabled population, hearing disabled persons constitute 5.76 per cent.

The proportion of disabled in hearing to total population was recorded the highest with 0.63 per cent in Sikkim to the lowest of 0.07 in Punjab among the states of India (Table 6).

Hearing is the second important cause of disability in Sikkim and Nagaland. In Arunachal Pradesh and Meghalaya, it is the third major cause of disability. In others like Haryana, Manipur, Mizoram, Tripura, Rajasthan, Orrisa, Madhya Pradesh, Chhattisgarh, Gujarat and Kerala it is the fourth major cause of disability.

Rural-Urban Distribution

More than half of the hearing disabled was in rural areas. The state of Bihar recorded 58 per cent of its total disabled in hearing in rural areas. On the other hand only 40 per cent of the total disabled in hearing in Daman & Diu in rural areas (Table 8).

Male-Female Distribution

Males in general in total hearing disabled population outnumbered the females. Males disabled in hearing to total disabled in hearing in Bihar shared 59 per cent. The smallest share of males was however recorded in Daman & Diu (Table 6).

MAJOR CONCLUSIONS

- There were 21.9 million disabled recorded in five different categories of disability in 2001 census of India. It was larger than the total population of Australia. The data on disability itself is disabled. The information gathered in one census is not comparable with the previous one due to inconsistencies in definition.
- The disabled population in India was recorded on the basis of information supplied to the enumerator by the head or any member of the household at the time of census. This information is never cross-examined by any agency. The degree of disability and the causes for them were never recorded.
- The disability in the population of Sikkim was most pronounced. Goa recorded the lowest percentage of disabled population. Numerically the largest number of disabled was recorded in Uttar Pradesh and the smallest in Goa.

- Spatially, more than half of the total disabled population were concentrated in six large size states: Uttar Pradesh, Bihar, West Bengal, Tamil Nadu, Maharashtra, and Rajasthan.
- Males appear to be more prone to disability than females. The state of Arunachal Pradesh recorded 67 per cent of its total disabled as males. In fact, except for Tamil Nadu, disabled males outnumbered the disabled females in all the states and union territories.
- The incidence of disability was more recorded in rural than urban. The state of Himachal Pradesh recorded 93 per cent of its total disabled in rural. In fact, except for Lakshadweep and Pondicherry, majority of the disabled in all the states and union territories was recorded in rural areas.
- The largest number of disabled were recorded in seeing followed by movement, mental, speech and hearing respectively. Only exception was Goa where disability in movement was the largest.
- Four major states of Indo-Gangetic plain i.e. Uttar Pradesh, Bihar, West Bengal and Rajasthan and the coastal state of Tamil Nadu account for more than half of the total disabled in seeing.
- The state of Arunachal Pradesh recorded 69 per cent of its total disabled population as disabled in seeing. Contrary to it, only 28 per cent of the total disabled in Goa were disabled in seeing.
- Majority of the disabled in seeing, except for Kerala, Tamil Nadu, and Lakshadweep, were recorded in rural areas. Similarly, majority of disabled in seeing was males except in Tamil Nadu and Lakshadweep.
- The state of Maharashtra recorded more than one third of its total disabled as disabled in movement. The disability in movement to the total disabled was the lowest in Arunachal Pradesh.
- A high proportion of mental disability in Goa, Kerala, Punjab, West Bengal and Maharashtra denotes an epidemiological transition.

SUGGESTED READINGS

1. Akhtar, Raiz " Scope of Geography of Health and Geomedical Research in India", in *Geographical Aspects of Health and Disease in India*, ed. Akhtar, Raiz and Learnmonth, A.T.A , Concept Publishing Company, New Delhi, 3–23, 1985.
2. Akhtar, Raiz "Spatial Aspects of Disease: A Survey", in *Annals*, National Associations of Geographers in India, Vol. XII, No. 1, December 1992, 39–54. 1992.
3. Biiussian , Punani and Rawal, Nandini " Demographic Details of the Visually Handicapped", in *Handbook: Visual Handicapped*, Ashish Publishing House, New Delhi, 39–54, 1993.
4. Chaterjee, L. "Spatial Distribution of the working age disabled in the U.S : 1998". Paper presented to the annual meeting of the Association of American Geographers; Chicago IL, 1995.
5. Dutt, A.K. and Dutta, H.M "Disease Dynamics in South East Asia with Special Reference to India" in *Geographical Aspects of Health and Diseases in India*, ed. Akhtar Raiz and Learnmonth, A.T.A., Concept Publishing Company , New Delhi, 38–47, 1985.
6. Foster, H.D. *Reducing the Incidence of Multiple Sclerosis Environments*, Routledge, New York, 19 and 14–34, 1998.
7. Garry, L. Albrecht and Verbrugge, M. Lois, " The Global Emergence of Disability" in *Social Studies in Health and Medicine*, Sage Publications, 193–205. 1994.
8. Gill, Valentine "Geography and Ethics: In Pursuit of Social Justice, Ethics and Emotion and Geographies of Health and Disability Research" in *Progress in Human Geography*, Vol. 27, No. 3, June 2003, 375–380, 2003.
9. Grant, R. "Transport for the disabled", in *Geography*, 77, 88–91, 1992.
10. Hahn, H. "Disability and the Urban Environment a Perspective on Los Angeles". *Society and Space*, 273–288, 1986.
11. Henderson, P. *Disability in Childhood and Youth*, Oxford University Press, London, 1972.

12. Hester, Parr " Medical Geography: Critical Medical and Health Geography" in *Progress in Human Geography*, Vol. 23, No.2, March 2004, 2004.
13. Hugg.l "A Map Comparison of Work Disability and Poverty Status in the United States", in *Social Science and Medicine* 13 D, 237–240, 1979.
14. Hunter, J.M and Akhtar Raiz "The Challenge of Medical Geography", in *Environment and Health: Themes in Medical Geography*, ed. Akhtar Raiz, Ashish Publishing House, New Delhi, 3–49, 1991.
15. Imrie, R.F and Wells, P.E., "Creating a barrier free environment", Town and Country Planning, 61, 88–91, 1993.
16. Imrie, Rob, "Albiest Geographies, Disabliest Spaces: Towards a Reconstruction of Golledge's "Geography and the Disabled", in *Transaction*, The Institute of British Geographers, N.S 21, !996, 397–403, 1996.
17. Jordan, E. Thomas "The Sources of Mental Retardation" in *The Mentally Challenged*, E. Merrill Books. Inc, Coloumbus , Ohio, 154–182, 1961.
18. Kelvyn , Jones and Moon, Graham " Concept and Issues in Mental Illness", in *Health Disease and Society*, Routledge and Kegan Paul, London and New York, 19–216, 1987.
19. Mayer, D. Jonathan, "Medical Geography some Unsolved Problems", in *The Professional Geographers*, Forum of the Association of American Geographers, Vol. 34, No. 3, 261–270, 1982.
20. Mc Glashan, M. D. "Blindness in Laupula Province" in *Medical Geography Techniques and Field Studies*, Methuen & Co. Ltd, London, 153–163, 1972.
21. Mehta, D.S, "The Darkness All Around" in *Handbook of Disabled in India*, Allied Publishers, New Delhi, 5, 1983.
22. Pandurkar , R.G, " Mortality Patterns and Trends of Poliomyelitis in Maharashtra", in *Transactions of the Institute of Indian Geographers*, Vol. 10, No. 2, July 1988, 15–20, 1988.
23. Park, C. Deobarh and Others "Disability Studies in Human Geography " in *Progress in Human Geography*, Vol. 22, No. 2, June 1998, 208–234, 1998.
24. Phllippa, Russel, " Some Causes of Handicap" in *The Wheelchair Child*, Condar Book Souvenir Press (E &A) Ltd., 11–30, 1978.
25. Rama, Mani D. *The Physically Handicapped in India: Policy and Programmes*, Ashish Publishing House, New Delhi, 1988.
26. Raza , Moonis and Nangia, Sudesh " Disabled Children" , in *Atlas of the Child in India*, Concept Publishing Company, New Delhi, 38–47, 1985.
27. Sen, Anima, *Psycho-social Integration of the Handicapped: A Challenge to the Society*, Mittal Publications, Delhi, 11–18, 1988.
28. Topliss, Eda, *Provisions for the Disabled*, Basil Blackwell, London, 1975.
29. Verma, M. Surinder, *Social Interpretation of Deaf*, Northern Book Center, New Delhi, 1999.

Index